The Emperor's
Four Treasuries

Harvard East Asian Monographs 129

*Published by
the Council on East Asian Studies
Harvard University
and distributed by
Harvard University Press
Cambridge (Massachusetts) and London
1987*

The Emperor's Four Treasuries

SCHOLARS AND THE STATE
IN THE LATE CH'IEN-LUNG ERA

R. Kent Guy

The Council on East Asian Studies at Harvard
University publishes a monograph series and,
through the Fairbank Center for East Asian
Research and the Reischauer Institute of Japanese
Studies, administers research projects designed
to further scholarly understanding of China, Japan,
Korea, Vietnam, Inner Asia, and adjacent areas.

Library of Congress Cataloging in Publication Data

Guy, R. Kent, 1948–
  The emperor's four treasuries.

  (Harvard East Asian monographs ; 129)
  Bibliography: p.
  Includes index.
  1. Ssu k'u ch'üan shu — History.   2. China — History —
Ch'ien-lung, 1736–1795.   3. Learning and scholarship —
China.   I. Title.   II. Series.
AC149.S73G89   1987   001.2'0951   87-495

For Christine Cordell,
With Love

Contents

Acknowledgments

This book was begun as a dissertation supervised by Professor Benjamin I. Schwartz at Harvard, and I am pleased to join the many who have made it their first order of business to acknowledge his guidance and inspiration. Professor Ying-shih Yü originally suggested the topic to me and proved an advisor of rare skill, allowing me to develop in my own directions, but offering encouragement, bibliographic and organizational advice at critical moments. Professor Philip Kuhn has served for me as for many a source of stimulation and support in thinking and teaching about issues of modern Chinese history. Professors Beatrice S. Bartlett, Ch'en Chieh-hsien, Paul J. A. Clarke, Jack L. Dull, Benjamin Elman, the late Joseph Fletcher, John K. Fairbank, William Kirby, Susan Naquin, Jonathan Spence, the late Suzuki Chusei, Wu Che-fu, Alexander B. Woodside, and Yin Chiang-yi have all made much-valued suggestions and comments, as have members of the History Research Group, the China Colloquium and the Department of History at the University of Washington and members of the History Department seminars at Yale, Harvard, Washington University and Bowdoin College. I have appreciated their criticisms even if I have been unable to implement all of their suggestions. The research staffs of the Chinese Library of the Humanistic Research Institute at Kyoto University, the Archives of the National Palace Museum in Taipei, the First Historical Archives in Beijing, the Harvard-Yenching Library and the East Asia Library of the University of Washington have graciously accommodated me and saved me countless hours of time. Thatcher Deane, Anthony Fairbank, John Herman and Tanya Ch'en Herman have assisted me in preparing the manuscript for publication. For clerical and research support, particularly during the last stages of manuscript preparation, I owe a debt of gratitude to the China Program of the Henry M. Jackson School of International Studies, University of Washington, and its successive chairmen, Jack L. Dull and Nicholas R. Lardy.

The *Ssu-k'u ch'üan-shu* took over twenty years and required the resources of an emperor, over two hundred scholars, and countless book collectors and clerks to complete. This project has taken rather less time, accomplished significantly less, but required almost as much support. It is a pleasure finally to thank all of those, named above and unnamed, teachers, colleagues, students, relatives, and friends who have borne me through the project, and to offer some small result of their many kindnesses.

1

Introduction

In the winter of 1772, the Ch'ien-lung Emperor initiated the compilation of the *Ssu-k'u ch'üan-shu* (Complete library of the four treasuries) by ordering local and provincial officials to search for, report, and make copies of all rare and valuable manuscripts held in libraries within their jurisdictions and to forward the results to Peking. At the same time, he urged private collectors to send their treasures to Peking voluntarily. In March of 1773 an administrative apparatus was created at the capital to receive the books and to evaluate their contents. The staff for the project grew, finally encompassing over seven hundred editors, collators, and copyists, including some of the most important figures at court and in the literary circles of the day. The final products, created and corrected over a twenty-two year period, were an annotated catalog of 10,680 titles extant in the empire grouped in four categories or "treasuries" (classics, history, philosophy and belles lettres), and an enormous compendium in which 3,593 titles were recopied filling 36,000 *chüan*. The darker side of the effort was a campaign of censorship undertaken by the imperial court in the late 1770s and early 1780s. By some counts as many as 2,400 titles were destroyed in this campaign, and another four or five hundred "revised" by official fiat. The products of the *Ssu-k'u ch'üan-shu* project were, of course, of enormous bibliographic significance. But in a sense the processes of collecting, editing, and censoring were even more significant, for they shed light on an issue and an era.

The issue is the role of the scholar in the imperial Chinese state,

1

a role which provided the foundation for much of the Chinese intellectual's creative endeavor. The social and political role of the man of knowledge has been explored in a number of studies of remarkable creativity and force, yet few general conclusions have been reached. In China, the scholar's role was one defined by an intricate network of assumptions, prerogatives, and institutions. Although Chinese thinkers throughout the imperial era, and Western thinkers since the Enlightenment, have regarded China as a land where knowledge and power were united, it is striking that neither Chinese, Japanese, nor Western scholarship has established the prerogatives of the Chinese state in the worlds of scholarship and literature, or explored the attitudes of Chinese intellectuals toward them. These prerogatives and attitudes were probably most clearly demonstrated in the book collections and censorship projects which nearly every Chinese dynasty sponsored. In the course of efforts like the compilation of the *Ssu-k'u ch'üan-shu,* intellectuals and rulers together defined the nature of imperial authority and explored its limits.

This process of definition and exploration took place in an institutional context unique to China. In theory, the only proper rule in China was rule by virtue, and the characteristics and mechanisms of virtuous rule were defined in the classical canon and elaborated in the body of commentary that had grown up around it. A dynasty's stewardship of the canon, and its invocation of classical sanction were vital bases of its legitimacy. The organs and mechanisms of this stewardship, therefore, occupied a far more important place in the apparatus of Chinese rule than they did in Western governments, and their histories were correspondingly more revealing of political dynamics. Not only were the cream of the Chinese bureaucracy involved in the preservation of classical texts, but important questions — of the relevance of textual and historical precedent in policy making, and the role of the emperor as creator, arbiter or model of sagely teachings — were addressed in discussions of classical texts. The relative absence, particularly in Western literature on China, of studies of institutions like the examination system, the Hanlin Academy or the Board of Personnel has left us with an image of traditional society as dominated by power-hungry despots and their machiavellian minions on the one hand, and idealistic and often eremetic Confucian scholars on the other, with little room for debate, discussion, or consensus between them.[1] Remedying this defect will clarify the quality and character of political life under a Confucian imperial system.

Institutions of imperial scholarship not only reflected ideas of government, they affected the government of ideas. Concepts such as freedom of conscience and expression and the separation of church and state have become so thoroughly a part of our Western political vision that it is difficult to imagine what forms intellectual creativity could take in a society where the ruler was seen as omniscient. Indeed, this obstacle has impeded our recognition that China could have had a genuine intellectual history of growth, vitality and change. Too often, the apparent consistency of Chinese scholars' pronouncements about human nature and the imperatives of government have blinded historians to the elements of change and argument within them. The chapters below will not deny that liberty of thought and speech are indispensable qualities of what we have come to regard as an ideal order; they will argue, however, that creativity and disputation were possible within the framework of Confucian government.

On the theoretical level, therefore, this book will examine a set of assumptions about the role of the scholar in the Chinese state. At the concrete level, it will examine the interaction of intellectuals and the Chinese government in the last quarter of the eighteenth century. This was a period of remarkable power but it was also a time when the fate of the dynasty appeared to be at a turning point. Modern historians have yet to examine all the dimensions of this transition, but they have noted its most vivid manifestation in the shifting attitudes and declining morale of China's political and social leaders. Within a few decades, the confident, almost arrogant leaders of the Ch'ien-lung era were replaced by a new group of troubled, beleaguered men whose writings were filled with images of decline and disillusionment.[2]

The broad question to be asked of the late Ch'ien-lung years is whether they should be viewed as the last years of Ch'ing greatness, or the first years of imperial decline. Was this the "glorious Indian summer of Ch'ing rule in China," a time when the Manchus' careful and on the whole successful governance of the Middle Kingdom brought forth fruit in the form of an age of wealth and harmony? Or was it an era of crumbling facades, when surface prosperity barely concealed the realities of corruption and decline?[3] The answer to this question surely lies in part in the attitude of the monarchy, bureaucracy, and social and intellectual elite toward each other, toward the regime under which they lived and toward the traditional goals of the Chinese state.

The goals of the monarchy are perhaps the most elusive, for although few institutions in world history have produced as much verbiage as the Ch'ien-lung monarchy did, the man behind the paper remained something of an enigma. Whether he was cruising down the Grand Canal to show the imperial flag in the southeast, reminiscing boastfully of ten great campaigns on the boundaries of his enormous empire, or presiding over a regime which produced voluminous descriptions of its goals, methods and philosophical foundations, the Ch'ien-lung Emperor seems to have been a man concerned almost exclusively with the image and style of the monarchy.[4] What impact, if any, did this pompous, posturing, and peripatetic monarch have on institutions and ideas in his own day? Probably the answer lies partly in the way he systematized and glorified relations between the Chinese elite and their Manchu rulers.

What were the attitudes of social elites of the day toward the institutions of government? The importance of changing elite attitudes and morale in shaping nineteenth-century Chinese history has long been assumed.[5] But little attention has been paid to the eighteenth-century elite, who were surely as critical to the success or failure of Ch'ing rule as their nineteenth-century successors. How did politicians and intellectuals of the eighteenth century, who were at once landowners and book owners, magnates, mentors and minions, view the times in which they lived and the regime which governed them? Two views of society have been traced among eighteenth-century elites, one predominant among those who served at court, the other prevalent among those outside the capital. Each world had its own priorities and perspectives. All who served at court shared an orientation toward the examinations: virtually everyone at court spent at least the first twenty-five years of his life studying for the examinations, and the next twenty-five years evaluating others' efforts. By the eighteenth century, the skills required in examination taking were limited, but they had to be so firmly mastered that the pursuit of them could become an obsession. As the examinations became more competitive with the growth of population in the seventeenth and eighteenth centuries, ever finer distinctions separated successful from unsuccessful candidates, forcing examiner and examinee alike to concentrate on the most trivial aspects of the process. A second concern of court officials was the ability to produce polished literary compositions for formal occasions. The name of the Hanlin Academy, where much of this competition took place, literally meant the "forest of quills" and the body was known unofficially as the "word commission." Finally, all at court had at least to pay lip-service to Sung dynasty interpretations of the classics, especially those by Chu Hsi (1130–1200) and his followers, for it was their work, repeatedly reprinted under official patronage, that formed the basis of state ideology and the examination system.

By contrast, a critical and questioning spirit prevailed among eighteenth-century intellectuals outside of government, an attitude which expressed itself in studies of the classics and histories. Texts long considered sacrosanct were challenged, unreadable passages clarified, and errors of transmission identified and rectified. The discursive and moralistic commentaries of the Sung were discarded in favor of more philological and precise explanations of earlier Han dynasty commentators like Cheng Hsuan (127–200). Many theories have been advanced about the origins of this spirit. In an era when the ranks of the educated were growing and the number of places in government remaining relatively constant, the new interests in texts may well have been part of a long term trend toward the creation of new roles for intellectuals outside, or on the periphery of, government service. Economic and demographic factors may explain how new intellectual roles were evolving, but they hardly explain the philosophical assumptions scholars brought to them. Recently, Yü Ying-shih has suggested that in addition to the "external factors" which impelled Ch'ing intellectuals toward pure scholarship, there may have been internal factors also at work. Specifically, he has suggested that scholars rediscovered the importance of intellectualism and study in the Confucian program in the course of their search for textual foundations of their metaphysical theories. Benjamin Elman has reconstructed the logic of Ch'ing intellectuals' passage from philosophy to philology and the social and intellectual forces that impelled it.[6]

The assumption of much recent scholarship has been that those who served at court collaborating with Manchu rule and those who did not were in conflict, separated from each other by an unbridge-able gap of suspicion, jealousy and mutual recrimination.[7] This has been a plausible assumption — particularly so for twentieth-century Chinese with visions politicized by the traumatic events of their own day — but it has been largely unexamined. The evidence of the following pages suggests that it must be modified. For, despite the competitiveness of the eighteenth-century examinations and the constraints of Manchu rule, the period offered Chinese a promise of unprecedented prosperity and security. So great was this promise that it seemed to many intellectuals outside the government and within it that the realization of the age-old ideals of Chinese society was within reach. It must have been frustrating for some to contemplate the realization of these ideals under foreign rule. But the history of the *Ssu-k'u ch'üan-shu* project affords many examples of scholarly factions successfully turning imperial initiative to their own ends, scholars and bureaucrats working together in occasionally uncomfortable but not unbearable tandem, and schools of learning in violent competition with each other for imperial sanction for their versions of truth. To be sure, there were differences between scholars and the state in eighteenth-century China, but perhaps our emphasis should be on tensions overcome rather than on irreconcilable conflict.

In particular, the pages of the *Annotated Catalog* of the imperial library bear testimony to the success of one group within the scholarly community, the group of intellectuals centered around Chu Yun and Tai Chen in Peking, in using the imperial compilation project for their own needs. Not only did the effort serve as a basis for their own textualist studies, the comments in the *Catalog* on specific books reflected their viewpoints. Moreover, the agenda and desiderata for intellectual life set forth in the *Catalog* were precisely those which Tai, Chu and their followers prescribed. The Chinese scholarly world largely followed this agenda for the next thirty years, until the pressures of domestic violence and foreign pressure turned Chinese attention to other questions in the mid-nineteenth century. So completely enthroned were Chu's views, in fact, that a reaction was provoked which has come to be known as the Sung learning. The debate between the followers of Chu Yun and the Sung learning was not over the propriety of government patronage of scholarship, for the scholars of the Sung learning were as willing to accept imperial patronage as their antagonists. It was over issues seemingly more important to eighteenth-century intellectuals, epistemological questions of the relation of classical texts, metaphysical insights and historical and philosophical truths. How could intellectuals best use the opportunities presented to them by material prosperity to reconstitute ancient truths? Chu Yun and his followers, sometimes styled the empiricist forerunners of a Chinese renaissance, felt this could be done through the examination of texts; their Sung learning opponents argued for a reaffirmation of traditional moral codes and metaphysical truths. The conflict between officially sanctioned textualism and Sung learning, which persisted far into the nineteenth century, can be seen in embryo in the deletions and revisions made in the draft reviews for the *Annotated Catalog*.

In view of the close collaboration of scholars and the state in some parts of the Ssu-k'u project, our traditional conception of the campaign of censorship known "as the literary inquisition of Ch'ien-lung" may need to be reconsidered. Existing interpretations of this campaign see it as part of a concerted campaign by Manchu rulers to establish firmer control over the lives and thought of their Chinese subjects.[8] But what of the many Chinese who participated in the "inquisition"? And why, in the midst of a reign of unchallenged power and prosperity, would an emperor so sensitive in other areas to the wishes of the Chinese elite embark on such a campaign? Archival sources suggest some of the answers. The Ch'ien-lung Emperor cannot be exonerated from the charge of trying to glorify his reign and those of his Ch'ing predecessors, or of trying to efface from the historical record the memories of Sino-Manchu unpleasantries in days gone by. But neither can the Chinese elite be exonerated from the charge of joining in the effort, as they saw before them a new mechanism for settling old scores. The contribution of both elite and the monarchy to the campaign of censorship must be recognized: ignited by Manchu sensitivities, the campaign acquired its malignant momentum as many of the tensions and conflicts within the Chinese elite were swept into it.

The history of the *Ssu-k'u ch'üan-shu* project and the subsequent campaign of censorship was thus mainly one of cooperation and collaboration, with only occasional conflict between the Chinese elite and their Manchu rulers. It is the purpose of this book to explore the motives of the various participants in this collaboration, the terms on which it was carried out, and its results for China's eighteenth-century history.

The interaction of scholars and the state is only one part of the eighteenth-century story. It is surely an important one, for gentry intellectuals formed the backbone of the Ch'ing state. But much more work must be done before our picture of the late eighteenth century is complete, or offers even a fraction of the richness of our view of the nineteenth century in China or of almost any European country during a comparable period. The outlines below are, therefore, as much hypotheses as conclusions, hypotheses offered in our continuing quest to understand the reality of life in China as it entered the modern era.

2

The Imperial Initiative

"A golden age of poetry and power . . ."[1]

The edict initiating what was to become the *Ssu-k'u ch'üan-shu* project was issued on 2 February 1772.[2] The emperor asserted that he had always looked to the writings of the past for inspiration in governing, and had labored daily and diligently to master and apply the precepts of the classical canon. Since the beginning of the reign, the edict noted, editions of the classics and the standard histories had been printed at government expense and distributed throughout the empire. But there might be other titles useful for guiding word and deed that had been overlooked. Therefore, governors-general, governors, and education commissioners were ordered to instruct subordinates to search the book collections in their respective jurisdictions and send lists specifying the author and contents of every valuable book they found to Peking for inspection by court officials. When provincial officials found books of particular value or significance, they were to borrow them and make copies, being sure to return the originals to their owners. By this means would the emperor's fondest wishes be fulfilled.

But what were those wishes? From the point of view of the modern reader, much was left unsaid in the edict, especially about motives and methods. Was this to be a project of thought control or literary patronage? What place was the effort meant to occupy in the long and not always harmonious history of interaction between Chinese scholars and their Manchu rulers? What special concerns motivated the emperor in beginning such a project, and why did he do so only in 1772, after he had been on the throne for thirty-six years?

To answer these questions one must look not so much to the edict itself as to the tradition of imperial scholarship and book collecting that it was meant to invoke, and the character of the emperor who sought to invoke it. This tradition has not been much discussed in the Western literature on China's political and intellectual history. Yet it was an important one, for the association of wisdom, kingship and the articulation of cultural identity was a fundamental element of legitimacy in traditional China. Emperors of China were not only political leaders, they were sages and stewards of the classical canon; as such they had very different prerogatives over scholarship and intellectual life than those to which we are accustomed in the West. Those prerogatives and the philosophical presuppositions on which they were based defined the parameters not only of thought control and literary patronage, but indeed of intellectual life itself.

Pre-Ch'ing Policies and Prerogatives

The closing lines of the edict of 1772 referred to book collection projects undertaken by Chinese governments during the reigns of Han Ch'eng-ti (r. 32–6 B.C.) and T'ang Hsuan-tsung (r. 713–756). But these were only the largest of the book collection projects sponsored by nearly every Chinese dynasty.[3] The scope of these efforts varied, according to the capacities and pretensions of the rulers who ordered them.

The First Emperor of China (r. 221–205 B.C.) adopted a policy toward the written word which all subsequent rulers would know but none would dare emulate. At the end of the second century B.C., as Hu Shih (1891–1962) has remarked, the "age of idle speculation had passed, and the problem of the day was how to govern the newly created empire."[4] To prevent intellectuals from criticizing "the government in the light of their own teachings," and casting "disrepute on the ruler," it was ordered in 213 B.C. that:

> ". . . all books in the bureau of history, save the records of Ch'in, be burned; all persons in the empire, save those who hold a function under the Bureau of Wide Learning, daring to store the *Shih*, the *Shu*, or the discussions of the various philosophers, should go to the administrative and military governors so that those books may be indiscriminately

burned. Those who dare to discuss the *Shih* and the *Shu* among themselves should be [executed and their bodies] exposed in the market
place. . . . Books not to be destroyed will be those on medicine and phar-
macy, divination by the tortoise and milfoil, and agriculture and arbor-
culture."[5]

The destruction which followed this order was probably not as great
as has commonly been alleged, since the edict was only in effect for
five years and several categories of books and book owners were
excluded. On the theoretical level, however, the edict made several
important assumptions. First, Ch'in Shih-huang's order showed,
albeit in a negative way, the importance of the written word for Chinese governments. Writings must have been perceived as posing
some threat to the government for such eminently practical men as
the First Emperor of Ch'in and his ministers to order a proscription.
Moreover, the rulers of Ch'in clearly respected certain categories of
books, those which could directly contribute to useful functions such
as curing the sick or divining the future; and were willing to permit
certain categories of officials, those attached to the Bureau of Wide
Knowledge, to see all books. The goal of Ch'in rulers was not to
destroy all writing, but to unify and harness knowledge in pursuit of
their own objectives. Scholarship was too important to be left to
scholars.

The policies of the Han dynasty, though often represented as
having developed in reaction to the Ch'in prohibitions, were based
on many of the same assumptions. In 191 B.C., the law prohibiting
private ownership of books was formally abrogated. Han Wu-ti (r.
141–87 B.C.) "set plans for the restoring of books and appointed
officers for transcribing them, including even the works of the various philosophers and the commentaries, all to be stored in the imperial library." Under Han Ch'eng-ti an imperial emissary was sent out
to collect books throughout the empire and four collators were appointed to review the texts received.[6] The most famous of these were
Liu Hsiang and his son Liu Hsin (43 B.C.–A.D. 25) who were in
charge of the classics and commentaries on them, the various philosophers, and poetry.[7]

Unlike the book burners of the First Emperor, Liu and his colleagues did not mean to stifle intellectual dispute. In fact there were
important debates throughout the Han, not only about the authenticity of various texts but also the lessons that could be drawn from
them.[8] But from the point of view of modern textual criticism or

even that of eighteenth- and nineteenth-century New Text Confucianism, the Lius' impact on scholarship was as pernicious as Ch'in
Shih-huang's.[9] For in correcting and evaluating texts in the name of
the emperor they had an enormous, perhaps determinative effect on
subsequent scholarship. While the policies of the Former Han and
the Ch'in dynasties had opposite effects, the one preserving books
and the other destroying them, they were based on a common
assumption: that the condition of scholarship—those arts by which
truth was preserved and proclaimed—was a proper and important
concern of the government. This assumption was, of course, the
mirror image of the Confucian view that the state of the government
was a proper and important concern of the scholar. But the notion
that the state had a role to play in the intellectual life was not solely
of Confucian inspiration for, as Ch'in Shih-huang's policies demonstrated, it was held by Confucians and Legalists alike. If, as has often
been suggested, the arts of ruling and writing developed together in
ancient China, then a sense of the basic unity of the two acts may
well have underlain both the Confucian and Legalist views of scholarship and government.

The Chinese conception of the unity of scholarship and government helps to explain the traditional evaluation of the scholarly
projects sponsored by various regimes which ruled China during the
ages of upheaval between the Han and T'ang dynasties. In 159,
under Han Huan-ti (r. 147–168), a special division of the Chinese
government known as the *pi-shu-chien,* was created to collect and collate books on a regular basis.[10] Since book collecting was by this time
seen as a legitimate activity of government, most rulers of north and
south China for the next several hundred years recreated this division of government, and undertook the publication of catalogs. After
the fall of the Han in 220, however, Chinese governments were short-
lived, seldom remaining in power long enough to complete a
thorough catalog. Furthermore, in the warfare and chaos of the era,
the imperial library was dispersed several times. Therefore the catalogs produced were usually short and unreliable. However, the criticism of these catalogs throughout Chinese history has been not that
they were inaccurate, but that they were merely catalogs, that is, the
regimes which sponsored them made no effort to assess or improve
upon the books they collected. Evidently, a government's ability to
collect books, correct texts, and pass judgments on them had become
a sign of its power, and a regime which failed to do this was somehow

inadequate. Writing and ruling had become, in some sense, opposite sides of the same coin, characteristic and interrelated expressions of landed literati dominance of imperial China.

The Sui (589–619), anxious to demonstrate its power in this as in so many other respects, embarked on a book collection project that ultimately involved three stages. In the first, undertaken before the dynasty had unified China, an emissary was sent out to collect rare books, offering one bolt of silk for each *chüan* collected; the originals were then copied and returned to their owners. Then, beginning in 589, scholars were summoned to collate and copy the texts collected. Finally, officials of the Sui dynasty judged the value of the books they had collected, divided them into three classes and issued a catalog entitled the *Ta-yeh cheng-yü shu-mu* (Catalog of the great enterprise era, compiled by imperial command).[11]

Another enormous book catalog, listing some 48,167 *chüan* was issued during the reign of T'ang Hsuan-tsung (r. 713–755) and was entitled the *Ch'ün-shu ssu-lu* (Catalog of the assembled books). As Yao Ming-ta points out, however, this catalog was not compiled in a day. T'ang bibliographers worked on the ample collection assembled by the Sui. Furthermore, the Department of the Imperial Library in T'ang times, which produced its first catalog under T'ang T'ai-tsung (r. 627–650), was staffed with one director, two subdirectors, one assistant, three secretaries, ten revisers of texts and four editors.[12]

The Sung dynasty produced two important catalogs of the imperial collections. The first, entitled *Ch'ung-wen tsung-mu* (Catalog of the imperial library at Ch'ung-wen gate) was edited by Wang Yao-ch'en (1001–1056) and issued in 1041.[13] It is one of the few imperial book catalogs available today, having been reconstructed by the Ssu-k'u editors from fragments preserved in the *Yung-lo ta-tien*. The Sung imperial library evidently functioned as something of a lending library, for the *Sung-shih* records several complaints by imperial librarians that books had been borrowed from but not returned to the collection.[14] The collection was largely destroyed by Jurchen invaders in 1126. Once reestablished in Hangchow, the Department of the Imperial Library tried to reconstitute its holdings "by making known its needs to bibliophiles in official circles, by sending directors on scouting trips, and by using the administrative hierarchy of the government, which extended throughout the empire." In 1177–1178, a catalog of the reconstituted library, entitled the *Chung-hsing kuan-ko shu-mu* (Catalog of the libraries of the restoration), was issued.[15]

Despite these achievements, the practice of cataloging and correcting texts in the imperial library declined after the Sung. This may be attributed in part to a lack of interest in things literary on the part of the Mongol rulers of the Yuan or the first rulers of the Ming. But other factors, including the changing social and geographical character of the intellectual community and the growth of printing were important in shaping the new pattern of relations between the scholar and the state.

According to the seventeenth-century historian Wan Ssu-t'ung (1633–1683), the Yuan imperial library was not at all inferior to those of preceding governments. Moreover, the Yuan government made a number of important innovations in intellectual life. Wan wrote:

> [The Yuan rulers were] particularly to be commended for their practice of allowing the writings of Confucians in the towns and prefectures to be handed up through the provincial administrations for the criticism of Hanlin academicians. The provincial governments of Chiang-che and other areas were ordered to print and circulate those worthy of publication. . . . The *Wen-hsien t'ung-k'ao* by Ma Tuan-lin rose to prominence by this means. In other cases, authors were rewarded with a rank or stipend, a custom unprecedented in ancient or modern times.[16]

The Mongol invasions and the severity of early Yuan rule hastened the move of many of China's intellectual elite from the capital to their native districts. The Yuan rulers of China, at least during the last years of their dynasty, were no less concerned than their predecessors about the state of scholarship in their realm, but they confronted an intellectual community far more diverse socially and geographically than any confronted by previous regimes. Scholarship was no longer the exclusive preserve of a small corps of aristocrats who looked to the capital for leadership, but the vocation and avocation of people of many different backgrounds and concerns spread over a wide geographical area.

Printing, which came into widespread use during the Sung dynasty, served as both a sign and spur of this development. Ch'ien Ta-hsin (1728–1804) noted in the Preface to his *Pu Yuan-shih i-wen-chih* (Supplementary treatise on bibliography for the Yuan history):

> Books before the T'ang were all produced by handcopying. During the Five Dynasties period printing first appeared. In the Sung, both private and public printing presses flourished. In addition to the standard editions of the classics collated and printed in the imperial library, there were Chekiang editions, Fukien editions, Szechwan editions and Kiangsi

editions. Scholars often boasted to each other of having bookshelves filled with both publicly and privately printed books.[17]

The growth of printing made imperial book collecting at once more difficult and less necessary: more difficult since many more versions of a given text had to be collected and checked before a standard version could be established; less necessary because many fairly accurate texts were now available. The Sung and post-Sung governments were no less concerned about the content of books than earlier governments, and practiced both pre- and post-publication censorship.[18]

The changing relationship between imperial authority and scholarly production was illustrated in the major scholarly work of the Ming, the compilation of the *Yung-lo ta-tien* (Grand encyclopedia of the Yung-lo era). Actually, this was an encyclopedia not a book catalog, but it was so often cited, both as precedent and negative example, by the Ch'ien-lung Emperor and his editors that its history deserves a brief consideration here. The compilation apparently began in the last year of the reign of Ming T'ai-tsu (r. 1368–1398), when the emperor ordered T'ang Yu-shen (fl. 1400) to prepare a classified compendium (*lei-yao*) of important language from the classics, histories, and philosophers. T'ai-tsu's death, however, touched off a succession struggle between his two sons, Chu Ti and Chu Yun-wen. While many Confucian scholars apparently sided with Chu Yun-wen, Chu Ti, having a larger army at his command, eventually won out. Shortly after he was enthroned as the Yung-lo Emperor (r. 1402–1424), intellectuals at court led by Hsieh Chin (1369–1415) memorialized proposing what was probably a continuation of the earlier project: "The past and present affairs of the empire are all recorded in books, but the chapters and volumes are so many that they cannot be easily consulted. We propose pulling out the information from each text and organizing it according to a rhyme scheme to facilitate its recovery." The Yung-lo Emperor accepted the proposal, hoping, as some modern historians have suggested, that the project would occupy the Chinese literati and defuse their opposition to his rule. Within a year, a draft entitled the *Wen-hsien ta-ch'eng* (Great compilation of source materials) was completed. But since this draft was found to contain many errors, a revision was ordered. In 1409, the revised text was presented to the emperor, who wrote a preface for the work and named it the *Yung-lo ta-tien*.[19]

Since the editors of the *Yung-lo ta-tien* copied into their compilation many works which were later lost, scholars since the fifteenth century have sought, often avidly, Yung-lo texts. But the original goal of the Ming effort was not so much to produce texts for a growing intelligentsia as it was to extract and condense the products of such an intelligentsia for use by the court. All imperial book collection projects, of course, had as one goal the collection of precedent and source material useful for the ruler. But the effort expended in earlier projects suggested, as has been argued above, that earlier governments were concerned with the state of scholarship and recognized that the products of scholarly effort had a value in themselves. The Yung-lo Emperor's rhyming reference guide, never actually printed, was clearly meant for narrower, more utilitarian ends. Thus, at the same time that private scholarship was becoming more independent and diverse, the government's scholarly efforts were becoming more narrow and limited.

The Ch'ing Context

The Ch'ien-lung era *Ssu-k'u ch'üan-shu* project thus represented a revival of the long tradition of centralized imperial book collecting, but it was a revival carried out in a new scholarly atmosphere, one in which scholars had developed a certain ambivalence toward public scholarship, "as careers in literature, the fine arts and private scholarship gained recognition as full-fledged alternatives."[20] The reasons for this revitalization of dormant traditions undoubtedly lay in the character of relations between the Manchu rulers and their Chinese subjects. The Manchus were a tiny minority who were ethnically and, at least for the first decades of their dynasty, culturally distinct from the vast population they ruled, and these facts very significantly affected the ideological justifications they fashioned for their rule. Ultimately, Manchus defended their dominance by pointing to Confucian texts which emphasized the right of the virtuous, regardless of ethnic extraction, to govern and by stressing their own political and moral qualifications. The Ch'ien-lung Emperor's reestablishment of the practice of imperial book collecting in the Ssu-k'u project was undoubtedly a part of this ideology of the reign of virtue. It would be ahistorical to argue, however, that the imperial initiative in the Ssu-k'u project reflected the goals or capacities of all Manchu rulers. As the work of Ono Kazuko, Thomas Fisher and Lynn Struve has shown,[21] Manchu ideological claims and Chinese scholars' attitudes toward them underwent considerable change during the early years of the dynasty. The Ssu-k'u project was the culmination of a long historical evolution of ideas and attitudes, during which many traditional assumptions about the relation of the scholar and the state were questioned, some discarded and some reaffirmed. The initiatives of Ch'ing emperors in the area of intellectual life reflected this complex evolution.

When the Manchus began to rule China proper in the mid-seventeenth century, they were confronted with three challenges from intellectuals. The first came from voluntary organizations of literati (*wen-she*) formed in increasing numbers in the late Ming, partly to influence government policy, but more importantly to influence civil service examiners and thus increase their members' chances in the competition for office. While literati associations had existed throughout Chinese history, the character and vigor of the late Ming groups was a relatively new phenomenon which reflected the growing independence of the scholar from the state in the late imperial era, the increasing competitiveness of the examination system, rising levels of popular education and a decline in the quality of the official school system.[22] Ming intellectuals repeatedly warned of the dangers of these societies and of literati infringement on the personnel and policy prerogatives of rulers, and these warnings were certainly not lost on the Manchus. A second challenge to the new regime from the intellectuals was explicit, ethnic anti-Manchuism. The difference in language, social manners, hairstyle and diet between the Manchu overlords and their Chinese subjects made Ch'ing rulers an easy target for Chinese, who had always had a strong sense of the identity and centrality of their agrarian lifestyle in the East Asian environment. Some Chinese were, no doubt, genuinely offended by the prospect of barbarians "stuffed with rudeness and recklessness" and "no different from birds and animals" ruling the Middle Kingdom. Others were probably using the fact of ethnic differences for their own political purposes. Almost constant Manchu efforts to suppress this racist idiom of political protest make it difficult to tell how widespread such thought was or what end it served, but there is no question that it represented an extremely powerful political weapon. Finally, Manchus were confronted with the challenge of having to learn to use effectively the abstract and literary language of Chinese government, including the proper ways of manipulating the classical canon to justify their own rule. Although these challenges presented themselves in different ways, from the Manchus' point of view the education, cliquishness, and ethnic hostilities of the Chinese literati were several dimensions of a single problem: how to dominate a literate and highly sophisticated Chinese elite? At least until the late eighteenth century, Ch'ing policies toward scholars and intellectual life must be evaluated against the background of this continuing concern.

Once established in Peking, the Manchus moved swiftly and forcefully to assert their control. Their first moves were designed to prevent literati from conspiring with each other and with local officials to influence government policies. In 1651, all officials were ordered to keep a record of all the visits which *sheng-yuan*, licentiates or holders of the first degree, made to their offices, and the business on which the *sheng-yuan* came. In the same edict, the formation of new academies, and holding meetings of more than ten literati were prohibited.[23] The next year, educational intendants (*t'i-hsueh-kuan*) were appointed in the provinces under Ch'ing control to lecture and exemplify the teachings of the classics and to oversee the studies and behavior of Chinese scholars.[24] The central government also ordered that a stone tablet be erected in the courtyard of every academy on which were to be inscribed eight prohibitions on student activity. Students were ordered not to "argue unreasonably with their teachers, organize fellowships or associations, publish their writings without authorization, exert pressure on local organizations, or hector or dictate to fellow villagers."[25]

The draconian spirit of early Manchu policies toward scholarship and literati was most clearly demonstrated in the regulations the new rulers formulated for publication and book-selling. The government issued in 1653 an edict that hardly boded well for imperial book collecting:

> Discussions of the classics ought to be based on the editions and interpretations of the Sung dynasty. In official documents, clarity and correct citation of classical precedents are most important. From now on, let those in charge of students charge educational officials with the responsibility of teaching students to know thoroughly and explicate carefully such works as *Tzu-chih t'ung-chien* and the *Tzu-chih t'ung-chien kang-mu*, the *Ta-hsueh yen-i*, the *Li-tai ming-ch'en tsou-i*, and the *Wen-chang cheng-tsung*. . . . Those works which stealthily advance strange principles or heterodox theories, or which self-consciously set themselves apart [from the classics], even though their prose is artful, should not be copied out by students. Bookstores should be allowed to print and circulate only books

of [the Sung school of] principle, works on government, and those books which contributed to the literary enterprise. Other works, with petty concerns or immoral language, and the publication of all literary exercise books or collections sponsored by literati associations are strictly forbidden. Those who violate this prohibition will be punished severely.[26]

Little changed in the early years of the dynasty's second reign. The regents for the young K'ang-hsi Emperor (r. 1661–1722) were, as Robert Oxnam has argued, survivors of the Manchus' "long march to power" whose "distrust of the Chinese elite blinded them to the possibility of using Chinese norms as a means of control."[27] The regents' hostility was manifested most clearly in a drastic reduction of the numbers of Chinese who were allowed to pass the triennial *chin-shih* examinations and in two famous prosecutions, the Kiangnan Tax Arrears case and the suppression of the works of Chuang T'ing-lung (d. 1600). In the first of these cases, the regents' ire was directed at the gentry of the prosperous, populous and often recalcitrant southeastern provinces of China who were resisting a reform in the tax collection system.[28]

The Chuang T'ing-lung case illustrated the atmosphere of suspicion and mutual incrimination which such rigid policies could engender. Some twenty years after his death it was discovered that Chuang, a literatus from Chekiang, had once acquired a private history of the late Ming and the loyalist regimes which followed its downfall and had proceeded to assemble a group of scholars to prepare a definitive edition of the book. The regents reacted to this discovery with a severity that sent waves of indignation and fear through the Chinese intellectual community: Chuang T'ing-lung and his father were both disinterred and beheaded, their works were banned, and Chuang Yun-ch'eng and over seventy printers, collaborators and relatives were sentenced to death or banishment.[29]

Such enmity between court and literati could not last for long. The political unity of traditional China rested on a delicate compromise between the interests of literate elites and those of the central government. The split between court and literati which developed during the early Ch'ing government, Hellmut Wilhem has observed, "had it been allowed to coalesce into a condition of permanency, might have undermined the Chinese governmemt and Chinese society as well."[30] As in so many other areas of Ch'ing history, the policies of the court toward intellectuals and intellectual life began to change in the years after the regency.

The best known of the K'ang-hsi period initiatives toward intellectuals was the *po-hsueh hung-ju* examination held in 1679. The most famous Chinese scholars of the day were invited to participate in this test, and those who passed were appointed to a commission to edit the Ming history. The young emperor demonstrated his interest in Chinese scholars and scholarship in other ways as well. Beginning in 1671, he attended daily tutoring sessions in the Chinese classics. In 1676 the first of a series of texts of lectures presented to the emperor on these occasions was published, with an imperial preface. In 1677, the *Nan-shu-fang* (Southern Study) was established where the emperor could meet with Chinese scholars on an informal basis and seek their advice.[31] The new spirit which infused Ch'ing policies was well illustrated when the regulations for publication and book selling, first issued in 1653 (quoted above), were reformulated in 1687 and 1714. Publishers were still forbidden to print or sell "salacious and immoral novels," but the references to publications of literary associations and works which "self-consciously set themselves apart from the classics" were eliminated. Furthermore, the new regulations were justified in terms of an emperor's duty to care for the hearts and minds of his subjects: "[In Our view] the customs and desires of Our subjects form the only basis for rule of the empire, and in order to purify those desires and strengthen custom, one must respect the classics. . . ."[32] With such language, the Ch'ing rulers proclaimed their wilingness to govern by the traditional rules of Chinese society.

The most evident results of the new imperial policies were a series of commentaries on the classics and original philosophical texts which were commissioned by the court, and set the philosophical tone of the capital. As the emperor reached out to the scholarly community, he turned to that branch of learning most favored by earlier Chinese dynasties, the school of "principle" or Neo-Confucianism.[33] Not only was the great synthesis of Neo-Confucian thought, the *Hsing-li ta-ch'üan* (Grand compendium of principles and morals) reprinted under imperial patronage, the emperor urged officials to read it and occasionally tested them on its contents. Another Sung text on government, the *Ta-hsueh yen-i* (Implications of the Great Learning) was translated into Manchu and distributed to all Manchu princes. Later in the reign, special homage was paid to the twelfth-century philosopher Chu Hsi, who was styled one of Confucius' most important disciples.[34] The Neo-Confucian flavor of official scholarship in the K'ang-hsi era served a variety of purposes;

probably, however, it did not reflect so much a conscious choice of philosophical tone as it did a commitment to certain individual scholars, men like Li Kuang-ti and Hsiung Tz'u-li and their vision of the appropriate principles for good government. Indeed, the personal tone of K'ang-hsi era imperial patronage represented both its strength and its weakness. For those scholars who received imperial patronage, the experience was exhilarating. Bridges were built between the court and famous scholarly families and communities, and a rather remarkable ambiance was created at the capital. The era was one in which "men rose by having their actual performance and practical proposals noted," and the premium was on productivity and creativity.[35] The buoyant atmosphere at court lingered long in the memories of those who lived through it, and served to inspire many, probably most importantly the K'ang-hsi Emperor's grandson who reigned as the Ch'ien-lung Emperor.

There were, however, at least two weaknesses of the K'ang-hsi era patronage system. The first was its limited extent: while many at the capital partook of the imperial largesse, scholars outside Peking, including some very creative minds, did not. Men like Yen Jo-chü and Hu Wei certainly did not actively resist the Ch'ing government, and had many official and semi-official contacts with court and officials. But they were not actively drawn into capital intellectual life, and their research lay in areas other than those being explored in Peking. When Mao Ch'i-ling heard of the attention being showered on Chu Hsi and Neo-Confucianism at court, he rather reluctantly destroyed his study of errors in Chu Hsi's commentaries on the Four Books, on which he had been working.[36] The limited scope of allegiance to imperial scholarly leadership was made graphically apparent in an episode which took place in 1686. Evidently impressed by the efforts of earlier dynasties to collect books, the K'ang-hsi Emperor sought to enlarge the scope of his own imperial library, reflecting that:

> Since ancient times rulers have exalted the written word in government. Even when all the laws and classics are assembled it is necessary to seek broadly the dispersed books in order to expand what we know and see. The proper citation of classical precedents is also an important matter. I have always devoted much [of my energy] to art and literature, and read day and night. Although the library of the imperial household is fairly extensive, the collection is not complete.
>
> It has occurred to me that many important manuscripts are probably

stored in the large cities of the empire. Moreover, how could there not be rare books stored in the countryside? A broad search ought to be made for all the classics, histories, and works of literature and philosophy, except for commonly available printed editions. The preparation of a catalog of the imperial library [in which is recorded the results of such a search], and the copying out of handwritten manuscripts are also tasks which ought to be accomplished. Let the Boards deliberate carefully and memorialize.[37]

A few days later, the Board of Rites memorialized, as ordered, suggesting methods by which the proposed aim could be achieved. But ultimately, nothing came of the proposal, no books were collected and no catalogs prepared. One tantalizingly brief reference to the project in the corpus of the seventeenth-century scholar Ku Yen-wu suggests one reason why. Book owners, Ku wrote, "tired of responding to orders and the increasing numbers of central directives, and reported that they had no books." Clearly, there were limits to the recognition of the K'ang-hsi Emperor's role as leader of the scholarly community, limits which K'ang-hsi's grandson would labor to eliminate.[38]

A second weakness of K'ang-hsi's personal system of patronage was the factionalism and favoritism that it fostered. The K'ang-hsi Emperor complained on several occasions of the grudges that various scholars bore toward one another, and the insidious effects these could have on scholars' political behavior. Even early in the reign, the seventeenth-century scholar Ku Yen-wu warned his nephew P'an Lei (1646–1708) against becoming involved in the world of court scholarship where so much time was spent currying favor and "the fulsome and the fragrant were mixed."[39] Toward the end of the reign, the scholarly factions came to be associated with imperial favorites or, more ominously, with political factions active in the succession struggle. The famous dictionaries *P'ei-wen yun-fu* (Repository arranged by rhyme of the studio of adorned literature) and *K'ang-hsi tzu-tien* (K'ang-hsi dictionary) were both edited by Chang Yü-shu (1642–1711), an imperial tutor who accompanied the emperor on his expeditions against Galdan and his fourth tour of the south. Scholars associated with the emperor's eighth son produced the famous encyclopedia *Ku-chin t'u-shu chi-ch'eng* (Synthesis of books and illustrations, past and present) and a collection of works on mathematics, music and astronomy entitled *Lü-lü cheng-i* (Oceans of calendrial and accoustic calculations).[40]

It was the problem of favoritism and factionalism which eventually brought the era to an end. The prosecution of Tai Ming-shih (1653–1713) in 1711 has been seen as signalling the twilight of the age.[41] Tai, a soured and withdrawn self-declared misfit, spent most of his life trying to pass the civil service examinations, succeeding only when he was fifty-six *sui*. After serving only two years as a Hanlin compiler, Tai was charged by a junior censor with writing a history of the late Ming dynasty in which the leaders of Ming loyalist regimes in the 1640s and 1650s were treated as emperors. The punishments were severe, including execution for Tai and his closest collaborator, and banishment and enslavement for many relatives and associates. Although the precise motives for the accusation against Tai remain a mystery, they probably had something to do with the vicious struggle then going on among the sons of the K'ang-hsi Emperor for the heir-apparency.[42]

The factionalism of the K'ang-hsi pattern of imperial patronage of scholarship would condemn it almost irredeemably in the eyes of the K'ang-hsi Emperor's son and successor, the Yung-cheng Emperor. The actions taken early in the Yung-cheng reign against Ch'en Meng-lei (b. 1651), compiler of the *T'u-shu chi-ch'eng* suggested some of the new emperor's concerns and motives. According to a tradition written down in the nineteenth century, Ch'en, who was then serving as secretary to the emperor's eighth son, Yin-chih, actually compiled the *T'u-shu chi-ch'eng* privately and presented it to the K'ang-hsi Emperor. The emperor, so the story goes, favored it with a preface and ordered it printed as a product of imperial scholarship. When Yin-chih lost out in the competition for the throne, and the emperor's fourth son was enthroned, Ch'en was accused of "overweening arrogance," and banished "in order to preserve harmony." Insofar as the *T'u-shu chi-ch'eng* was a product of the previous emperor, it was praised as a "prodigy of literary compilation." However, a new commission of scholars was appointed to correct the mistakes and add the "requisite finish" to Ch'en's work, so that the deceased emperor's sublime intention would be fulfilled.[43]

The condemnations of the last years of the K'ang-hsi reign and the first years of the Yung-cheng reign marked an important change of emphasis of the part of Manchu rulers. The Yung-cheng Emperor, veteran of the bitter succession crises of the early 1700s, was particularly suspicious of those scholars who had supported his rivals and of Chinese factions in general. Instead of extolling the virtues of individual Confucians, the new emperor expounded Confucian virtues to the empire as a whole. Projects of imperial scholarship, which usually involved patronage of individuals, declined.[44] In the first year of the reign, the membership of the Hanlin Academy, where most imperial scholars were formally employed during the K'ang-hsi reign, was cut in half.[45] In the place of the imperial patronage system, a system of stipends paid to all students in official academies, regardless of their productivity, grew up. At the same time, a new network of government school inspectors (*hsueh-cheng**), personally chosen and supervised by the emperor, was established to judge the recipients of the stipends.[46]

Under the Yung-cheng Emperor, scholars were supported by the state as they had never been before; but they were also lectured to as never before. In 1724, for instance, the famous edict on factions was issued. The emperor reflected that:

> . . . just as Heaven is exalted and the Earth is low, so are the roles of prince and minister fixed. The essential duty of a minister is to be aware that he has a prince. For then his dispositions will be firmly disciplined, and he will be able to share his prince's likes and dislikes; hence the saying, "One in virtue, one in heart, high and low are bound together."

The emperor took issue with the Sung scholar Ou-yang Hsiu's view that factions were evidence of a scholar's commitment to higher principles:

> "Superior men [Ou-yang Hsiu had written] form cliques because they share the same *tao*." But how can there be *tao* among them when they deny their sovereign and work for their own interests? Ou-yang Hsiu's *tao* is simply the *tao* of inferior persons.[47]

Yung-cheng's argument that, in effect, it was the ruler's prerogative to define the *tao* was quite traditional, but his energy in proclaiming the principle demonstrated one of the dominant emphases of his reign. This edict was made required reading at the monthly meetings of the officials academies in which all holders of the first official degree were enrolled.

The sedition trial of Tseng Ching afforded the Yung-cheng Emperor another opportunity to address Chinese scholars on the subjects of Confucian virtue and Manchu legitimacy. In this case, Tseng Ching (1679–1735), an undistinguished first degree holder from Hunan, attempted to incite a rebellion using the works of the seventeenth-century Ming loyalist scholar Lü Liu-liang (1628–1683) as his text. The attempt met with little success, and soon came to the attention of provincial governor O-erh-t'ai and eventually of the emperor. A massive investigation was launched. Two long refutations of Tseng and Lü, the *Po Lü Liu-liang ssu-shu chiang-i* (A refutation of Lü Liu-liang's commentary on the Four Books) and *Ta-i chueh-mi-lu* (Record of the righteous way for enlightening the misguided) were prepared, evidently under the emperor's personal guidance, in what Thomas S. Fisher has termed the most extensive "media campaign" ever launched in a non-modern society. Both refutations asserted the Manchus' right to rule, and reminded Confucian scholars of their duty to family and ruler. The *Ta-i chueh-mi-lu* was the more successful, directly attacking Lü's ethnic anti-Manchuism and proclaiming that: "the virtue of loyalty to the state was foremost among all human relationships."[48]

By the end of the Yung-cheng reign, therefore, the outlines of Ch'ing policies and dilemmas toward intellectuals were fairly well established. The question of how a Manchu emperor could express his respect for Chinese literati without surrendering any of his authority had been given several answers. The stern measures of the early years had preserved Manchu authority, but at the cost of excluding from government the very people who seemed to have most expertise in governing, the Chinese literati. The initiatives of the K'ang-hsi years represented an effort to woo these literati, but the informal and personal character of his wooing brought with it almost inevitable factionalism and heightened political tensions. The Yung-cheng Emperor tried to systematize the procedures for literati participation in government, sacrificing the K'ang-hsi era creativity and buoyancy for political security. While some of the actions of the Ch'ing government certainly had a repressive character—the prosecutions of Chuang T'ing-lung or of Tai Ming-shih for instance—Ch'ing policies as a whole cannot be described in these terms. Even these, the most famous literary cases in early Ch'ing history, were brought about as much by squabbling among the literati as by imperial absolutism. On the whole Ch'ing policy must be described as having been characterized by a growing recognition of the importance of Confucian doctrine and the unity of *cheng* (governing) and *chiao* (teaching) in ruling China. The continuation and, to anticipate the argument, the exaggeration of this trend during the Ch'ien-lung reign are the subjects of the next section.

Court Publication and Censorship in the Ch'ien-lung Reign

The Ch'ien-lung Emperor drew upon the precedents of patronage and court publication that his father and grandfather had laid down, but he adapted them in ways that reflected the political environment of his court and conditions in the empire which he ruled. When the new emperor ascended the throne in 1736, the times seemed to call more for the stabilization and extension of the existing order than for any new policy initiatives, and he seemed determined to keep it that way by the only method endorsed by Chinese statecraft — careful attention to ritual requirements, historical precedents and classical principles. The emperor argued in one of his premonarchical essays that a ruler's duty to instruct his people was a critical part of the mandate of heaven. His argument was a philosophical one, resting on the goodness of human nature and the weakness of man. Human nature was basically good, but since man dwelt in a world of temptation, the common man might easily fall into evil ways. Heaven had therefore created rulers and teachers to purify man's emotions and restore the fundamental goodness of the universe. The ruler accomplished this by "separating the noble and the base, establishing a social hierarchy, clearly prescribing rewards and fines, and distinguishing the honors due those of various ranks," and by promoting the rituals of community feasting, sacrifices to the altars of the soil, marriage, capping, mourning and ancestral sacrifice. In essence, every act of government had as its purpose the enlightenment of the people.[49]

Such an approach to government seemed to suit the new ruler's education, for the Ch'ien-lung Emperor came to the throne with a far more sophisticated grasp of Chinese history and the classics than any of his Ch'ing predecessors. But it also was an approach well-suited to the task before him: ruling an enormous empire through a fairly small group of bureaucrats whose main common attribute was their classical education. Like most Chinese emperors, Ch'ien-lung seldom celebrated the pursuit of knowledge for its own sake — he saw scholars and scholarship as adjuncts of government — but he clearly respected men of learning. In a very early edict, he rejected the idea that "studious" and "bookish" men were unfit for office, claiming that a knowledge of historical precedent was precisely what would be required of government servants during his reign.[50]

Initially, however, the emperor seems to have been intent to demonstrate that the austerity of the Yung-cheng reign had ended, and that the buoyancy of the K'ang-hsi reign could be recaptured. There were probably two reasons for this policy emphasis. The first was Ch'ien-lung's deep admiration for his grandfather; the future Ch'ien-lung Emperor had been the K'ang-hsi Emperor's favorite grandson, and the old emperor had personally supervised Ch'ien-lung's education. Perhaps as important, the twenty-five year old monarch found himself surrounded by men who had risen to power during the K'ang-hsi reign. Chang T'ing-yü (1675–1755), who had served in the Nan-shu-fang and the Hanlin Academy in the late K'ang-hsi years, became regent for the Ch'ien-lung Emperor and chief grand councillor. At least according to the court gossip of the period, Chang was able, from this position, to appoint many of his old friends and proteges to key positions.[51] The imperial patronage system began to work almost immediately on Ch'ien-lung's accession. In the first year of the new reign, the emperor ordered provincial treasurers to print and distribute copies of all the works of government-sponsored scholarship to bookstores within their jurisdictions. Two years later, an edict was issued encouraging private printers to reproduce and distribute scholarly works produced under government auspices.[52]

While these edicts theoretically applied to all publications of the Ch'ing, the vast majority of state-sponsored scholarly works were compiled during the late K'ang-hsi years. Only one philosophical work, a rather stern tract on the obligations of filiality entitled *Hsiao-ching chi-chu* (Collected interpretations of the *Classic of Filial Piety*) was issued during the Yung-cheng reign. The Ch'ien-lung Emperor's attitude toward this work, or at least toward the text on which it was based, was evinced in an edict of 1737. Rejecting a proposal that the *Hsiao-ching* be formally declared equal to the Four Books, the emperor accepted his councillors' review of the evidence that the text was a forgery, and their view that it did not supplement the principles of government.[53] This remarkably straight-forward pronouncement contrasted sharply with the emperor's vigorous efforts to print and distribute other government publications. Several other actions provided further evidence of the new imperial attitudes toward scholars and scholarship. In 1736, a second *po-hsueh hung-ju* examination was held, on the model of the K'ang-hsi examinations. In one of the most startling reversals of the era, Tseng Ching, who had been pardoned by the Yung-cheng Emperor, was abruptly executed, and the *Ta-i chueh-mi-lu* recalled and banned.[54]

Although he emulated his grandfather's policies in many respects, the Ch'ien-lung Emperor was just as concerned as his father had been to prevent debilitating factional conflict. Several of his pre-monarchical essays dwelt on the necessity, so often stressed by Yung-cheng, that the ruler and his ministers be of one mind on policy and morality. The Ch'ien-lung Emperor issued several edicts on factions, displaying a sensitivity to talk of party divisions. The emperor's sensitivity suggested that he knew something of the reality of factional conflict, and indeed, suggestions of such conflict can be found throughout the history of his reign. Furthermore, the emperor and his advisors were well aware of the connection between literary projects and factional politics. When Chang T'ing-yü and O-erh-t'ai, leaders of opposing factions, fell from power in 1745 and 1750 respectively, their disgrace was signalled by the implication of their followers in literary suppression cases involving Hu Chung-tsao and the works of Lü Liu-liang. Later in the reign, an edict was issued praising the contents of the *T'ung-chih-t'ang ching-chieh* (Classical commentaries from the T'ung-chih Hall), which had been compiled by Hsu Ch'ien-hsueh (1631–1694) when he was tutor to the K'ang-hsi emperor's fourth son, but decrying the partisan conflict which had surrounded its issue.[55]

In 1741, the Ch'ien-lung Emperor tried to revive the book collecting project which his father had begun:

> Since ancient times, rulers have revered literature and sought to collect lost books, and the imperial library is at present largely complete. In recent times, however, the amount of writing has grown daily. Among the scholars of the Yuan and Ming, and those of our dynasty, there are more than a few who, having studied the Six Classics and immersed themselves in the study of metaphysics, have produced work that is pure and faultless. Although these authors reside in the countryside and have not yet been discovered at court, provincial governors and education commissioners ought to devote themselves to finding them.[56]

Nothing came of this proposal, however. Most notably, intellectuals who had lived through the factionalism of the K'ang-hsi reign and the austerity of the Yung-cheng era needed more encouragement and reassurance before embarking on a project of imperial sponsorship. If so, this was what they got. In 1749, the emperor wrote:

> The members of the Hanlin Academy serve us with their talent for literary composition, and recently we have posed questions on examination involving poetic composition, requiring that much energy be devoted to the study of rhetoric.

But there are no doubt many who immerse themselves in the classics, exult in their hidden glories, and research their sublime implications. Is it that those who study the classics properly are rare, or that we seldom hear of them? Of course, reading the classics is not comparable to carrying out their principles. But knowing how to put principles into practice, and respecting classical techniques is an important part of understanding man's essential nature and the way of the world. . . .

Now the world is at peace. Among the scholars there must be some who have attained insight into the basic meaning of the classics and who have spent their lives in diligent contemplation of former Confucians. Why is it that they grow old sitting at home by their windows while at court those with a mastery of classical studies are so few?

In fact, several individuals who were nominated on this occasion rose to high positions in the Hanlin Academy.[57]

Imperial rhetoric changed little in the course of the reign, but the response to it did, as scholars became more sure of the court's intentions and more confident of their own futures. The imperial trips to the south in 1757, 1762, 1765, 1780 and 1784 were probably both a cause and a symbol of the relationship the Ch'ien-lung Emperor wanted to cultivate with the scholarly community. In these massive and well-publicized expeditions, the emperor, surrounded by courtesans and courtiers, would cruise along the Grand Canal, stopping at major points of historic or scenic interest, where he would exchange poems, books, and samples of calligraphy with the influential land holders. The trips could not have been unpleasant for anyone involved, except the hapless souls who had to pay for entertaining the imperial party along the way, but their purpose was political, to demonstrate the emperor's concern for the elite of China, and the harmony of court-elite relations. K'ang-hsi had, of course, made similar trips, but those expeditions had at least had the ostensible purpose of inspecting water conservancy projects and testing the mood of the people following the rebellion of the three feudatories. Ch'ien-lung's trips seemed to have no purpose other than celebrating the wealth, commercial prosperity, and intellectual attainments of the southeast.[58]

By the 1750s, young scholars with ability and potential were coming into the government in increasing numbers — the cohort who passed the *chin-shih* examinations of 1757 were a truly remarkable group (see chapter 3 below) — and a career in the capital bureaucracy without service on one of the imperial publication projects became unthinkable. The history of court and elite preoccupations during the Ch'ien-lung era could be told in terms of the products of imperial patronage. In the 1740s, the first products of Ch'ien-lung patronage, works on court music, painting, and geography, appeared.[59] Also among the earliest products of the era was the *Ch'in-ting ssu-shu-wen* (Examinations essays on the Four Books, compiled by imperial order) a primer in which the late K'ang-hsi period teacher and Neo-Confucian partisan Fang Pao (1668–1749) selected and annotated a group of Ming and early Ch'ing examination essays to serve as a standard for the empire.[60] At the same time, commissions were formed to prepare elaborations of the existing official commentaries on the classics. In the 1750s, some of the first new commentaries appeared: works on the *Rites* texts were published in 1754, and on the *I-ching, Shih-ching* and *Ch'un-ch'iu* in 1758.[61]

The Neo-Confucian tone of these works suggested that they represented a continuation of the K'ang-hsi pattern of patronage. In the 1740s, however, the focus of Ch'ien-lung publications began to shift from classical to historical texts. In 1747, court scholars were set to work on continuations of the three famous Chinese encyclopedias, the *T'ung-chih* (Comprehensive treatises), *T'ung-tien* (Comprehensive institutes), and the *Wen-hsien t'ung-k'ao* (Comprehensive examination of historical source materials), a project which was to continue until the end of the reign. At least part of the goal of this project was to compile, in a form acceptable to Chinese scholars, a version of Ming and Ch'ing history acceptable to Manchu rulers.[62] Circumstances during the eighteenth century accelerated the trend toward imperial history writing. As the century progressed, the central government became more concerned with military campaigns along its borders, and a new group of imperial relatives began to take the place of the aging K'ang-hsi era courtiers in the highest councils of the central government. New sorts of scholarly works were commissioned which focused on the Manchus' history and martial achievements. These included a series of publications on the history, language and customs of Mongolia, designed perhaps to celebrate Ch'ing triumphs in central Asia.[63] Beginning in the late 1750s, a cluster of works on Manchu language and history were commissioned. The *Ch'ing-wen-chien* (Dictionary of Manchu), a dictionary of the Manchu language in Manchu, was reissued in 1771, systematizing and in some sense dignifying a language which was already fading at court. The grandiloquent but often deceptive *K'ai-kuo fang-lüeh* (An account of the founding of our nation), commissioned in 1773, provided not only a heroic dimension but a Confucian patina for early Manchu history.[64] In 1749, the first of a long series of military campaigns, the *P'ing-ting Chin-chuan fang-lueh* (An account of the pacification of the Chin-chuan rebels) appeared. Here, the Ch'ien-lung Emperor seems once again to have taken a leaf from his grandfather's book, for the only *fang-lueh,* accounts of military campaigns, to appear before 1749 were the K'ang-hsi record of the suppression of the revolt of the three feudatories, and the records of K'ang-hsi campaigns against the Dzungars. A division of the Grand Council known as the Fang-lueh-kuan (Military archives office), which had been created on an *ad hoc* basis in the K'ang-hsi period, was made a permanent office in 1749 and charged with the responsibility of preparing military histories; it also bore some responsibility in the other editorial projects of the period. With the changing subject matter of imperial compilations, the locus of publication work shifted from the Board of Rites and Hanlin Academy to the Grand Council, where it was to remain until the end of the reign.[65]

To be sure, the shift from classical commentary to military history signalled a change of tone in court patronage, but the fact that the best and the brightest among Chinese scholars were employed preparing works which celebrated the collaboration of Manchus and Chinese suggests that the purpose of court literary activity had not changed. Combining the intent of K'ang-hsi patronage, to demonstrated Ch'ing commitment to Confucian rule, with the standardizing impulse of the Yung-cheng years, Ch'ien-lung court scholarship reflected a maturing of Manchu policy toward Chinese intellectuals. K'ang-hsi's publications, issued early in the dynasty, had represented a promise of present and future fidelity to Confucian principles; for the Ch'ien-lung Emperor, on the other hand, publication and patronage of scholars was an existing dynasty policy, meant to glorify the ruler and serve as a sign and pillar of the prosperity of the era. Unlike the K'ang-hsi Emperor, the Ch'ien-lung Emperor did not patronize individuals; eighteenth-century court scholarship was meant to express attitudes toward a class, and was administered through regular institutions of Ch'ing government like the Grand Council, Board of Rites and Hanlin Academy, rather than through *ad hoc* arrangements in the emperor's private study. Ch'ien-lung publications represented a commitment, not so much to scholarship in an abstract sense, nor to specific individuals, but to scholars in general and the world they represented.

Ch'ien-lung's censorship also served political as well as literary ends. An emperor as concerned as the Ch'ien-lung Emperor was with the relationship between the court and the scholarly community could be a vigilant and ferocious censor, but at least until the 1770s Ch'ien-lung censorship was tempered by a sense of political reality. Two cases early in the reign illustrated this. In 1741 Hsieh Chi-shih, a grain attendant from Hunan who had by his political obstinacy annoyed both the Ch'ien-lung Emperor and his father, was found to have written some classical commentaries which challenged the Chu Hsi interpretations endorsed in the official commentaries. The emperor ordered the works destroyed but carried the matter no further, being reluctant, as he put it in an edict on the case, to make utterances *per se* a crime. Thirteen years later, the emperor reacted quite differently when he read Provincial Education Commissioner Hu Chung-tsao's line "My emotions ponder on the corrupt and the pure," in which the word corrupt (*cho*) was placed next to the dynastic name pure (*ch'ing*). But more was at stake in the Hu case than word choice, for Hu Chung-tsao was a student of the grand councillor and imperial tutor O-erh-t'ai, and his poetry contained oblique and partisan references to the conflict then going on between O-erh-t'ai and Chang T'ing-yü. Hu Chung-tsao's poetry was felt to constitute an attack on Chang, the powerful chief councillor to the emperor, and ultimately on the emperor himself. This was intolerable; Hu was summoned to the court, interrogated, and executed. The outcomes of the two cases reflected the different combination of literary and political concerns in the cases. Hsieh Chi-shih was condemned, but ultimately pardoned and returned to official life, for focusing on what was "between the lines" in the commentaries, as opposed to what he should have been doing in the emperor's view, exerting himself to promote correct behavior in the world. Hu Chung-tsao's poetry, regardless of its literary content, was read as a political attack, and hence brought its author to grief.[66]

The two cases set informal parameters for the censorship of the reign. Whenever the emperor saw himself attacked, he did not hesitate to use all the police power at his command to prosecute rapidly and demonstratively. The cases of Ting Wen-pin in 1752 and Liu Chen-yü in 1761 provoked just such reactions. Ting, a rather unstable scholar from Shao-hsing who styled himself Master Ting (*ting-tzu*), was sentenced to death by slow slicing for presenting to a descendant of Confucius two works of his own composition. One, entitled *Wei-shih hsien-shu* (A calendar for a false age) proposed new reign titles for recent Chinese rulers; the other, titled rather more ominously *Ta-Hsia ta-Ming hsin-shu* (A new discussion of the great Hsia and great Ming dynasties) was said to be full of anti-Manchu language. A similar fate befell Liu Chen-yü, a seventy-year-old *shang-yuan* from Kiangsi who published a plan for apprehending bandits as well as a work entitled *Tso-li wan-shih chih-p'ing shu* (Toward the eternally peaceful government of the realm), which discussed the sumptuary laws and ritual systems of the dynasty, and presented them to the examiners at the 1753 Kiangsi provincial examinations. Remarkably, both the *Ta-Hsia ta-Ming hsin-shu* and the *Tso-li wan-shih chih-p'ing shu* had undergone review by provincial officials—the Ch'ing equivalent of pre-publication censorship—before the cases in question. It seems that it was not so much the content of the books which were offensive, but the claims that Ting and Liu were implicitly making in presenting their works, respectively, to a descendant of Confucius and to an official at the politically charged examinations.[67]

Where no such political claims were present, and no attacks were apparent, the Ch'ien-lung Emperor was inclined to take a much more tolerant view of works which others at his court considered misguided. In 1757 the Hunan *sheng-yuan* Ch'en An-p'ing was indicted for writing two commentaries, one on the *Ta-hsueh* (Great learning) and the other on the *Chung-yung* (Doctrine of the mean), in which he disagreed with Chu Hsi. The Hunan provincial governor was concerned with possible parallels with the case of Hsieh Chi-shih. But the emperor was more concerned with the parallels with the case of Hu Chung-tsao, that is, with whether there was any direct or implied political attack in the works. Finding none, he freed Ch'en and ordered the governor and provincial education commissioner not to pursue the case further. The emperor delivered a similar verdict on the case of Yü Teng-chiao. Yü was evidently a scholar of some repute; he was nominated to take the second *po-hsueh hung-ju* exam in 1736. He failed these exams but earned his *chin-shih* degree nine years later, rising to the post of secretary at the Board of Punishments before his retirement. Yü evidently had a talent for making enemies, and in 1761 a kinsman brought a volume of Yü's poetry to the attention of provincial authorities, who relayed it to the emperor. The emperor reviewed the volume, but found no cause for further prosecution. If the court monitored poetry too strictly, the emperor remarked in his edict on the case, not only would the authors of the empire have no way to express the concerns of their hearts, but poets would not dare to write a word.[68]

Thus, while the Ch'ien-lung Emperor never hesitated to destroy books that he saw as dangerous, until the 1770s he used his powers of censorship cautiously, for his relationship with the book-owning community was too valuable to be cast aside lightly. Censorship, like patronage, was not a capricious act for the Ch'ien-lung Emperor, but one of the institutional expressions of his relationship with the writing and book holding community. It was perhaps not until the Ssu-k'u's thirty-six thousand *chüan* of testimony to the legitimacy of his dynasty almost rested securely on his shelves that he felt confident enough to undertake a search for the writings of those who had once questioned his rule.

The Edict of 1771

In February of 1771, the Ch'ien-lung Emperor made a second effort to organize a book collecting project. Two factors probably precipitated the decision to embark on a book collection project in this year. The first was the simultaneous celebration in 1771 of the emperor's sixtieth birthday and his mother's eightieth. The event was celebrated with much festivity, the building of monuments and temples in Peking, and the Manchus' summer residence at Jehol, a spate of new official publications, a series of special examinations and *ad hoc* increases in the numbers of people allowed to pass the regular examinations. Contemplating his own arrival at sixty, and his dynasty's successful rule of over two hundred years, the Ch'ien-lung Emperor seemed to want a summation of the principles on which Ch'ing rule had been based, a final testimony to his own and his ancestors' commitment to civilized rule. The emergence of a new group of Chinese leaders at the Ch'ien-lung court following the death of Fu Heng in 1770 may also have been a precipitating factor. Certainly, these men were to be very closely associated with the project and they very likely were active in its initiation. The self-congratulatory tone of the edict suggested, however, that it was meant to celebrate the success of Ch'ien-lung policies toward intellectual life rather than impose new themes in the document; its dominant tone of imperial benevolence and openness to all scholars was characteristic of the Ch'ien-lung era. The document was carefully drawn to express these themes, to celebrate the achievements of Manchu policy and to point to continuing goals.

Pompous as they seem, the opening lines of the edict of 1771 were hardly an idle boast, they were an assertion of the emperor's adherence to one of the fundamental tenets of emperorship in China: "[In our rule] We have always been mindful of precedent, revered the writings of the past, relied on the brush to govern and ruled in accordance with principle. We have been diligent from day to day in our study." For the Ch'ien-lung Emperor, as for intellectuals and rulers throughout most of the history of imperial China, the special relationship of knowledge and power had a corollary in the duty of an emperor to maintain a catalog of the best books in his empire: "So I often think of the great imperial libraries of the past and the literati attendants who recorded their contents. The greatest of these catalogs became commentaries in themselves, [commentaries which] being passed from generation to generation, served as mirrors for a thousand years of history." Of course, as the Ch'ien-lung Emperor knew well, historians would judge his dynasty not only by its achievements and the quality of its institutions, but by the character of intellectual life that they fostered. A compilation completed under his sponsorship would not only assist the ruler, but it would assemble and advance learning in an increasingly specialized and geographically dispersed empire: "Those who know only a small subject and compile narratives so detailed that every last fact or phenomenon is enumerated will be unified and each will be fit into his proper tradition. How could there be any who would not make discoveries? By such means are the arts advanced and minds nurtured."

The emperor's policies since the beginning of his reign, he asserted, had amply demonstrated his respect for the classics and histories as the ultimate sources of moral and political authority: "For this reason, when we first ascended the throne, we called for lost books to be collected and ordered scholars to edit the thirteen classics and the twenty-one histories. [We further commanded] that their works be widely disseminated as a benefit to future scholars. We established commissions successively to compile the *Kang-mu* for the Sung, Yuan and Ming dynasties, the *T'ung-chien chi-lan* (Imperial comments on the comprehensive mirror), and the three encyclopedias." But much more remained to be done if the traditional goal of Chinese rulers, a comprehensive repository of the truth of man's nature and the principles of government, was to be accomplished. The emperor offered the *Tu-shu chi-ch'eng*, compiled under his grandfather, as an example of what could be done: "But the only goal of study is to obtain guiding principles and know more of the words and deeds of the past in order to gain virtue. Only if book collections are broad and encompassing can research be fine and precise. For instance, in the K'ang-hsi era, the *Tu-shu chi-ch'eng* was collected and annotated, and provided a great view of policy making [through the ages]. Using various compilations, and arranging according to category materials from such a vast number of sources that no one reader could expect to understand them all, the compilation enabled its user to trace the sources [of various ideas] and to examine one by one their points of origin."

Just as the interaction of state and private scholarship during the K'ang-hsi reign had resulted in this remarkable compilation, a massive project of state-sponsored scholarship could leaven the intellectual life of the present reign: "Now, to organize the books stored in the imperial household cannot but be a good thing. Similarly, the books of past and present authors, regardless of their number, who perhaps still live in the mountains and have not ascended to the ranks of the distinguished should also be collected from time to time and sent to the capital. By such means, the unity of scholarship past and present can be made manifest."

The second half of the edict described the sort of books the emperor sought, and outlined procedures for collecting them. The emperor distinguished between works of enduring truth useful in government and mere ephemera. "Let the provincial governors be ordered to collect all books," the edict went on, "except for editions of examination essays (*shih-wen*) prepared for sale by bookshops, genealogies [which circulate] uselessly among the people, collections of letters, and samples of calligraphy. [Governors need not send] works by authors who have little real learning, or who write only to implicate or to frighten, or who prepare collections which contain only flattery, praise, and trivia. Other than these, any preserved book which clarifies the essential methods of government or which concerns human nature ought to be purchased. As for commentaries, explications, and booknotes, if they are of real benefit, they ought to be included." In this listing there were, to be sure, ominous echoes of the early Ch'ing prohibitions of works with "petty concerns or immoral language" or literary exercise books. However, the Ch'ienlung Emperor was quick to point out that he did not mean to exclude any individual of genuine ability. "Furthermore, the scholars and worthies of the present dynasty, like the famous men of old, have their literary collections. Those of recent times who have immersed themselves in classical studies seeking the original elegance, also have literary collections. These cannot be compared with plagiarized words or dangerous ideas and ought to be sought thoroughly."

The procedures to be used in collecting books were dictated by the nature of scholarship and book production in the empire: "The books for sale in bookstores should be bought at the going rate. Prints of woodblocks stored in private homes should be made at government expense. Officials should order copyists to duplicate those books which have never been printed and exist only in handwritten manuscripts, being careful to return the originals to their owners." Above all, this project, which had been conceived as a benefit to scholarship, should not become a nuisance to scholars: "Governors should order their subordinates to manage everything carefully. In no case should yamen underlings make the project into a pretext to disturb the populace." Finally, the emperor proposed a practical measure to simplify and rationalize the enormous task of book collecting he had ordered. "Since the books to be collected are manifold, if we do not carefully compare them, but simply send everything, there will inevitably be duplication and waste. Therefore, let the governors and governor-generals first make a list of all titles collected, and annotate it with the names of authors, dynasties, and the general content of the works. Write these lists down simply, and memorialize. After [the lists have been received], I will order court officials to compare the lists carefully. A list of those which seem worthy of further inspection, will then be prepared and sent around."

The final lines of the edict reminded the readers that the project proposed was one of a great tradition of such efforts dating back to the Former Han dynasty: "By this means the glories of the *Ch'i-lueh* [prepared by Liu Hsiang during the Former Han] and the *Ssu-k'u shu-mu* [of the Sui] will be exceeded, and my wishes fulfilled."[69] Ironically, this perhaps greatest book collection project in Chinese history would also be the last. But before it ended, emperor, bureaucrats, and scholars would be brought together for a final affirmation of the unity of knowledge and power in Chinese history, an affirmation which would demonstrate the strengths of the great tradition, its weaknesses, and the delicate balance that existed between scholars and the eighteenth-century state.

3

The Scholars' Response

The edict of February 1772 was not the first call by a Chinese, or even by a Ch'ing emperor, for a book collecting project. It led to what was probably the largest book collecting effort in Chinese history because it struck a responsive chord among the social and political elite of the day. In part, of course, this response was an indication of the success of Ch'ien-lung era policies toward the Chinese elite, but it was also a reflection of the character of eighteenth-century society: the growing interest of intellectuals in collecting, collating, and correcting ancient texts; a flourishing economy which made it possible for poor scholars to find wealthy patrons for their relatively specialized studies; and the resulting integration of social and intellectual elites, particularly in the lower Yangtze Valley. None of these trends was new in the eighteenth century; some had been several centuries in the making. But their simultaneous maturing in the last years of the eighteenth century rendered the elite of this day a different group from the rather fragmented community, leery of central government initiative, which had allowed the K'ang-hsi and early Ch'ien-lung book collecting orders to become dead letters. The characteristics of late eighteenth-century society not only made possible the Ssu-k'u project but shaped, perhaps decisively, its products.

Chinese society in the eighteenth century was, in many respects, a difficult society to reconstruct. Benjamin A. Elman's recent application of the notion of "discourse" to eighteenth-century society is most apt, for the tone of the era was better expressed in the shared vocabulary and common concerns of the day than in any single document or event.[1] Eighteenth-century intellectuals seldom made broad programmatic statements; they expressed their intellectual commitments in the increasingly detailed and self-confident studies of ancient texts and ideals, the commentaries and glosses, prefaces, colophons, letters, essays, and biographies which they produced. Their changing political commitments were expressed not in manifestos, but in their steady gravitation toward the centers of power and wealth in Chinese society. Many elements of the social life of the century were epitomized in the semi-official circles of patronage and pedagogy which grew up particularly in the latter half of the century. In these circles, contacts were made, personal bonds were cemented, research goals were articulated, research projects formulated, and sources of financial support secured. Also, and perhaps as important to the participants, the skills of textual criticism and philological investigation were passed from master to disciple in these groups, and the results of the application of these techniques — critical editions of the major classical texts and essays about their interpretation — found their promoters and publishers. One of the first of these circles to develop coalesced around the leadership of the eminent philologist and Hanlin academician Chu Yun (1729–1781) in the late 1750s. Well-connected at court and among the social and scholarly elite of the day, Chu expressed the concerns of and guided the careers of many scholars, and when the Ch'ien-lung edict on book collecting was promulgated in the winter of 1772, it was Chu who formulated the scholars' response and urged his wealthy book-owning friends to cooperate with the imperial summons in the cause of *k'ao-cheng* scholarship.

K'ao-cheng *Scholarship in the Eighteenth Century*

The emergence of a community of intellectuals committed to the task of preserving ancient texts reflected developments in the social, political and intellectual history of late imperial China. In intellectual history, the turn toward textualism represented a turning away from Sung metaphysicians to earlier, specifically Han, sources of insight and intellectual legitimacy. But it would be a mistake to characterize *k'ao-cheng* scholars as mere antiquarians. For, sharing in the universal Chinese belief that the golden age of the past could teach valuable lessons for the present, eighteenth-century scholars saw themselves as active and creative, laying the foundation for a better future by forging new pathways into the past. Ironically, the future for which they were laying the foundation was one few in the eighteenth century could have foreseen. The skills scholars were developing in such fields as epigraphy, textual criticism, philology and historical investigation, when applied in a more critical age, would transform the classical canon into a museum piece, an object for investigation rather than reverence. The age was still some years in the future, however. In the eighteenth century the goal was not the destruction of classical learning, but its reestablishment on a firmer foundation.

The textualist movement has been called by various names — *han-hsueh* (Han learning), *p'u-hsueh* (unadorned learning), and *k'ao-cheng* (rectification through investigation).[2] By most reckonings, the movement began in the early seventeenth century. Twentieth-century Chinese historians have advanced several theories to account for its origins, but in doing so they have all too often read the concerns of their own politicized eras into the eighteenth century. Scholars writing at the turn of the century, influenced by the heat of anti-Manchu feeling in their own day, tended to portray *k'ao-cheng* scholarship as a response on the part of the Chinese elite to Manchu despotism. In this view Ch'ing scholars, recognizing that they had no more strength vis-à-vis the Ch'ing state than "so much fish roe or so many fleas," sought refuge in safe, apolitical studies of ancient texts and expressed their opposition to the state, if at all, only by means of an aesopian language.[3] The writings of Hu Shih and his followers in the May Fourth Movement have been equally influential. Writing in the late teens and twenties of this century, Hu analyzed *k'ao-cheng* scholarship from the standpoint of his training in western philosophy, and found in it an empiricism and objectivity which, he felt, heralded the beginning of a Chinese scientific revolution. In making his case, Hu was more concerned with the methods of Ch'ing scholars than with their results or explicit concerns. He saw the principle goal of *k'ao-cheng* scholarship as overthrowing Sung metaphysics and reconstructing Chinese learning on a new foundation.[4] A third view has associated the rise of new sorts of learning in the eighteenth century with changes in material and economic life occurring simultaneously, among them the growth of cities, commerce, and commercial wealth.[5]

Each of these views has its truth, but none is wholly satisfying; in particular none explains the dynamism of the *k'ao-cheng* movement in the eighteenth century, or its peculiar combination of intellectual iconoclasm and social elitism. Whatever the seventeenth-century origins of the *k'ao-cheng* movement, its growth reflected eighteenth-century developments: the gradual accumulation of knowledge and perspective about the classics, which both reinforced and subtly recast the goals of the movement; the resolution, or at least abatement, of tensions between Chinese scholars and the Manchu state; and finally the increasing wealth of Chinese society, which provided a new financial foundation for scholarly life in the century.

The evolution of intellectual concerns from the seventeenth to the eighteenth centuries was subtle. In almost every field of concern to them, eighteenth-century scholars built upon the work of their seventeenth-century predecessors. But as Benjamin Elman has compellingly demonstrated, *k'ao-cheng* scholars brought new methodological sophistication, new resources and new perspectives to bear on traditional problems. And as they did so, the problems themselves assumed new forms. This process was most evident in *k'ao-cheng* studies of the Five Classics. The core of classical studies during the Han dynasty, the *I-ching* (Book of changes), *Shang-shu* (Book of documents), *Shih-ching* (Book of poetry), *Ch'un-ch'iu* (Spring and autumn annals) and the *Li-chi* (Record of rituals), were thought to have been edited by Confucius, and to contain the most authentic vision of the Chinese past. The Five Classics had been eclipsed, during later imperial times, by the Four Books, somewhat more philosophical texts emphasized by the Sung scholar Chu Hsi (1130–1200). It was characteristic of the *k'ao-cheng* movement that its adherents should seek to resolve the controversies that had developed around the often corrupted texts of the Five Classics, and restore them to a central place in the corpus of Chinese learning.

The disputes on the texts of the Five Classics to which *k'ao-cheng* scholars addressed themselves were of considerable duration and significance. Perhaps the most important issue in the study of the *I-ching* was the relationship of the charts in the work to its text, a question inspired by the importance which Chu Hsi and other Neo-Confucians attributed to the charts as sources of Confucian wisdom. Followers of Wang Yang-ming (1472–1529) and his early Ch'ing disciple Huang Tsung-hsi (1610–1694) sought to attack Chu on this matter by contending that the charts were forgeries of Taoist inspiration. Chu's defenders countered that the Sung master stood on perfectly valid scholarly grounds in his emendations, and that he preserved a rich and valuable traditional understanding of the text.[6] The debate engaged some of the most important figures in Ch'ing intellectual history. Following Huang Tsung-hsi were Hu Wei (1633–1714) and Mao Ch'i-ling (1623–1716); Chu's defenders included Ku Yen-wu (1613–1682), Tai Chen (1724–1777) and, from a slightly different point of view, Fang Tung-shu (1772–1851).[7] As both sides brought more erudition to bear on the issue, the arguments became more sophisticated and lengthy. What Huang Tsung-hsi had suggested in a work of 6 *chüan* in 1661, Hu and Mao proved in works of 10 and 30 *chüan*, respectively. Similarly, what Ku Yen-wu suggested in a footnote to his *Jih-chih-lu*, Tai Chen showed much more conclusively in *Ching-k'ao*.

Scholarship on the *Shang-shu*, as Benjamin Elman has clearly demonstrated, displayed a similar evolution from concern with a metaphysical issue to concern with a text. In this case the issue was the relationship of the "moral mind" (*tao-hsin*) to the "mind of man" (*jen-hsin*), and specifically the way this issue was formulated in the "Counsels of the Great Yü" chapter of the *Shang-shu*. Chu Hsi and his followers had interpreted this passage to imply that the source of moral truth lay outside the human heart, and had to be sought by a laborious process of refinement and self-cultivation. Late Ming scholars responded that there was only one mind, in man as in the universe, and that moral truth and human desires coexisted in it. The debate was a heated one, and was almost certainly one of the inspirations for Yen Jo-chü (1636–1704) in writing his *Shang-shu ku-wen shu-cheng* (An examination of the ancient text version of the *Shang-shu*), published in 1745, which demonstrated not only that the passage in question was drawn from the Taoist canon, but cast doubt on the entire ancient text version of the *Shang-shu*.[8] This work drew an angry retort from Mao Ch'i-ling, and an even angrier response from Fang Tung-shu. The conflict became a *cause célèbre*, stimulating classicists and historians until well into the nineteenth century.[9]

The *Shih-ching* did not inspire such metaphysical disputes. It did, however, provide a foundation for a branch of learning that was to become a backbone of Ch'ing textualism, phonology. Ch'ing scholars were concerned above all with the matter of how to prevent corrupt texts and the mistaken inferences that might be drawn from them. One of the most serious problems they faced was that characters with the same sound could easily, either through inadvertence or by design, be exchanged for one another. It therefore was vital from the textualist point of view to establish families of characters with the same sound, and to show how these sounds had changed over time. The *Shih-ching,* a collection of ancient poetry with a fairly predictable rhyme scheme, seemed an ideal starting point for such research. Ch'en Ti (d. ca. 1617) revived the study of phonology and the *Shih-ching* with his *Mao-shih ku-yin-k'ao* (A study of ancient pronunciations in the Mao recension of the *Shih-ching*), published in the early 1600s. But, at least according to eighteenth-century scholars, Chen's work was nothing but "chaff and weeds" and it was not until Ku Yen-wu's work on phonology, probably completed fairly late in his life, appeared that phonology began.[10] Chiang Yung's (1681–1762) *Ku-yun piao-chun* (A primer on ancient phonology) corrected a number of the factual errors and idealistic assumptions of Ku's work, and through Chiang's friend and student, Tai Chen (1724–1777), the discipline of phonology was passed to scholars of the late eighteenth century. A number of phonological studies were produced in the last years of the century, but one of the most careful and ardent students of the field was Tuan Yü-ts'ai (1735–1815), whose *Liu-shu yin-yun piao* (Classification of the phonology of the Six Books), appeared in the late 1760s.[11]

On one level at least, authors on each of these three texts seemed to be following in the footsteps of their predecessors, posing traditional questions in more and more refined ways and offering ever more sophisticated answers. But, as scholarly works became longer and more detailed, their emphasis seemed to shift from questions of faith to matters of proof. Tenets which previously had been assumed or asserted were now proven, and elements of tradition that had previously been left unexamined were now elaborated and enumerated. Scholarship on the ancient texts on rites illustrated this phenomenon. The earliest Ch'ing writers on the rites texts were all, apparently, inspired by the legacy of Chu Hsi. The great Sung Neo-Confucian had been at work on a commentary on the *I-li* (Ceremonies and rites) at the time of his death, and shortly before he died, he professed the need for more work on the rites. Two of his disciples undertook to complete his work after his death, but they were apparently unsuccessful, so that the task of properly explaining various rites remained on the intellectual agenda.[12] In 1696 Hsu Ch'ien-hsueh produced an index to the various mourning rites in 120 *chüan;* some years later, Chiang Yung compiled an 88 *chüan* index to the rites, and in 1761, Ch'in Hui-t'ien produced his 260 *chüan Wu-li t'ung-k'ao* (A study of the five texts on rites). Once these compendia had been accomplished, the way was open for more detailed study of various facets of the rites texts. Both Chiang Yung and his friend Tai Chen produced studies of one chapter of the *Chou-li* (Rites of Chou) dealing with ancient vehicles and weapons. Others worked on land tenure systems, enfeoffment practices and uniforms.[13]

There is ample evidence that each of the early students of the rites texts was inspired by the prospect of fulfilling one of Chu Hsi's goals. The authors of later studies may have been similarly inspired. But, with the shift to limited topics and the scholarly disputes that often accompanied this shift something of the original Neo-Confucian faith was lost. At the verbal level, evidence of this loss was not hard to find. It had become routine by the middle of the eighteenth century for increasingly precise and scholarly philologists to condemn the more religious of Neo-Confucian writings as "empty metaphysical speculation." At the level of philosophers' conceptions and self images evidence of a decline, or at least a transformation, of Neo-Confucian faith could be found in such passages as Ling T'ing-kan's (1757–1809) explanation of the relationship of his exegesis of ritual texts to Chu Hsi's traditional goal of "investigation of things" (*ko-wu*). Chu's goal, of course, has been variously defined and conceptualized, but at base it represented the search for moral universals. Ling had no complaint with this; the problem was how to accomplish it. As Ling saw it, the key to morality lay in individual feelings and these seemed not susceptible to the kind of analysis necessary to establish moral certainty. Ling's solution to this problem began with his perception that human feelings were best expressed in rituals and one could, by investigating rituals and ceremonies, accomplish Chu's goals. The best way to investigate rituals was to study the texts which recorded them. Exegesis of ritual texts (*k'ao-li*) was, therefore, morally equivalent to Chu Hsi's investigation of things.[14] Textual scholarship had replaced, for Ling at least, self-cultivation as the center of the intellectual's life.

The potential for fairly radical intellectual and political change was implicit in the textualists' redefinition of the purposes of Chinese scholarship, as work on the *Ch'un-ch'iu* demonstrated. This was not a particularly controversial text in early Ch'ing times, though a number of exegeses were prepared. The most prominent student of the text in the eighteenth century was Chuang Ts'un-yü (1719–1788). Chuang's interest in this text led him beyond the traditional Tso commentary on it, into other early exegeses, including the Kung-yang and the Ku-liang commentaries. In the first half of the nineteenth century Chuang's students, Liu Feng-lu (1776–1829) and Wei Yuan (1794–1856) among others, found in these neglected commentaries a powerful sanction for social and political activism.[15] Significantly, however, it was not Chuang who drew these implications, and it was not until the economic and social crises of the 1820s and 1830s that they were developed. In the eighteenth century, scholars seemed largely unaware of, or uninterested in, the political implications of their work. Chinese thought in the late eighteenth century seemed thus to be at a turning point: poised between a Neo-Confucian past and a more activist future, committed to the rigorous examination of texts but not yet emancipated from the concerns that had traditionally surrounded the texts.

One factor which had impelled Ch'ing scholarship to this point was undoubtedly the self-conscious pride that Ch'ing scholars took in their own methodology. Tai Chen was one scholar whose methodological awareness was a particularly important element in his self-image. Commenting on the publication of his *K'ao-kung chi-t'u,* in a letter to Yao Nai (1732–1815) written in 1775, Tai remarked:

> Some of my works are based on thoroughly conclusive evidence, others are not yet so based. By conclusive evidence I mean that the text must be verified by antiquity in every particular, the work must be in such complete accord with the truth as to leave nothing debatable, all the major and minor points must be traced, and fundamental as well as secondary points considered.
>
> If we rely on hearsay to determine the meaning of a text, use the various interpretations to point out its strengths, merely express its arguments with empty words, and use random proofs to verify it, then although we are moving upstream to determine the source, we don't see the original spring with our own eyes.[16]

In Renaissance Europe, of course, such methodological awareness was associated with the rise of science, but the implication that Ch'ien-lung China was on its way toward an intellectual transformation comparable to the scientific revolution in Europe, must be approached with caution. For Tai and his colleagues meant not to move in new intellectual directions but to use the tools they were evolving to examine old texts and render ancient questions solvable. It was precisely because new research techniques were useful in solving old problems that eighteenth-century scholars took pride in them. The methodological awareness of eighteenth-century intellectuals probably signalled a growing confidence in the powers of the intellect, a confidence which in turn eroded many traditional beliefs. But science, and the systematic exploration of natural phenomena, taken as an end in itself, was still a long way off.[17]

Economic and social growth also conditioned intellectual development. Eighteenth-century scholars' attitudes toward the world and their role in it must have been shaped in part by what they saw around them. Like the late Ming, the period of the eighteenth century was a time of growing cities, increasing commercial wealth and slowly rising prices. Unlike the late Ming, however, the Ch'ien-lung era was a time of relative peace; although there were wars along the frontier during most of the reign, the resources of the government and social elites were not excessively drained by the demands of defense. The resulting accumulation of wealth may well have served to convince intellectuals that the realization of classical ideals was possible in their own times. In fact, in a sense, scholars' very careers were founded on this wealth. The best philological scholarship required not only a significant investment in books and editions, but the freedom from the demands of livelihood that only patronage, or very extensive personal resources provide. Certainly the commercial wealth of the eighteenth century underlay the era's textual scholarship.

Relations between scholars and the wealthy differed in the various centers of *k'ao-cheng* scholarship. By the middle of the eighteenth century, the Yangchow area was fast becoming one of the wealthiest areas in China owing to the influence of area merchants who were licensed in the government salt monopoly to sell to the prosperous lower Yangtze delta. Some *k'ao-cheng* scholars were descendants of merchants, but even when they had no scholarly relatives, merchants' concern to demonstrate a commitment to Confucian values led them to engage in a lavish philanthropy.[18] In the last half of the eighteenth century, salt merchants endowed three academies in the Yangchow area, and three more in nearby Huichou, and some of the most prominent scholars of the day were attracted to teaching positions at these or to research positions sponsored by individual merchants.[19] Collectively known as the "Yangchow school," scholars of northern Kiangsu and southern Anhwei led the way in the development of new philological techniques in the eighteenth century, and particularly in research on the Rites texts.[20]

Intellectuals from the Soochow area tended to be from families of older wealth, and were often most interested in book collecting, bibliography, and participation in the state examination process. A twentieth-century descendant of the Yangchow school labelled these concerns of "decadent wordsmiths." But in the eighteenth century, Soochow scholars' interests rendered their city one of the empire's major centers of book production and trade.[21] Although one Soochow academy was founded with merchant capital, most of the educational resources of the district were more closely tied to landed wealth. The early twentieth-century publisher and bibliophile Yeh Te-hui has traced the fate of large private libraries first assembled by Ch'ien Ch'ien-i (1582–1664) and Mao Chin (1599–1659) in the late Ming and passed along through students and descendants, largely landholders of the Ch'ang-shu region northeast of Soochow, to Huang P'ei-lieh (1763–1825).[22] From the midst of this gentle labyrinth of land and libraries came some of the great titans of *k'ao-cheng* scholarship — the three "phoenixes" of the Ch'ien family, Ta-hsin (1728–1804), Ta-chao (1744–1813), Tung-yuan (d. 1824) and Wang Ming-sheng (1722–1798).

In the Hangchow area older wealth, mostly of mercantile origins, had made possible the assembly of a vast network of scholarly resources by the middle of the eighteenth century. Not only were the Hangchow book owners men of very considerable means, they were close personal friends.[23] The intellectual tradition which drew its sustenance from this assemblage of scholarly resources had a few special characteristics, notably its interest in late Ming history and its rather activist emphasis, but it participated in all the developments of Ch'ing scholarly life. The Hangchow libraries formed a major source of books for the Ssu-k'u Project.[24]

The growth of Peking as an intellectual center paralleled the growth in the capacity of the Ch'ing state to attract men of wealth and education to its service. From rather modest beginnings in the seventeenth century, the Liu-li-ch'ang, or bookseller's quarter, had grown by the mid-eighteenth century to be a market "justifiably famous throughout China." Its merchants regularly sent to Soochow for books and baubles to amuse their wealthy customers.[25] Among these customers were sinified Manchu princes and successful examination candidates who stayed on in the capital to occupy court sinecures. But there were also several families who stayed for several generations in Peking, forming a kind of local elite. Most famous of these were the families of Chu Yun and Weng Fang-kang. Regularly infused with new blood from China's heartland, the eighteenth-century scholarly elite in Peking was attuned to all the intellectual developments of the empire. Its particular contribution probably was the study of stone inscriptions which dotted the North China plain.

Yangchow, Soochow, Hangchow, and Peking were not the only centers of *k'ao-cheng* scholarship in the eighteenth century, but they were the major ones. The close connection between wealth and scholarship in these centers suggested the importance of the new prosperity of the century for textual scholarship. But, one of the frustrations for the historian of the eighteenth century is that the quantitative changes of the era did not always lead to qualitative ones. The new wealth of the period served more to support the fulfillment of traditional goals than to inspire new social ideals. To be sure there were some, like Tai Chen and Chiao Hsun, who formulated new ethical conceptions in which human feelings played a greater role than in earlier conceptions.[26] But these works could hardly be considered typical of the period, either in terms of content or influence.

It is similarly difficult to see *k'ao-cheng* scholarship as a kind of statecraft thinking turned in upon itself in response to Manchu despotism. *K'ao-cheng* scholars were fond of quoting statecraft thinkers, and they did so even in the Ssu-k'u catalog itself, but they quoted only the passages that interested them. They rejected, or simply ignored, the more political writings of seventeenth-century thinkers.[27] In a fundamental sense, of course, early twentieth-century historians were right; Manchu government created the climate in which *k'ao-cheng* scholarship flourished, and this fact imposed some constraints on Chinese scholars. The Ch'ing court reserved some positions of real political authority for Manchus only, and the great Chinese court favorites of the century—men like Hang Shih-chün (1696–1775), Shen Te-ch'ien (1673–1769), Ch'ien Ta-hsin and Chi Hsiao-lan (1724–1805)—were all writers and poets, rather than strategists or planners.[28] When Hang Shih-chün complained in 1743 that only Manchus could occupy the highest offices of the realm, he was banished from court with a speed which belied the Manchus' often proclaimed desire to rule without regard to ethnic distinctions.[29]

But what more would *k'ao-cheng* scholars have sought? They were or represented the wealthiest segments of Chinese society. While the Manchus reserved for themselves the right to make appointments to court sinecures and to control border affairs, they quite consciously did not interfere with the rights of landholders or the prerogatives of wealth. In an era when economic growth almost guaranteed the continued dominance of those who owned land or controlled resources, *k'ao-cheng* scholarship, which tended toward an acceptance of established authority and contained at best an impulse to gradual reform, may well have suited Chinese elites' interests and predilections as well as the political opportunities open to them. The speed with which many intellectuals began to abandon the *k'ao-cheng* approach once events made glaringly evident the need for economic and social reforms in the early nineteenth century, suggests that Ch'ing intellectuals had not forgotten the role they could play in policy-making. But the eighteenth century was a time when the opportunities for realizing traditional intellectual goals seemed to elites more compelling than the need for social or political reform. Rather than seek political reform, scholars of the eighteenth century seemed content to function within the existing order, turning court initiatives to their own ends wherever possible. Such at least, was the apparent aim of Chu Yun and his friends as they set out to formulate a response to the imperial book collecting order of 1773.

The Circle of Chu Yun and Its Significance

Western scholarship has not yet probed deeply enough into the eighteenth century to realize Chu Yun's significance, but among his contemporaries and among modern Chinese and Japanese historians, Chu's importance has long been recognized. Yao Ming-ta wrote in his 1937 biography of Chu:

> If one were to read through the writings and biographies of the Ch'ien-lung and Chia-ch'ing periods, one would realize Chu's unparalleled influence on the scholarly climate of the day. On the one hand, he proposed the establishment of a government bureau to collate books, and created the environment for such a project. On the other hand, he repeatedly received and nurtured scholars and created a climate for the encouragement of scholarship. He was truly the founder and leader of the "unadorned learning" (*p'u-hsueh*) movement in the Ch'ien-lung and Chia-ch'ing periods.[30]

Kawata Teiichi finds some of Yao Ming-ta's language, particularly his characterization of Chu as "founder and leader of the unadorned learning movement," a bit excessive, but still sees Chu as a figure who enriched enormously "not only the intellectual but the human side of the eighteenth-century life."[31]

Chu's family history illustrates the increasing attractiveness of life in Peking for the eighteenth-century elite. Although they were originally from the Hangchow area, the Chus had resided in Peking since Chu Yun's grandfather chose to retire there, in the late K'ang-hsi era. Except for a seven-year term as a magistrate in Shensi, Chu Yun's father spent his entire life in the capital, serving at court and teaching Peking students for over twenty-five years.[32] Chu Yun and his younger brother Chu Kuei attracted the attention of the city when they ranked first and second, respectively, on the Shun-t'ien *hsiu-ts'ai* examinations of 1745.[33] Chu Yun's rise to court prominence, however, began with his selection in what was, from the point of view of intellectual history, one of the most distinguished *chin-shih* classes in the eighteenth century. In the second and forty-second places, respectively, of the class of 1754 were the two distinguished eighteenth-century historians Ch'ien Ta-hsin (1728–1804) and Wang Ming-sheng (1722–1798). The bibliographer and future editor-in-chief of the *Ssu-k'u ch'üan-shu,* Chi Hsiao-lan (Chi Yun) ranked sixth in the examinations, and the great epigrapher and classicist Wang Ch'ang (1725–1806) ranked ninth. Chu Yun himself ranked thirty-eighth.[34]

All except Wang Ch'ang were appointed to the Hanlin Academy. Working and socializing together, they soon became a tightly knit group whose opinion and scholarly style came to dominate the thinking of the capital. Their social circle was soon joined by the philosopher and classicist Tai Chen and the epigrapher Weng Fang-kang (1733–1818), who was a neighbor of Chu's in Peking. Even Yao Nai, who would eventually dissent from their views, was temporarily drawn into their midst.[35]

For about ten years Chu served with this group at court, employed on such Ch'ien-lung era historical compilations as the *P'ing-ting Tsun-ko-erh fang-lueh* (An account of the suppression of the Dzungars).[36] In the mid-1760s, however, Chu began to make his own particular contribution to intellectual life, the recruitment and training of younger scholars. Two events in Chu's life marked and hastened this development. The first was his move in 1764 into a new home across from that of fellow Hanlin Academician Chiang Yü-ts'un (d. 1770). Chu named his study in the new quarters, a room which became famous as the scene of many parties and discussions, the "Pepper Blossom Humming Boat."[37] In the same year Chu's father died, and the three-year mourning period of enforced retirement from official business gave Chu the leisure to begin cultivating younger talents on a larger scale.[38]

Chu's contemporaries claimed that he had between five hundred and one thousand students (*men-jen*) in his lifetime; Yao Ming-ta has located references to about one hundred in Chu's collected writings.[39] The larger figures were probably inflated by the fact that candidates who passed the *chin-shih* examinations usually referred to their examiner as "master" even if they had not had any previous contact with him, and by the tendency of eighteenth-century literati to form "belletristic friendships" of the sort described by James Polachek, that is, relationships in which aesthetic commonalities are asserted for essentially political purposes. Clearly some of Chu's students were of the latter type, coming to him shortly before or after the *chin-shih* examinations in the hope that some last minute coaching or a chance meeting with a senior courtier might stand them in good stead. But many of Chu's students, like Chang Hsueh-ch'eng (1738–1801), Wu Lan-t'ing (*chü-jen*, 1774) and Ch'a Pi-ch'eng (*chin-shih*, 1778) stayed with Chu for a number of years and actually lodged at his house.[40]

Congeniality was a hallmark of Chu's household. In an epitaph for Chu's neighbor Chiang Yü-ts'un, Chang Hsueh-ch'eng described the atmosphere that prevailed at Chu's home:

> Student and teacher would exchange presents, pitchers of wine or platters of vegetable delicacies, then sit and eat them together. Or, student and teacher would recall together an old story and laugh over it. Morning and evening passed so. In discussing writing or questioning the origin of words, there was not a day of leisure. If there was wine, it was never left undrunk.[41]

Understandably, when the term of mourning for his father was finished, Chu was loath to leave the Pepper Blossom Humming Boat. Fortunately, after a brief service at court, Chu was appointed to a post which allowed him to continue his lifestyle and calling. The post of provincial education commissioner had been created by the Yung-cheng Emperor to supervise the holders of the *hsiu-ts'ai* degree who received government stipends. It was unranked and formally outside the bureaucracy, but it carried enormous prestige, especially as the appointees were supposed to have been personally selected and instructed by the emperor.[42] Although they were based in provincial capitals, provincial education commissioners traveled from county seat to county seat in order to interview students and give exams.

Chu was appointed education commissioner of Anhwei in 1771. He took with him a rather large retinue that included Chang Hsueh-ch'eng, Shao Chin-han (1743–1796), Hung Liang-chi (1746–1809) and Huang Ching-jen (1749–1803), and they traveled through the province on a regular basis. For instance, in the spring of 1772, they examined students at Wu-hu in the end of March, then journeyed south to conduct examinations at Hui-chou in April. On the four-teenth of May, they conducted examinations at Hsiu-ning, and on the thirteenth of July at Ning-kuo, before returning to T'ai-p'ing for a brief respite. The following fall they embarked on a tour of the northern half of the province.[43]

The retinue which Chu brought with him to Anhwei constituted the nucleus of his circle. Although Chu's was the first group of this type to coalesce, such circles became a fairly common feature of the era's intellectual life. Among the more important successors of Chu's circle in this regard were the groups which Pi Yuan (1730–1797) formed in Shensi in the late 1770s and 1780s, and the group that Juan Yuan formed at the Hsueh-hai-t'ang in Canton in the first decades of the nineteenth century.[44] The later circles were more active in publication: Pi Yuan underwrote the publication of many monographs and critical editions, and Juan Yuan was responsible for the fourteen-hundred-volume compilation of classical commentary entitled *Huang-Ch'ing ching-chieh* (Ch'ing period commentaries on the classics). But all circles were springboards for scholarly talent, and probably also bridges between Manchu political power and Chinese intellectual life. Life in these circles illustrated some of the ways in which Chinese intellectuals adapted themselves to the political and economic realities of the eighteenth century. Three sorts of ties seem to have bound together the members of Chu's circle — economic, social and intellectual.

The economic aspects of Chu's circle were obviously important, especially for poorer members of the group like Wang Chung (1745–1794) and Chang Hsueh-ch'eng. A *chin-shih* degree was a guarantee at least of economic security in eighteenth-century China and Chu Yun's advice on the examinations, if only because he was so well con-nected with those likely to serve as examiners and so well attuned to scholarly trends in the capital, was valuable. Thus, Chu could advise Chang Hsueh-ch'eng that he had no talent for writing the *pa-ku* essays required on the examinations, but reassure him: "Why should passing the examinations be difficult? And why should this require you to study *pa-ku?* Follow your own way, trust to your own nature, and a degree will not necessarily be unobtainable."[45] It was not that Chu could guarantee his students success on the exams, but he had a far clearer sense of what was required of them than most if not all of his contemporaries.

Chu was also in a position to find temporary employment for needy students. Li Wei (fl. 1770–1800), a student of Chu's in the late 1770s, once pointed out to Chu that his habit of recommending for employment students who had failed on the examinations had earned him the scorn of many in Peking who said "Master Chu's recommendation that a student has exceptional talent and unusual abilities only means that the student has failed." Chu sighed and remarked: "I, too, have doubts on this point. Yet if a man whose talents are incomplete has come from a poor family and traveled for over a thousand *li* . . . and has certain abilities, I praise those abilities. This may not completely accord with righteousness, but what harm can it do?"[46]

Chu's capacities in this respect were not unlimited. In 1769, one of his favorite students, Jen Ta-ch'un (1738–1789), passed fourth on the list in the *chin-shih* examinations. Although this did not guarantee Jen a place in the Hanlin Academy, most people assumed he would be appointed, both because of the breadth of his reading and his particular expertise in the three books of rites. Surprise and consternation attended his appointment in the Department of Ceremonies at the Board of Rites. Shortly after this appointment was announced, Jen asked Chu to intercede on his behalf to secure a transfer to a less burdensome post. Jen pointed out that, since his house was close to Chu's, he would be able to borrow books more often from his patron's library if he had more leisure time: "Of ten years spent in office, at least five could be spent reading." But Chu was unable to effect the transfer.[47]

Less tangible, but real nevertheless, were the social ties that bound Chu and his students. The frequent references to feasts and outings in Chu's writings and those of his students attest to the active social life the group led together, and the descriptions by Chu's students of the warm and supportive atmosphere of his circle are too frequent and sincere to be dismissed as sycophancy. Chu himself discusses the importance of social ties between students in a preface he wrote in 1766. In that year Ch'eng Chin-fang (1718–1784) and Feng T'ing-cheng (1728–1784) stopped at his house on a visit to Peking. Several students were invited to join the party. As was customary, Feng assembled the poems written during the evening into a small volume entitled *Chiao-hua chin-fang hsiao-chi* (A small collection from the Pepper blossom humming boat).[48]

In his preface Chu wrote that while he was delighted that Feng had made the collection, he was afraid that people outside his circle might take his group for a band of dissolutes who gathered because they were dissatisfied with the world. Why, he asked, did men of old give the name "friends" (*yu*) only to the drinking groups who gathered for wine and amusement, isolated themselves from the world, and were of no use to society? If these were the only kind of "friends" that existed, then friendship would depend on a person's employment in the world which in turn depended on fate (*ming*) and the times (*shih*). Chu was at some pains to distinguish the relationships among the members of his circle from political friendships which depended on a person's employment in the world and, in turn, on "fate" and the "times."

> He who asks nothing of heaven can be faithful to himself, and he who keeps faith with himself is a *chün-tzu*. One who keeps faith with himself and asks nothing of heaven can be a friend, and through such friendship there is a genuine and tranquil happiness. Meeting on one occasion, two people become friends. Afterwards when they have parted they know that the friendship will not be betrayed . . . A *chün-tzu's* friendship ought to be ceaseless like the waters.[49]

In Chu's vision, his students were cementing bonds which would transcend physical distance and social and intellectual differences as they explored scholarly issues together at his home. His preface, of course, reflected an elite lifestyle, but there was also an unarticulated sense that in an increasingly fragmented society, circles like Chu's provided important social and personal support for their members. It was important, however, that the unity which Chu saw himself as fostering was not in opposition to established authority: these were not drunken dissolutes who rejected social and political convention. Neither were they concerned patriots; the personality ideal of the eighteenth century was quite different from the image of ardent and passionate commitment which dominated the nineteenth century.[50] The style of the eighteenth century emphasized success, going along and getting ahead. It had, of course, all the classic components of collaborationism; but it is difficult to imagine how eighteenth-century scholars could have evolved any other.

Finally, Chu's circle was held together by the fact that he stood for something in the intellectual world. Chu published little himself, but by virtue of his many terms as civil service examiner he was in a position to enforce a view of learning on his contemporaries. Narrowly conceived, his cause was the advocacy of certain Han dynasty texts and techniques. More broadly viewed, he offered a general approach to learning, an epistemological conception with ramifications for all branches of scholarship. Chu believed, as did virtually all traditional Chinese scholars, that China's ancient sages had actually achieved insight into the truth. The problem was how to recapture those insights. Ch'ing scholars came to believe that the best way to accomplish this was to study their actual words. For men of Chu Yun's generation, this meant studying Han dynasty commentaries, since they were the earliest extant explications of the ancient classics.

Chu's interest in Han commentaries combined the concerns of the Peking and Chiang-nan scholarly communities. The deciphering of stone steles was a particular pastime of the Peking elite; outings to the sites of steles were an important part of capital social life. Many of the sites Chu visited were of Han date, and reminiscences of the outings and reflections of Han history fill Chu's early writings.[51] As his intellectual world grew, so did Chu's concern with the philological methods prized by scholars from the south. He sponsored works on dialects and classical dictionaries, as well as a study of variant characters in the classics. While education commissioner of Anhwei he ordered that copies of the *Shuo-wen,* an etymological dictionary of the Han period, be distributed to all examination candidates.[52]

In an essay entitled "Ch'üan-hsueh-pien" (An exhortation to study), which was written as a preface to a collection he had compiled of the best examination essays he had read during his term as education commissioner in Fukien, Chu said:

> What I seek in examination essays may be called "interpretation of the classics," that is, explanations of the meaning of the Four Books and the Five Classics. Explaining the classics began in early Han times when various texts were preserved and protected by the elders from the fires of Ch'in Shih-huang. The *Han shu* claims that from the time Han Wu-ti established the five erudites and their disciples until the end of the Former Han dynasty, over a million words could have been written in explanation of a single classic. There were over a thousand masters, and scholarship was a road of wealth and influence . . . The explanations of the classics were both numerous and earnest. But now they are regrettably lost. . .

Our present government is also anxious to attract men of merit, and so also dispatches officials to give examinations. The goals and methods are the same as in Han times. As Han Yü of the T'ang said: "If a student does not know the classics, he is not worthy of holding office." Or, "In order to write one must understand words."

The texts of Han Confucianists which have been distributed to all the academies include the writings of Mao Ch'ang, Ho Hsiu, Chao Ch'i and Cheng K'ang-ch'eng.[53] Also in ready circulation are the dictionaries of Master Hsu Shen.[54] If the student has not read Hsu's *Shuo-wen,* he will not know the meaning of words. If they do not read Mao Ch'ang, Ho Hsiu, Chao Ch'i and Cheng Hsuan, they will not be able to penetrate the meaning of the classics. Students who take the examinations without this preparation will find it very difficult to conform to the standards of elegance and clarity demanded. Now, clarity is not simply achieved through empty words, nor is elegance a matter of useless expressions. If the student wants to avoid useless expressions and empty words, and conform to examination standards, he must know the classics and the meaning of words.[55]

The contentions of Chu and his colleagues have often been misrepresented. They argued not that Han texts were sacrosanct — after all even the earliest Han commentators were several centuries removed from the golden ages of the sages — but that the philological methods of Han commentators were more likely to be productive than the discursive methods of later commentators. The focus was on the method rather than on the results of Han commentators. As Chu put it in another essay, if modern students couldn't even distinguish between similar but distinct characters like *ch'an* (flattery) and *t'ao* (doubt), how could they possibly comprehend the classics or express the intent of the sages?[56] Chu expressed well the emotional force that underlay the movement when he wrote to a friend in 1755 that what he feared most was the self-satisfaction of those who read texts only perfunctorily.[57]

Chu Yun's Memorial

If philologists can ever be described as giddy with excitement, that surely was the mood in Chu Yun's yamen at T'ai-p'ing in the fall of 1772 as Chu and his students pondered the possibilities of the imperial edict on book collecting. In December, Chu memorialized, submitting some seventeen books for the emperor's scrutiny.[58] Not surprisingly, these books represented the major contribution of each district of Anhwei province to Ch'ing scholarship in general and the *k'ao-cheng* movement in particular. From Hui-chou came books by Chu Yun's secretary Tai Chen and Tai's teacher Chiang Yung (1681–1762).[59] Representing An-ch'ing Prefecture were Fang I-chih (d. 1671), one of the pioneers of the *k'ao-cheng* movement in the late Ming, and his two sons Fang Chung-te and Fang Chung-lü (fl. 1700).[60] The works of the historical geographer Hsu Wen-ching (1667–after 1756) were submitted from T'ai-p'ing; and selections by the poet Mei Ting-tso (1549–1618), the prose-stylist Shih Jun-chang (1619–1683) and Wu Hsiao-kung (ca. 1700?) were selected from Ning-kuo Prefecture.[61] This first submission of Chu Yun's demonstrated the degree to which regional loyalties and philosophical allegiances could still shape scholarly discourse, despite the many concerns that all eighteenth-century scholars shared.

On more careful consideration, Chu saw potential in the imperial edict on book collecting for more than an assertion of localism or intellectual loyalism. "This," he was said to remark of the edict, "is an extraordinary statute." If collators of the proper skills were selected then "knowledge could be traced to its origins, the outline of the ancient sages' vision established and the interactions of heaven and earth examined."[62] The reference to collators of the proper skills was probably not to specific individuals. Chu was not so much contending for control of the personnel in the project as he was exploring ways in which the community he represented might be able to accomplish through the effort some of its fundamental goals.

After discussions with his staff, Chu submitted a palace memorial in early January proposing that four projects be undertaken in conjunction with the compilation of the imperial library. This was a rather unusual action, for at least two reasons. First, although provincial education commissioners had the right to submit palace memorials, they usually did so only to convey examination results and to submit the required annual reports on their staffs. It was also rare, in the late Ch'ien-lung era, for any official to use the secret memorial system to engage the emperor in discussion of an edict already issued. Given these considerations, the emperor's rather unusual response to his counselors' comments on the memorial (see chapter 4 below), and the fact that there were apparently no other responses to the book-collecting order, it seems likely that Chu was addressing the emperor not only in his formal role as provincial education commissioner, but also in his informal role as representative of the scholarly community well positioned at court. The four projects Chu proposed were (1) the preservation of handwritten texts, particularly those of the Sung and Yuan dynasties, (2) the reconstitution of texts lost since the Ming dynasty from fragments preserved in the *Yung-lo ta-tien*, (3) the preparation of a master catalog of all the stone inscriptions in the empire, and (4) the insertion in each book of a brief evaluation of its content and bibliographical significance.[63] The discussion below will examine each of these projects in detail, and consider their relation to ideas circulating among Chu's colleagues and friends.

Handwritten texts. Chu's first concern was that adequate attention be paid to handwritten texts of the Sung, Liao and Chin dynasties:

> The collection of older printed and handcopied editions is particularly urgent. Extant Han and T'ang books are, of course, extremely rare. Book collectors have more copies of classical commentaries and literary collections from the Sung, Liao, Chin and Yuan dynasties, but unless there are printed editions, copies of such books in circulation will become rarer day by day. Other works, of philosophy and history, often do not exceed one or two *chüan* in length. But it would be appropriate to concentrate first on acquiring the best of them, ordering officials to make copies of them and returning the originals to their owners. These works can usefully supplement the bibliographical treatises in previous dynastic histories, and fill the gaps in Our Dynasty's collection, so that what is written [about these periods] will have a foundation.

As a scholar of Han learning, Chu was hardly likely to be advocating the preservation of Sung scholarship for the ideas it expressed, and while the idea of supplementing the bibliographical treatises of previous dynasties or providing a foundation for what was written about the Sung, Liao, Chin, and Yuan dynasties might have appealed to the emperor, Chu's proposal went somewhat further than this.

On one level, Chu was inviting the government to engage in a competition that had fascinated the wealthy book collectors of Soochow for about a century. The collection of Sung editions had become such a vogue by the late eighteenth century that some collectors identified their libraries by the number of Sung volumes they possessed, while others prepared special catalogs of their Sung holdings. Huang P'ei-lieh, for instance, named one segment of his library the Pai-Sung i-ch'an (The room with one hundred Sung editions). The passion for Sung books was partly a fad, but many owners also prized the Sung editions for their supposed textual accuracy.[64]

More fundamentally, what Chu was proposing involved not so much new goals and procedures as a new rationale. Where the emperor had ordered that books be assembled because they contained insights valuable to the ruler and his ministers, Chu was suggesting that books be collected regardless of their content, simply because they were in danger of being lost. In the emperor's plan, scholarship was to be encouraged in the interest of better government; in Chu's proposal, the government was being invited into the scholar's world in the interest of better scholarship.

Similar arguments had been advanced by Chu's student Chang Hsueh-ch'eng in his writings on bibliographical classification and local historiography.[65] But the most forceful and memorable plea for government action to preserve books was an essay entitled "Ju-ts'ang-shuo" (A plea for Confucian libraries), by an associate of Chang's and Chu's, the Shantung bibliophile Chou Yung-nien. The essay took the form of a dialogue between Chou and various real and imagined opponents; in the course of these debates, Chou anticipated many of the arguments that would be used against Chu's project. One spoilsport was made to argue in Chou's essay that "the body of ancient and modern books is as vast as the fog on the seas, and your plan to collect them would be as futile as the efforts of the foolish old man to move the mountain [stone by stone]."[66] Chou responded that the task of collecting all the books in China for a Confucian tripitaka would be nothing compared to the effort made during the T'ang and Six Dynasties to assemble, annotate and translate the Buddhist tripitaka. All that was required, Chou asserted, was a leader willing to issue a summons for the books, and a patron willing to sponsor their collation.

Another imagined opponent, this time the historical Cheng Chiao (1088–1166) was quoted as arguing that: "There are specialized books and specialized scholars to maintain them. Therefore, although individuals come and go, traditions of scholarship will be preserved." This argument had a certain resonance with the Han learning view that specialized traditions of learning were important vehicles for the preservation of sagely teachings; indeed, Chang Hsueh-ch'eng particularly admired Cheng Chiao's pronouncements on this subject. Chou Yung-nien's response to this argument was both practical and effective: "Why not store the products of specialized scholarly traditions away," he asked, "so that the public will protect them?" Moreover, Chou pointed out that specialized traditions of learning would be more numerous if a central library were established, since the books to support them would be easier to obtain.

Cheng Chiao was also represented as asserting that "essays are as common as dew in the morning, . . . but principle, though as deep as a mountain ravine, may not always be attained by the seeker." Why, in short, collect so many books when only a few have the truth? Because, answered Chou, if all kinds of books were collected, everything the scholar could need would be assembled. When all knowledge could be reviewed, what appeared to be useless studies might well prove to be useful.[67]

The premise of both Chu's proposal and Chou's essay, one very common in the eighteenth century, was that all books had some value, if not as repositories of truth at least as historical evidence. Today's trash could become tomorrow's treasure. It behooved a government intent upon playing a role in scholarship to assemble a library worthy of the vast financial and organizational resources it could command. Chu's proposal would have turned the government's book-collecting project into the scholar's bibliography of last resort. It would also have set the government in competition with private collectors, those scholars who had access to the imperial library being the ultimate beneficiaries.

The Yung-lo ta-tien. Chu's second proposal dealt with the books stored in the inner court library:

> The current holdings of the inner court library ought to be made public, in order that the gaps in the collection can be filled. There are very few books which are not held either by the imperial library (in the inner palace) or by the Hanlin Academy (in the outer palace). I recall that in the Han minister Liu Hsiang's method of collation, books outside the palace supplemented those inside the palace, and those inside the palace were collated against those from the outside. If your majesty were to order prepared a catalog of the library of the imperial household and make it available to scholars of the outer court, then later you could order each official to submit any volumes he possessed which were not on the list. By this means, the holdings of the inner court library could be made even broader.

The division between the inner court library, presumably for the emperor's personal use, and the outer court library was an old one in Chinese history, and there had been many efforts to consolidate the two holdings. As the remainder of Chu's proposal made clear, however, Chu was not as concerned with the completeness of the emperor's personal library as he was with making the contents of both court libraries available to scholars. He was particularly concerned with one text in them.

> When I served in the Hanlin Academy, I often perused the *Yung-lo ta-tien*. There is little order among the books and articles in this work; in some cases, texts have been divided up (under more than one heading). But there are many books preserved in the collection which are never seen outside of it. I would urge that your majesty order several of the complete works contained in the encyclopedia (which are not available elsewhere) copied out and included in your collection. By such means lost texts would be restored and scholarship benefited.

The *Yung-lo ta-tien* fascinated Ming and Ch'ing scholars. When the work was completed in 1409, two copies were made — one for storage in Peking, the other to be stored in Nanking. The Peking copy was partially destroyed by fire in 1562, but the burned volumes were later recopied from the Nanking set. When the Ming dynasty fell, the Nanking set was destroyed, but the Peking copy was left intact, and it was said that the Shun-chih emperor browsed in it during moments of leisure. The order of the topics in the work was indeed unique. One character was chosen from the name or description of each event or idea discussed. These characters were then grouped according to a rhyme scheme devised at the court of the first Ming emperor. Passages drawn from classical texts were quoted under the character representing the event they described. Although the passages quoted were sometimes quite long, occasionally, as Chu Yun noted, passages from the same book could appear under different headings. The form of the encyclopedia made the task of recovering individual titles a daunting one, particularly in view of the length of the whole. The original work comprised eleven thousand string-bound volumes, probably well over a million lines, although perhaps as many as two thousand of these volumes had been lost by the eighteenth century.[68]

Very few scholars had actually seen the encyclopaedia. Liu Jo-yü (fl. 1630), a late Ming thinker, did not even know where it was stored. In the late K'ang-hsi years, Hsu Ch'ien-hsueh applied to use the *Yung-lo ta-tien* in his compilation of the *Ta-Ch'ing i-t'ung-chih* (Comprehensive gazetteer of the Ch'ing empire), but permission was denied. One of the most elaborate attempts to make use of the *Yung-lo* encyclopedia was made by Li Fu (1675–1750), a senior court official, and the historian Ch'üan Tsu-wang (1705–1755) when Ch'üan was living at Li's home in Peking in the early 1730s. Li, perhaps in his capacity as editor of the *Pa-ch'i t'ung-chih* (Gazetteer of the eight banner system) was able to borrow some volumes of the encyclopedia which the two proceeded to examine together. Seeing the importance of the work, they formulated rules for copying out texts from it and even hired copyists. The effort was halted, however, when Ch'üan was transferred to a magistracy in the first year of the Ch'ien-lung reign.[69]

The failure of this effort did not stop other scholars from trying to obtain *Yung-lo* texts. Fang Pao (1668–1749) tried to use it in the 1740s in his commentaries on the three books of rites, and Tai Chen, one of Chu Yun's secretaries in Anhwei tried to use it in 1767. Tai Chen described his efforts to use the work:

> I have tried to reconstruct the *Chiu-chang hsuan-shu* (Nine essays on computation) for over twenty years to no avail. I discovered that the text might have been copied into the *Yung-lo ta-tien,* which is currently stored in the Hanlin Academy. In 1767, I went to the Academy together with the Hanlin compiler Ts'ao Wen-chih who was from my native district (to look into it). Although there were errors in the text, and it was dispersed in several entries, I thought it could be reconstructed.[70]

Chu Yun's proposal of 1774 that the reconstruction of texts from the *Yung-lo ta-tien* be made a part of the *Ssu-k'u ch'üan-shu* project was hardly a novel idea. Perhaps the only novelty of the situation was that Chu felt comfortable enough with the centers of power in Peking to propose the project formally, and Manchu rulers were favorably enough disposed toward Chinese scholars and scholarship to underwrite the project. The suggestion and subsequent effort afforded, at any rate, a measure of both the growing textualism of eighteenth-century scholars and the changing attitudes of Ch'ing rulers toward them.

On cataloging. Chu's third proposal was to lead the imperial compilation project into yet another area of special interest to eighteenth-century scholars:

> Making lists and collating texts are both important. Officials who collated books under previous dynasties, like the officers of White Tiger Hall and the Pavilion of Heavenly Emoluments,[71] assembled the various editions, compared, and produced a new one. At the T'ang and Sung academies, officials were chosen especially for this. Liu Hsiang, Liu

Chih-chi and Tseng Kung were (therefore) all engaged in a specialized enterprise.[72] Through the ages, catalogs like the *Ch'i-lueh, Chi-hsien shu-mu* and the *Ch'ung-wen tsung-mu* all reflected traditions of learning passed from Master to disciple. Your servant requests that your majesty order that Confucian officials be chosen especially to collate books, which can then be organized either according to the *Ch'i-lueh* or according to the four division system.[73] Each book presented to your majesty ought to be preceded by a brief note setting forth its accomplishments and failures and summarizing its main points. These notes can then be presented for your approval. Your servant observes that if able directors-general, revisers and collators are selected to work at the Wu-ying Throne Hall on this task, then each day there will be accomplishments, and every month there will be progress, and soon the task will be complete.

While many imperial book catalogs had some annotation, very few had such complete annotations as Chu Yun was proposing. One that did was the Han dynasty work *Ch'i-lueh,* by Liu Hsiang, which however was known only through its surviving fragments. The frequent references to this work, together with the discussion of "passing learning from Master to disciple," strongly suggested that the impetus for Chu's proposal was the admiration that Chu and his colleagues felt for Han dynasty institutions of imperial scholarship, which they saw as largely responsible for the productivity of Han dynasty commentators. In particular, they felt that imperial bibliographers like Liu Hsiang had been able, through their annotations, to reestablish traditions of learning and the chains of discipleship through which these traditions were built.[74]

A second and rather different concern may also have influenced Chu Yun's proposal. As I have argued in chapter 2, the rise of printing in the late Sung and Yuan had resulted in a proliferation of editions; by Ch'ing times, the library builder was confronted with a bewildering variety of texts, with very different editorial standards. Lamenting this state of affairs, one late eighteenth-century intellectual wrote:

Since the Sung, printed editions have become common. Yet if you compare two different editions of the same book, they are as different as a wide hall and a narrow path. When you look at book collectors' catalogs, you see entries for so many volumes of the classics, so many volumes of history, etc., etc., but you can't tell whose editions they are. Therefore, you really don't know whether a book is in fact what it purports to be and whether its text is reliable. How can you talk about the merits or significance of the collection?[75]

Chu's memorial did not mention the problems of editions directly, but in calling for annotations, in the manner of Liu Hsiang, he may well have had this issue in mind. Certainly, in a number of reviews drafted by Chu's colleagues for the imperial catalog, this question was dealt with at some length.[76]

Stone inscriptions. Chu's fourth proposal was that the government undertake to compile a master catalog of stone inscriptions in conjunction with its master catalog of books:

> Stone inscriptions and rubbings must be recorded. The Sung official Cheng Chiao made two surveys of the stone rubbings missing from earlier listings. Ou-yang Hsiu and Chao Ming-ch'eng prepared lists of stone inscriptions, while Nieh Ch'ung-i and Lü Ta-lin made compilations of rubbings.[77] These books can be relied upon in studying the past. Your servant urges that a collection of rubbings be made in addition to the collection of books. Rubbings should be made of all the steles and grave stone inscriptions of each province, sent to the capital, and collated and listed in a complete form.

Interest in epigraphy was especially great among Chu's students and contemporaries. In late October of 1771, Chu and some students had paid a visit to the Liu-li-ch'ang, or book-selling district of Peking, where they had unearthed a series of stone inscriptions which had enabled them to piece together a map of Liao dynasty Peking. The episode was a famous one, recorded in many contemporary writings, and evidently served to inspire many who heard of it. In the next quarter-century, many of Chu's students and friends published their own catalogs of stone inscriptions. In 1787 a volume of comments on the inscriptions by Ch'ien entitled *Chin-shih-wen pa-wei* (Comments and colophons on stone inscriptions) was printed. Wang Ch'ang's massive work *Chin-shih ts'ui-p'ien* (A compendium of stone inscriptions) was published in 1805. Weng Fang-kang produced in 1786 a study of Han stone inscriptions, *Liang-Han chin-shih chi* (Notes on Han stone texts). Sun Hsing-yen's catalog, titled *Huan-yü fang-pei-lu* (Stone inscriptions of the empire), was evidently a collective effort, produced by the collaboration of many of Chu Yun's students and friends in 1802.

The compilers of these catalogs saw themselves as part of a long, if occasionally broken, tradition. The oldest stone inscriptions in China dated from the Chou dynasty, but the first extant catalogs were the *Chi-ku lu* (Catalog of antiquities) of Ou-yang Hsiu (1007–1072) and the *Chin-shih-lu* (List of stone inscriptions) of Chao

Ming-ch'eng.[78] These two pioneers had a number of followers during the Sung dynasty, but during the Ming, stone inscriptions came to be regarded as objects of aesthetic rather than historical interest and the nature of catalogs changed accordingly. Ku Yen-wu (1613–1682) had "rediscovered" the importance of stone inscriptions in the early Ch'ing, using them extensively in his monumental *Jih-chih-lu* (Record of knowledge accumulated day by day), and it was to Ku that eighteenth-century scholars looked for inspiration. Chu and his friends were particularly anxious to collect stone inscriptions from the Yuan and Ming dynasties, and before the Liang dynasty, since neither of these periods had received adequate treatment in previous catalogs.[79]

It was characteristic of eighteenth-century thought to have regarded the collection of stone inscriptions as a natural complement to the collection of books; both projects were seen as establishing with certainty the words and deeds of the past. "Without stone inscriptions," Wang Ch'ang asked, "how can one check the language of the classics?"[80] Weng Fang-kan especially valued Han inscriptions because they provided the most reliable samples of the prose style of the Han, a style he regarded as more "grand and imposing" than that of any other era.[81] One could, of course, find records of stone inscriptions in such sources as local gazetteers but relying on secondary records would, Sun Hsing-yen observed, defeat the epigrapher's primary purpose. Unless one had made a rubbing or transcription with one's own hand, how could one be certain that a careless scholar had not erred in recording a name or date, introducing just the sort of error that stone inscriptions were meant to be proof against?[82] It was natural for Chu and his colleagues to look to the government, with its far-flung personnel network and capacities for standardization, for leadership in an effort to collect all surviving stone inscriptions in the empire.

With these four proposals, Chu signaled the willingness of those around him to participate in the imperial book collecting project, and outlined some of the benefits they hoped to realize from such an effort. That Chu was expressing the attitude of the scholarly community as a whole has been one of the principal arguments of this chapter. Chu was by no means typical of eighteenth-century scholars or scholarly patronage; the resources he could command, the talent he could attract, and the depth of his commitment to the philological method were all exceptional. But his circle embodied in their most advanced form the social, economic, and intellectual trends which would shape late eighteenth- and early nineteenth-century thought. Chu's proposals were not, therefore, the *pro forma* response of a bureaucratized literatus to imperial initiative; they reflected the genuine desires of a creative intellectual in touch with the hopes and concerns of his contemporaries.

The ends Chu sought may be both broadly and narrowly conceived. The influence that a project of the sort Chu was proposing, sanctioned by imperial authority, could have had over scholarly life was enormous. The prospect of proclaiming his own principles from the practically unimpeachable imperial pulpit was pleasing, no doubt, for Chu. The need to find employment for intellectuals outside of or on the peripheries of government service in an era when population was increasing and the number of places in government remained constant may also have lain behind Chu's proposals.[83] As David Nivison has written: "It may be that one of Chu's motives for initiating the Ssu-k'u project was the hope of finding employment for some of his more promising proteges. A man like Chu, who had the means to play the patron on a small scale, would probably attract more students than he could support."[84] Certainly, the Ssu-k'u project afforded significant opportunity for the employment of out-of-work philologically inclined intellectuals. It should be noted, however, that most of the staff of the Ssu-k'u project came from the ranks of men who already held high official degrees and positions.

Important as these considerations may well have been, they hardly accounted for the fervor or consistency that can be found in so many of the writings of Chu and his friends. The fact, for instance, that generations of Ch'ing intellectuals expressed the same desire to see the *Yung-lo ta-tien,* the immense amount of time and energy spent in collecting and classifying stone rubbings, and the remarkable dedication of eighteenth-century intellectuals to Han dynasty scholarly institutions suggested a deeper commitment. One of the hallmarks of Confucian thought has always been the belief that a golden age had actually existed in the past. *K'ao-cheng* scholars believed sincerely that by studying ancient texts they could recapture ancient truths. Therefore, in addition to the delights and economic rewards of scholarship, *k'ao-cheng* scholars had the ethical satisfaction of knowing that they had done their duty as Confucians and contributed to reestablishing truth. They believed that their cause was one that both merited and required government support.

4

Scholars and Bureaucrats at the Ch'ien-lung Court:
The Compilation of the *Ssu-k'u ch'üan-shu*

The process of compiling the *Ssu-k'u ch'üan-shu* was above all an exercise in coordination and compromise. That the court and the scholarly community both saw value in the effort evidenced a certain area of agreement between them. Both saw as desirable a coordinated project to collect, collate, and comment on China's literary and scholarly heritage, and both recognized the propriety of government leadership in such an endeavor. But there were areas of disagreement as well, disagreements apparent more often in the practices than in the pronouncements of emperor and scholars. Often implicitly at stake, for instance, was the question of whether the enterprise was to be undertaken for its own sake, or whether it was to serve larger political and ideological purposes. The Chinese bureaucracy also brought its own peculiar biases, limitations and capacities to the tasks assigned it, those of collecting books, selecting editors and detecting treason, negligence, and error. The compilation effort succeeded only to the extent that areas of agreement between emperor, bureaucrats and scholars were broad enough to support the intellectual and institutional edifice built upon them; areas of disagreement could be expressed and resolved; and bureaucratic pattern and precedent could be harnessed for the realization of editorial goals. This chapter will examine how the work of compiling the *Ssu-k'u ch'üan-shu* was organized and managed, and how imperial, scholarly, and bureaucratic interests were expressed and furthered in the process.

To assess the significance of bureaucratic interests in the Ssu-k'u

project, one must address the larger issue of roles in the Chinese government.[1] While all Chinese officials were to some extent scholars by
training and politicians in practice, some were clearly more inclined
toward scholarly and literary pursuits, and others more interested in
attaining and wielding political authority. The difficulties involved in
defining and differentiating the two groups, however, would horrify
even the least fastidious analyst. The differences between them were
not wholly expressed in institutional allegiances—even a relatively
"literary" institution like the Hanlin Academy could accommodate
men of both literary and political inclinations.[2] Nor were the individuals themselves likely to articulate their stance in relation to this
dichotomy for, in Confucian theory at least, "cultivating the self" and
"ordering society" were "two parts of an indivisible whole."[3] Furthermore, the composition and political weight of the two groups would
likely have shifted with changes in the political fortunes of the ruling
dynasty.[4]

Yet differences of outlook between scholars and bureaucrats did
fundamentally affect the workings of the Ssu-k'u Commission. Try
as he might to represent his circle as a group of apolitical "friends"
in search of scholarly truth, once Chu Yun sought imperial patronage for projects of his own design, he had entered an intensely competitive realm where few favors were granted easily and even the
most trivial matter could occasion violent dispute. The reason for
this had to do with the nature of political and social authority in late
imperial China. Despite the complexity of eighteenth-century society, there existed within it only one center of legitimate authority,
the emperorship, and all who sought to legitimate their projects had
to do so through his office. The emperor himself, of course, was the
ultimate arbiter, but much of the capital bureaucracy was established
to guide and implement his decisions. In a long reign like the Ch'ien-
lung period, the pathways to patronage were at once well-known and
well-worn. Any effort by one group to utilize them for its own ends
would be carefully watched by all who took interest in the process of
Ch'ing government.

The process began in the case of Chu Yun's project when his
memorial reached the capital on 3 January 1773. In the initial
consideration of Chu's memorial and at each subsequent stage in the
process—shaping the institutional form of the project, selecting and
administering the editorial staff, and reviewing the final product—

compromises were made which reflected the interests and concerns of those involved. The texts of the *Ssu-k'u ch'üan-shu* reflected these compromises, and the often critical assessments of the work have served to highlight their nature.

Decision-making at the Ch'ien-lung Court
The Establishment of the Ssu-k'u Commission

The bureaucratic environment which conditioned the Ssu-k'u project was first apparent in the official consideration of Chu Yun's memorial. It was an environment shaped both by institutional factors and individual personalities although, to be sure, the extent to which the two interacted has been obscured by imperial rhetoric and two centuries of often biased history writing. The formal procedures, at least, were clear. When it was received at the capital, Chu's memorial was first referred to the Grand Council for deliberation. A memorial summarizing the Council's deliberations was submitted to the emperor in late February, to which he responded in edicts of 26 February and 3 March. The last of these edicts named and formally established the Ssu-k'u Commission.

Thus the first members of the Chinese bureaucracy to pass judgment on Chu's proposal were the grand councillors, who were situated at the pinnacle of officialdom in the Ch'ien-lung era. Originally a group of imperial friends convened informally to assist the ruler in administration and military matters, the group had become by the late eighteenth century an organ of policy formation worthy of its rather imposing title which exercised control over most of the functions of government.[5] It was composed of senior statesmen and prominent representatives of the major interest groups in Ch'ing politics. Usually the council was dominated by Manchus. For over twenty years during the Ch'ien-lung reign, the dominant figure was Fu-heng (d. 1770), the emperor's brother-in-law, president of the Board of Revenue, and mastermind of most of the military campaigns of the era.[6] His death in 1770, however, had left the council leaderless, and the Manchus consequently were represented only by relatively junior figures of little influence.

Indeed, as Chu Yun and his friends undoubtedly saw, the 1770s were the first time in the eighteenth century when the council was dominated by Chinese. But the senior Chinese statesmen were men of very different personality. Liu T'ung-hsun (1700–1773), Fu-heng's successor as chief grand councillor, had made a remarkable career for himself instigating and investigating charges of corruption against imperial friends and relatives. In 1741, Liu, then a newly appointed president of the Censorate, indicted Chang T'ing-yü, who was then chief grand councillor, president of the Board of Personnel and unofficial regent for the young Ch'ien-lung Emperor, on charges of corruption and nepotism. This charge was followed in subsequent years by indictments of the imperial in-laws and powerful governors and governors-general. Strikingly, his charges were usually upheld by the emperor, a fact which suggested that at the very least Liu kept a close watch on the emperor's own moods and perceptions. Nonetheless, Liu earned for himself a reputation for honesty and incorruptibility, and his own career was untouched by scandal.[7]

Yü Min-chung (1714–1780), the second senior member of the Council, had a reputation almost the opposite of Liu's. Liu had come from a relatively humble northern China family, but Yü was a representative of the Chiang-nan literati who had been so assiduously cultivated by the emperor in the early years of his reign. Yü's career was founded on his literary accomplishments as well as his several terms as civil service examiner and President of the Board of Personnel. The place he had carved out for himself between Manchu rulers and Chinese subjects was a vital one in an administration so self-consciously organized on Chinese principles, but it could also be a precarious one. Yü was accused of corruption, probably with reason, several times during his life, but was usually saved by the emperor's personal intervention until his posthumous demotion for revealing state secrets in 1782.[8]

The unusually long time which elapsed between 3 January, when Chu's memorial was referred to the Council, and 26 February, when the emperor acknowledged receipt of the Council's views, suggests dispute over the proposal, and this is confirmed in eighteenth-century sources.[9] Yao Nai, an associate of Chu's well connected in the senior bureaucracy, reported in a biography of Chu that Liu argued strenuously against Chu's proposal during Council discussions, remarking that the project was not necessary for government and would become an empty nuisance. Support for Chu's proposal was said to have come from Yü Min-chung.[10] Perhaps because of Yü's subsequent disgrace for revealing state secrets, eighteenth- and nineteenth-century sources were discreetly silent about his contacts

with literati and his remarks on state policy. But his advocacy of
Chu's project would have been very much of a piece with his role as
defender of Chinese literati interests at the Manchu court.

In view of this configuration of forces, what factors ultimately
determined the fate of Chu's proposals? If personal contacts alone
had done so, there probably would have been little problem, for Chu
was well connected with all the centers of power and authority in
Peking. This fact, perhaps as much as his amiable nature and love
of parties, accounted for his prominence in scholarly circles and the
pivotal role he played in the inauguration of the Ssu-k'u project. Chu
Yun and Liu T'ung-hsun, for instance, had known each other for
over twenty-five years and were to all appearances quite close.[11] In
1750, Liu invited Chu Yun to his house to help in compiling the
Sheng-ching-chih (History of the capital). A contemporary noted that:

> When there was a discussion of law or public affairs at court and Liu
> had a question, Chu often volunteered the information. . . . When their
> opinions differed, they argued forcefully. T'ung-hsun was quite stern and
> even of those in his generation, few dared to question him lightly. Chu
> alone dared speak freely with him.[12]

Liu came to feel an almost fatherly affection for the younger Chu.
When Chu passed his *chin-shih,* Liu was said to remark: "You can no
longer reprimand me with stories of old officials. I am old and can
no longer do anything. You must now be energetic."[13]

Chu Yun also had even higher connections. He had accompanied
the emperor on a tour to Mu-lan in the summer of 1758, and had
worked on literary projects close to the ruler's heart. Chu Kuei reported that during an imperial audience the emperor asked specifically after the whereabouts and occupation of Chu Yun.[14] Later,
when Chu Yun was returning from Anhwei in 1773 to take up a post
as reviser for books submitted from the provinces at the Ssu-k'u
Commission, his friends advised him to return quickly, rather than
sightsee in Shantung as planned, since the emperor inquired about
him frequently.[15] The very fact that Chu memorialized on the book
collection project at all suggests his remarkable position, for it was
most unusual for an education commissioner to memorialize on any
but routine matters.

Chu, it may be surmised, would not have memorialized unless he
had reason to believe his proposals would meet with a favorable
response. As close as Chu's relations were with Liu and the emperor,

however, they did not guarantee smooth sailing for his memorial when it reached Peking. That Chu's ties did not pave the way for his proposals suggests just how important apparently subtle differences of perspective could be in the highly political environment of the Chinese court. The differences between Liu and Chu were of priorities and experiences. Liu had served for over twenty-five years in senior substantive posts in Chinese government, defending central interests against the spendthrift tendencies of senior officials and imperial friends. Chu, though thoroughly familiar with court life, would never hold a substantive post and seemed consciously to avoid government service. Yü Min-chung, on the other hand, had founded his career on his proximity to the emperor and his ability to sponsor Chinese literati and their projects. Each of these perspectives was probably typical of the strategies of many Chinese trying to succeed in a Manchu court, and each would shape attitudes toward the tasks of compiling the imperial library.

Precedent and public opinion both sanctioned the various stances of politician and literati toward the Ch'ing court. For instance, grand councillors were not supposed to receive calls, except on public business. When Chu was in the Hanlin and Liu was a councillor, they met at court one day. Liu, noting that he had not seen Chu for some time, asked "Have you forgotten me?" Chu answered: "I wouldn't dare to visit you except on public business. People would gossip." Liu sighed and agreed.[16]

Two other incidents during the period under review, though they did not directly involve Liu and Chu, demonstrate the fairly rigid separation between individuals of the two orientations. In the fall of 1773, some of the editors of Ssu-k'u tried to elevate one of their number to the Grand Council. Yü Min-chung, a councillor, wrote back to them informing them quite firmly that what they requested could not be done, since "The Grand Council and the Hanlin are two different roads."[17] Some years later, in 1796, the names of three Ssu-k'u editors, Liu Lun, Chi Hsiao-lan and P'eng Yuan-tuan, were being considered for positions on the Grand Council. The emperor rejected all three of them: "Liu Lun has never shown himself willing to work really hard, P'eng Yuan-tuan is not attentive to his conduct, and Chi [Hsiao-lan] though he has read many books, does not really understand *li*."[18] Although there were many personal and social bridges joining the worlds of the scholar and politician in Chinese officialdom, there were important differences in the outlooks, priorities and career patterns of the two groups. The significance of these differences were quite apparent in the Grand Council's comments on Chu Yun's memorial, which were received by the emperor on 26 February 1773.

The Grand Council memorial was neither as idealistic as the emperor's original edict, nor as committed as Chu Yun's responding memorial. It was an even-handed discussion, which balanced considerations of ideological significance, historical precedent and administrative practicality. The memorial considered the merits and feasibility of each of Chu's proposals separately. The generally negative tone of the document which probably reflected Liu T'ung-hsun's influence was suggested by the order in which the recommendations were offered. The first two proposals considered were flatly rejected; the third and fourth were accepted only with significant modifications.[19]

Addressing first Chu's proposal that the government concentrate initially on acquiring and copying handwritten texts, especially commentaries and *wen-chi* of the Sung, Liao and Chin dynasties, the councillors asserted that:

> There are certainly enough wood-block printed copies of the best works in circulation for broad benefit and wide dissemination. While the works of some unknown authors have, perhaps, never been printed, handwritten copies [of these books] can be stored and utilized for research when needed.

The councillors then pointed out that the emperor's edict originally called for handwritten as well as printed books to be forwarded to the capital, and concluded that there was "no need to establish special procedures" for Sung, Liao and Chin books. In fact, the councillors were probably wrong: in view of the heavy demand for Sung and Yuan editions, there was probably a need for special procedures in handling these books (see chapter 3 above). The question, however, was not so much about the state of bibliography as about the extent of government responsibility for book preservation. Chu had urged that the government take on a new obligation in this regard; the councillors objected.

The councillors adduced both historical precedent and an assessment of the difficulties involved to oppose Chu's suggestion that the government compile a master list of all the stone inscriptions in the empire. They acknowledged the value of such an effort: "Since antiquity, inscriptions and [recorded] history have formed the warp and woof of scholarship. . . . Stone inscriptions are the oldest historical source, can be relied upon in examining the past, and never lose their truth." However, neither Juan Hsiao-hsu (479–502), in his *Ch'i-lu* (Seven Summaries),[20] nor Ma Tuan-lin, in the bibliographical section of his *Wen-hsien t'ung-k'ao*[21] had established special categories for stone inscriptions. Furthermore, the work of compiling stone inscriptions belonged to specialists in epigraphy, and was not for all scholars. While it might be useful to assemble all the catalog of stone inscriptions in one place,[22] the government could hardly undertake a master catalog: "There are numerous stone inscriptions that can be used to examine the truth of the classical canon, and many of them are located in out of the way places. If local officials are ordered to copy them all out, we fear that in some area the work will be managed improperly and there will be trouble." It is not clear whether by the phrase "there will be trouble" (*chuan-tzu fen-jao*) the councillors envisioned a distortion of the meaning of stone inscriptions, the difficulty of getting at bronzes which were located in many cases in private collections, or simply the disruption of ordinary and official life which would occur if magistrates were commanded to comb the countryside hunting for steles. In any event, the councillors rejected Chu's proposal.

Considering Chu's suggestion that books in the imperial household, including the *Yung-lo ta-tien,* be made more accessible, Liu and his colleagues took some pains to point out that the analogy Chu had drawn between the Ch'ing and Han imperial libraries was inaccurate. Whereas in Han times it was important that a catalog of the imperial library be prepared since commoners were forbidden to enter the imperial library, the Ch'ing had made a special effort to make the learning of the sages available to scholars of the empire, publishing standard editions of the thirteen classics and twenty-one histories and distributing them free of charge. The councillors were thus more concerned with establishing the principle that government-sponsored works were accessible to the public than with the issue of how widely they were diffused. It was practical difficulty, not principle, that inhibited further work on the *Yung-lo ta-tien.*

The councillors pointed out that the *Yung-lo ta-tien* was so large that Hanlin compilers were not allowed to see it, therefore Chu Yun could only have a vague notion of what he was proposing. To ascertain the true situation, Grand Council clerks had been sent to inspect the Yung-lo encyclopedia. The clerks found, first, that only nine thousand of the original eleven thousand volumes were extant and, second, that the plan of the encyclopedia, quoting from diverse sources on each topic, made it difficult to reconstruct original texts and therefore inconvenient to use. The councillors concluded that there were many books in the Yung-lo collection that could fill gaps in the libraries of the empire. But, they remarked, "such an effort was not the original intent of the project." They recommended a compromise procedure. Instead of commissioning members of the Hanlin Academy to copy out texts from the *Yung-lo ta-tien,* the councillors suggested that the Hanlin members copy out only the titles of works not in circulation. Once the list was finished, a final decision could be made about which works were to be copied out in full.

Finally, the councillors discussed Chu's proposal that a short comment on the value and significance of each work be appended to the first *chüan* of the text in the Imperial Manuscript Library. As in the case of Chu's proposal for a catalog of stone inscriptions, they acknowledged the value of such a procedure, quoting the *Han-shu* view that Liu Hsiang's "summaries of the chapters and sections of the work he collated, and excerpts of their major conclusions" performed an important service for the reader.[23] Nonetheless, they observed:

> There are so many more books in circulation now than there were in Han times, and since the emperor has decreed that this collection of books be broader than has ever been undertaken before, if we append a comment at the beginning of every book, the resulting set of comments will be as broad as the seas.

The councillors then cited the examples of the *Ch'ung-wen tsung-mu* (Catalog of the imperial library at the Ch'ung-wen Pavilion) by Wang Yao-ch'en,[24] and the *Ch'ün-chai tu-shu-chih* (Ch'ün-chai's notes on reading) by Ch'ao Kung-wu (fl. 1144),[25] annotated catalogs which were compiled after the collections of books they were based upon had been assembled. After the work of assembling books had been completed, the coucillors suggested, some court officials could be assigned to prepare a catalog listing the title and author of each book in the collection.

The councillors' position was a complex one. As men of learning, they recognized the value and historical precedent for much of the work that was proposed. As senior officials of the government, however, they would ultimately be responsible for carrying out the proposals adopted, and could judge the commitments of time and personnel necessary. The second of these considerations was stressed throughout their memorials, and apparently shaped the council's recommendations. In view of the administrative problems that arose during the course of the project, the councillors' hesitations were perhaps well-founded.

But ultimately, of course, the emperor's was the final voice in the decision-making process, and his relation with the proposed book collection project was a unique one. Unlike the scholars, his concern was not so much to determine the truth as to proclaim and defend it. Unlike the bureaucrats, his position rested not on his ability to perform the tasks of government, but on his ancestors' achievements, and ultimately on his subjects' perception that he was carrying out the mandate of heaven. The Ch'ien-lung Emperor himself, insofar as we can tell, was a man of excellent classical education and some literary and scholarly inclination. But he was also a very political man, supremely concerned with establishing and upholding the power and prestige of his unique position.

The importance of these considerations led him to take a rather unusual step when confronted with his councillors' memorial. Most of Grand Council deliberation memorials were, if not binding on the emperor, at least documents to be taken very seriously.[26] In the winter of 1773, however, the emperor selectively took issue with his councillors' recommendations. In the edict of 26 February, the emperor accepted their recommendation that the government not undertake the compilation of a master catalog of stone inscriptions, or concentrate particularly on books of the Sung, Chin and Yuan dynasties.[27] However, he objected to the methods proposed for work on the *Yung-lo ta-tien* and overruled them on the matter of short summaries. In both cases, the emperor argued that the merit of Chu Yun's proposals overrode the practical considerations advanced by the councillors.

The work of recovering books from the Yung-lo encyclopedia, the emperor conceded in his first edict, could go on for months, even years, without completion. Yet it was important work, "particularly because the search for books at the time [the encyclopedia was compiled] was so broad, and there could be rare works in it which" were seldom seen in the eighteenth century. In order to insure that the work was carried out properly and as expeditiously as possible, the emperor ordered grand councillors themselves to serve as editors-in-chief for the project. Under their direction, Hanlin members were to serve in rotation as collators for the *Yung-lo ta-tien*, determining which texts could be reconstructed, and comparing *Yung-lo ta-tien* versions of texts to those in the *T'u-shu chi-ch'eng*. The results were to be reported regularly by memorial.

Considering his councillors' opinion that writing short summary comments at the beginning of every work collected would be too troublesome, the emperor remarked that: "If we follow Liu Hsiang's rules, then the complications will be endless." However, he noted that "a brief summary and description was inserted in each work stored in the imperial library during the K'ang-hsi period, and these summaries are indeed useful in examining the books." Therefore, the emperor ordered that when all the books were assembled, those serving as collators should prepare short summaries for insertion in the texts.

Four days after this edict was issued, the councillors sent to the emperor the sixty *ts'e* of the table of contents of the *Yung-lo ta-tien* on which they had based their recommendations. The following day, 3 March, two edicts were issued, one commenting on the Ming encyclopedia, the other naming the proposed book collection project.

The emperor pronounced himself unimpressed by the *Yung-lo ta-tien*.[28] Although the preface was very broad in scope and full of high sounding phraseology, he asserted that it was really nothing more than a reference book, an exercise in intellectual covetousness (*t'an-to wu-te*). Its use of rhyme as a principle of organization distorted and disordered classical teachings, which were the origin of all learning. For instance:

> Of the hexagrams in the *I-ching*, the *Yung-lo ta-tien* recorded the *men* first; of the sections in the *Shih-ching*, it records the "Ta-tung" first; and of the sections in the *Chou-li*, it considered the "Tung-kuan" first.[29]

all of which was contrary to the order established by the sages. Ch'ien-lung remarked that his own book collection would be organized on the immutable four treasuries system. This system, in which books were grouped into four categories — classics, histories, philosophies and belles lettres — had first been used during the Chin dynasty (265–313). Although quite popular during the T'ang and Sung dynasties, it had gone out of use after the Sung, so that the Ch'ien-lung era use of the system represented a historical revival.

The purpose of this revival was never made explicit, but fairly clearly the emperor preferred to associate himself with the glories of T'ang and Sung rule rather than the ignominies (in his view, at least) of Ming disorganization. Although the Ch'ien-lung collection would draw upon the Yung-lo encyclopedia, it would clearly supersede it, both in scope and in organization.

The same day, the book collection was officially named *The Complete Library of the Four Treasuries (Ssu-k'u ch'üan-shu)*.[30] In so naming his collection, the Ch'ien-lung Emperor was making a statement about bibliographical classification, but, perhaps as important, he was also making a statement about the nature of the project he envisioned. As Yao Ming-ta has pointed out, while imperial book catalogs and encyclopedias compiled under imperial sponsorship had names, book collections did not. In Yao's view, the Ch'ien-lung emperor was trying to combine the name of an encyclopedia with the reality of a book collection.[31] It may also have been that the opportunity of affixing a name of his own devising to the corpus of Chinese learning appealed to the emperor's sense of his own role in Chinese government and letters.

These edicts, transparent but certainly not unprecedented examples of imperial breast-beating, reflected some of the emperor's motives in the Ssu-k'u project. Clearly, the idea of superseding a predecessor, particularly a Ming predecessor, was important to him. Equally apparent is the assumption that an emperor should defend the truths of classical teaching, and could affect by edict the organization of knowledge in his realm. Above all, however, these edicts reiterated the notion, first expressed in the initial edict on the Ssu-k'u project, that the project was to represent a celebration of the unity of knowledge and power achieved in the Ch'ien-lung reign. That such concerns should have led the Ch'ien-lung Emperor to order his most senior councillors to supervise the recovery of books from a four-hundred-year-old encyclopedia and the preparation of a comment on every book extant in the empire (and this at a time when the nation was at war in the southwest and soon to be in rebellion in the northeast) was characteristic of the emperorship, the man who held it, and the era.

Institutional constraints and individual predilections both shaped the attitudes that emperor, bureaucrats and scholars brought to the *Ssu-k'u ch'üan-shu* project. The circle of Chu Yun, at one with other scholars who contended that only through the most thorough and meticulous canvassing of the classical canon could truth be established, saw in the project an opportunity to accomplish scholarly goals beyond their own resources. Overall, the Grand Council's reaction was one of concern with the strains such a project could put on the fabric of personnel and interests that governed China, although there is evidence that some individual councillors, like Yü Min-chung, favored the project because it seemed likely to enhance their position in court politics. The emperor, though perhaps personally inclined toward literary matters, was also concerned with assembling a collection of unprecedented scope and proving himself worthy of occupying his position as leader of the literate community. In a later time, when military and political necessities would galvanize a reluctant court into action, these differing perspectives could harden into factional allegiances. In the late eighteenth century, however, they remained differences of perspective, which in a leisured era could be expressed with confidence and security. They did, however, have a profound impact on the junior men who would be appointed to the Ssu-k'u Commission.

Collators and Commitments

The errors in Chinese texts — errors which had arisen through centuries of transmission in an orthography which, even to those most familiar with it, offered seemingly endless possibility for both meaningful and meaningless variation — explained, at one level at least, the need for an effort like the Ssu-k'u project. But the long and sometimes tormented history of Chinese texts hardly explained why so many aspiring bureaucrats of the 1770s and 1780s were anxious to embark on a task of such staggering scope. Strikingly, some of those who were to be employed as editors on the project did not have substantial reputations as scholars at the time the project began, and few went on to careers in the eighteenth-century equivalent of academe. They were, on the whole, men bound for political office, and they saw in the Ssu-k'u project an opportunity to advance their bureaucratic careers.

The balance of literary ability, personal connections and administrative competence which determined a candidate's chances for employment and promotion in Ch'ing China was always a delicate one. But evidence suggests that it was changing in the last years of the eighteenth century, with significant consequences for the way in which government was to be conducted for the next several decades. In particular, the influence of individual grand councillors over this and many other processes of government was growing. Because the changes in personnel procedure affected the Ssu-k'u project, and because the project affords a unique opportunity to view changes with major impact on the subsequent structure of government, some attention will be devoted here to selection and administration.

The emperor's main concern was that responsibility in the Ssu-k'u project be clearly delegated and delineated. Sometime in the spring of 1773, the Grand Council evidently suggested to the emperor that members of the Hanlin Academy serve in rotation on the Ssu-k'u Commission without specific appointments. This was evidently something of a rear-guard action; having failed in their efforts to limit the scope of the project, councillors now sought to diffuse responsibility, hoping that in doing so they might limit the workload. The emperor opposed this, ordering the council to select a definite number of men for the Commission, and assign them specific responsibilities. The councillors dutifully memorialized on 2 April that this had been accomplished.[32]

Formal responsibilities brought with them formal titles, but not impersonal recruitment. The methods for selection of Ssu-k'u revisers parallelled the procedures used for the selection of provincial education commissioners, and examiners for the *chü-jen* and *chin-shih* examinations. The councillors relied on examination results, personal knowledge of the candidates and the recommendation of friends in making the appointments. Friends' recommendations were particularly important in selecting more scholarly collators. Tai Chen and Yü Chi were, for instance, appointed after being recommended by their teacher Ch'iu Yueh-hsiu, then serving as president of the Board of War, to Grand Councillor Yü Min-chung.[33] Apparently many at the Hanlin Academy tried to prevail on Chu Yun to put in a kind word on their behalf. But Chu chose not to hear their pleas, characteristically scorning the rewards of government service: "What the wealthy and powerful compete for is inconsequential." He added, in what must have seemed cold comfort to those anxious to begin their careers with an appointment at the Ssu-k'u Commission, "self-satisfaction can be obtained from heaven alone."[34]

No complete list of the men appointed by the council to the Ssu-k'u Commission in 1773 has been preserved. However, a list of those who served on the Commission, dated 1782, is published in the first *chüan* of the *Ssu-k'u ch'üan-shu tsung-mu t'i-yao.*[35] The work of collation and recovery was entrusted to four categories of revisers (*tsuan-hsiu*):[36] revisers for the collation of books submitted from the provinces, revisers for the collation of books from the *Yung-lo ta-tien,* a group of troubleshooters known as "yellow tally research revisers," and revisers for astronomy (for their relative ranks, see fig. 1). Fifty names were listed in the four categories. Forty of them had served in the Hanlin Academy prior to their appointment; twenty-three were Hanlin bachelors at the time of their appointment.

They were a relatively young group; the ages at the time of appointment, of those whose ages can be determined, ranged from 26–48, and the median age was 31.[37] The old men of the group were Tai Chen (48), Li Shou-chien (in his forties), Chu Yun (43), and Chou Yung-nien (42), all men appointed from outside the Hanlin Academy and known primarily for their scholarly ability. The majority of the revisers, men appointed directly from the Hanlin Academy or related institutions at court, were men at the beginnings of their official careers.

A modified version of the method devised by Adam Yuen-chung Lui in his book, *The Hanlin Academy: Training Ground for the Ambitious,* can be used to characterize the subsequent career patterns of the men appointed. Lui assesses the importance of various groups of Hanlin bachelors by determining how many of the bachelors in a given group reached the third rank or above, and finds that, for the Ch'ien-lung period as a whole, 21.2 percent of the Hanlin bachelors reached this rank.[38] The men appointed as Ssu-k'u revisers did significantly better than this average. Of the forty men appointed to the Commission who had served in the Hanlin Academy prior to their appointment, twelve, or thirty percent reached the third rank. Among men who were relatively young at the time of their appointment, men admitted to the Hanlin from the *chin-shih* examinations of 1769, 1771 or 1772, the success of those who served at the Commission was more striking. Twenty-five of the one hundred men in these three classes were appointed to the Ssu-k'u Commission; from this pool came sixty-three percent of the senior officials in the three classes. Clearly, the young men appointed at the Commission were an elite, even among their Hanlin colleagues, destined for career success.[39]

The critical question is what these statistics meant. Did the selection process merely identify members of a pre-existing elite? Or was

Figure 1. The Staff of the Ssu-k'u Commission,
Arranged by Bureaucratic Rank

Rank*	Title (Number of Appointees)
1A	Directors-General (16)
1B	Assistant Directors-General (6) Senior Readers (15)
2A	
2B	
3A	Chief Editors (3); Chief Collator (1)
3B	
4A	
4B	
5A	
5B	Revisers for the collation of books submitted from the provinces (6)
6A	Hanlin Proctors (20)
6B	Revisers for the collation of books from the *Yung-lo-ta-tien* (38)
7A	Assistant Editors for the *Annotated Catalog* (7)
7B	Wu-ying Throne Hall Proctors (9)
8A	
8B	
Unranked	Copyists (212)

*Service on the Commission did not carry an official rank. These figures represent the ranks Ssu-k'u personnel had when they were appointed to the Commission staff.

subsequent career success a result of service on the project? If the latter, how direct a result was it, and how could the prestige acquired by working on an imperial project manifest itself as career advancement? At issue in these questions, which must have been as apparent to young men contemplating service at the Ssu-k'u Commission as they are to us today, was the character of bureaucratic loyalty. Although many of the Ssu-k'u recruits were men at the beginnings of their careers, they must have been aware of the many scholarly projects and institutions which constituted precedents for the Ssu-k'u effort, and their impact on the lives of those who had served in them.

Career success was probably not a direct result of service at the Commission. Ostensibly, the reward for meritorious service at the Commission was the privilege of *i-hsu*, advancement on the waiting list for appointment to office. On two occasions, in March of 1778 and November of 1780, this privilege was awarded to Ssu-k'u revisers, and a total of eleven revisers were so rewarded.[40] As the subsequent careers of the men rewarded suggest, the significance of the privilege was not great. Wang Tseng (*chin-shih,* 1771), rewarded in November of 1780, was demoted from his post as compiler and assigned to a post in the provinces three years later.[41] Ch'en Ch'ang-ch'i (1743–1820), rewarded in 1778, was found to be a third-rate official in the metropolitan personnel evaluations of 1785, and demoted.[42] Chou Ping-t'ai (1746–1821), placed at the head of the list for appointment in 1778 and 1780, was not promoted to substantive office until 1788, and this promotion appears to have resulted more from favorable imperial reaction to his proposal that the *pi-yung,* a Han dynasty institution in which the emperor lectured on the classics, be reinstituted than from his Ssu-k'u service.[43] A fourth reviser, Wu Sheng-lan (d. 1810), was also recommended for advancement to office twice, but had to wait eleven years for his chance; when it came, his promotion was a big one, however, from assistant secretary of the Supervisorate of Imperial Instruction (rank 6b) to vice-president of the Board of Works (rank 2a).[44]

On the other hand, the prominence of Ssu-k'u revisers in government was probably not coincidental. Appointment to the Commission meant that one had attracted the attention of the grand councillors, either through personal achievement or the recommendation of friends and patrons. This attention was a vital factor in Chinese official careers, probably expediting one's first appointment out of the Hanlin Academy, and certainly necesssary for promotion into the highest ranks of government service.[45] Councillors could even make life more pleasant for those serving in the Hanlin by recommending them for the lucrative commissions as *chin-shih* and *chü-jen* examiners that supplemented the modest stipends paid the bachelors.[46] In fact, attention from on high played a decisive role in several of the examples discussed in the previous paragraph. Wu Sheng-lan was a student in the school of the Palace of Universal Peace for children of bannermen at the same time as Ho-shen and became his protégé. Wu rose very rapidly at court after Ho-shen became a grand councillor, holding four commissions as chief *chü-jen* examiner and two as education commissioner in spite of the fact that he did not possess a *chin-shih* degree. He was appointed vice-president of the Board of Works in 1788, and vice-president of the Board of Rites and grand secretary in 1792. When the Ch'ien-lung Emperor died and Ho-shen fell, Wu fell too, being demoted overnight to the rank of compiler at the Hanlin Academy.[47]

Ch'en Ch'ang-ch'i's career illustrates the negative impact a councillor could have on an official career. Appointed to the Ssu-k'u Commission in 1773, and commissioned chief *chü-jen* examiner twice in the next four years, Ch'en appeared to be on his way to a distinguished career. Ho-shen, at the time a councillor, evidently noticed Ch'en and passed word that he would like to have Ch'en call. Ch'en refused, asserting that "since he is not head of the Academy, there is no etiquette (*li*) for visiting him." This was taken, probably correctly, as an insult by the councillor and as a result Ch'en was found to be a third-rate official in the next metropolitan personnel evaluation.[48] Ch'en, a feisty Cantonese, persisted in his course of opposition to the leadership, submitting a memorial some years later proposing that the education commissioner of Chihli be empowered to undertake an investigation of corruption among senior officials at the court. It received no official response. Long after Ho-shen died, Ch'en regained imperial favor with a series of courageous and knowledgeable memorials on defense against pirates along the Fukien and Kwangtung coasts.[49]

Although recommendations were an important factor in personnel administration throughout the Ch'ing period, there were important changes in their significance in the very years that the Ssu-k'u project was underway. Yü Min-chung may well have opened the door for informal influence when he abolished the rule, to which Liu T'ung-hsun had adhered strictly, that a grand councillor could not meet other members of the bureaucracy except on official business. It was perhaps in light of this change in the social ground rules of government service that Ho-shen took offense at Ch'en Ch'ang-ch'i's refusal to visit him, although it was possible that Ch'en was protesting the rule change as well as snubbing his superior. In fact, during his term of office, Yü Min-chung was said to have "controlled the destinies of all the young officials of the empire."[50] Yao Nai, who was appointed a reviser on the Ssu-k'u Commission on the recommendation of Liu T'ung-hsun, evidently attracted Yü's attention during his service. Yü asked Yao to "become his follower" (*ch'u ch'i men*), and slated him for service at the Censorate. Yao, however, refused, and eventually returned to his ancestral home in Anhwei.[51] To be sure, the most egregious examples of nepotism occurred during the tenure of Ho-shen as grand councillor, but he did not create the system. The Ssu-k'u revisers, at least, had had ample experience of it before the young Manchu's rise to power.

This changing situation probably affected morale and performance at the Commission. The members of the Ssu-k'u staff who advanced to high office did so in large part because of their ability to attract and hold the attention of senior officials, and their appointment to the Commission merely represented the first evidence of such ability. Many of those appointed to the Commission had, of course, distinguished themselves by their literary and scholarly achievements prior to their appointments. It goes without saying that such men had an aptitude for the kind of work they were to be involved in at the Commission. However some, or perhaps all in differing degrees, saw the project as an opportunity to gain access to the corridors of power in Peking, and were thus committed to book editing as a means rather than as an end.

Working at the Ssu-k'u Commission may have been prestigious, but it was not meant to be profitable. In their memorial on the organization of the Commission, the grand councillors asserted, perhaps pointedly, that there was no need to pay the appointees from the Hanlin or other institutions in "food, firewood or silver."[52] Within a month, the emperor reversed this decision, prescribing that the appointees be provided with food and drink according to the precedent of the employees at the Wu-ying Throne Hall. According to a memorial of 1783, this provision was converted into silver payments. The rate for collators was not given, but supervisory personnel were paid 4.8 ounces of silver per month.[53] At irregular intervals, the emperor rewarded each member of the editorial staff of the Commission with a small gift, in one case a bolt of satin, in another case a cantaloupe.[54] Revisers had another perquisite which, as the case of Sun Ch'en-tung (1737-1781) demonstrates, could be profitable, that of nominating one person of suitable education and calligraphic ability to serve as a copyist. The copyists so appointed were not paid anything, but were rewarded with the privilege of advancing in the waiting lists for appointment to office. Sun Ch'en-tung passed third on the *chin-shih* examinations of 1772 and was appointed reviser on the Ssu-k'u staff. His nineteenth-century biography portrayed him as politically naive but of extraordinary integrity. Dutifully nominating a copyist recommended to him by the senior editors, he was astonished and angered when the man offered him money in return.[55]

As the number of books to be inspected grew, the physical locations of Ssu-k'u activity multiplied. In the early spring of 1773, the collators were based in a set of rooms adjoining the west side of the Hanlin Academy. By April of that year, a storeroom of the Imperial Printing Press and Bindery was converted for use. Evidently, more book storage space was necessary; a memorial of June 1778, reported that the premises of the Military Archives Office were so full of books that they would hold no more, and that the stacks of books were interfering with the daily work of the Office. By 1783, collation was taking place at four sites, each with its own supervisory and support personnel, and special budgetary provisions were made for maintenance and security at book storage warehouses.[56]

The arrangements for supervision were important both to the revisers, for whom the chance of meeting senior statesmen was an important reward for work at the Commission, and for the grand councillors who ultimately bore responsibility for the work. Quite early, the councillors urged that several of their clerks be established as proctors (*t'i-tiao-kuan*) whose responsibility was to manage the smooth flow of books and oversee the work in general.[57] The title "proctor" was ubiquitous in Ch'ing administration, appearing over thirty times in Brunnert and Hagelstrom's catalog of Ch'ing official titles. Originally coined for capital police officials in the Yuan dynasty, by Ch'ing times it seems to have had the sense of executive officer, one who "managed documents and oversaw clerks."[58] In the Ssu-k'u Commission the proctors' role seems to have been administrative rather than editorial. At least seven of the men appointed to the post had served as Grand Council clerks, three of whom did not hold a *chin-shih* degree. Five of the appointees were Manchus, and this was the only post in the Commission, other than copy clerk, to which Manchus were appointed. By 1783, and perhaps earlier, archivists and librarians as well as copy clerks were appointed at each collation site.[59]

Responsibility for management and editorial direction of the project was vested in three layers of officials. The lowest ranking of these, and perhaps the most directly involved in day-to-day operations, were chief editors Chi Hsiao-lan and Lu Hsi-hsiung (1734–1792) and chief collator Lu-fei Chih (d. 1790). Their specific responsibilities cannot be certainly established, but Lu-fei Chih seems to have been in charge of the copying of manuscripts,[60] and Chi Hsiao-lan's name was most frequently associated with the editing of the *Annotated Catalog*. Lu Hsi-hsiung had purview over a wide range of matters, but was probably most directly involved with the collation of texts. Senior to these three but with much less direct responsibility for the production of manuscripts were fifteen chief readers. They were, as a rule, presidents of boards and old political associates of the emperor or grand councillors, and served as final proofreaders of texts prepared for inclusion in the *Ssu-k'u ch'üan-shu*. At the top of the Ssu-k'u Commission were sixteen directors-general and ten assistant directors-general. Mostly Manchu princes, military leaders or grand councillors, the directors-general did not all play an active role in the Commission.[61]

Some were quite influential, however. When Yü Min-chung and the emperor were both in residence in Peking, Yü evidently conveyed imperial instructions and intentions to the editors orally. When the emperor went to Jehol for the summer months, Yü accompanied him, communicating by letter with chief editor Lu Hsi-hsiung. These letters were edited and published by Ch'en Yüan (b. 1880) in 1938.[62] Yü was succeeded in his role as interpreter by Ho-shen (1755–1799), whose activities will be detailed more fully below. The emperor evidently took a sincere and continuing interest in the project. In one of the first letters Yü Min-chung wrote to Lu Hsi-hsiung, Yü related that the emperor had asked to be consulted on "all matters requiring discussion." Yü, surprised and probably chagrined, noted, "I had no choice but to agree."[63]

Collection and Evaluation

Although the work accompanied at each stage of the process will be examined in some detail below, the steps of the process may be outlined here. One advantage the Ch'ien-lung government appeared to have over private collectors, and no doubt part of the reason Chu Yun and his friends were so enthusiastic about the Ssu-k'u project, was the empire-wide network of personnel and connections which could be enlisted in the search for rare and valuable books. The book collection process illustrated the character and capacities of this network, and the nature of the court's relations with the various strata of people who owned books in eighteenth-century China. Clearly, for a project like the Ssu-k'u compilation to succeed, it had to appeal to the wealthiest collectors who owned rare editions, as well as collectors of more limited means who might only have one or two volumes of interest to the imperial collators. The emperor entrusted provincial governors and governors-general, initially at least, with the responsibility of making contact with book collectors. Many governors, and particularly those in prosperous, book-owning provinces, established in their provincial capitals *ad hoc* committees known as book-bureaus (*shu-chü*) to oversee the process. The terminology here was not novel. Groups of prominent local leaders created by provincial governors to perform semi-official tasks were often called *chü* in the nineteenth century. *Chü* created by Tseng Kuo-fan and Li Hung-chang in antebellum Chiang-nan oversaw the provision of relief and administration of taxes, for instance; Tseng in particular also created provincial *shu-chü*, which reprinted many important classical texts, and it seems likely that modern private publishing companies have borrowed the usage of the term to fashion their own names.

These committees were generally led by expectant officials and subordinates of the governors; occasionally they included natives of the province on home leave, or famous book collectors.[64] But the basic link with the book-owning populace was a group of functionaries known as "local educational officials." These were the directors of the state academies in which first degree holders (*sheng-yuan*) were enrolled. They were generally men who had passed the regional examinations for the *chü-jen* degree, but failed in the metropolitan examinations for the *chin-shih*. Since the rule of avoidance was not applied to these positions, local educational officials were often natives of the districts in which they served, and their appointments were controlled by provincial governors. Working together with centrally appointed provincial education commissioners, local educational officials were responsible for guiding the intellectual lives and supervising the political activities of students in their districts. Despite these apparently awesome responsibilities, they were among the lowest ranking Ch'ing officials, and also were among the last groups of officials to receive "nourishment of virtue" salaries, which were not extended to them until the early Ch'ien-lung reign.[65]

Local educational officials' status did not prevent governors from placing a fairly heavy burden on them. According to governors' reports, after being notified of the nature of the Ssu-k'u compilation by local education officials, book owners took their valuable books to their provincial book bureaus, where the books were inspected and evaluated. To alleviate the necessity of sending to Peking all the books collected, the book bureaus prepared lists of the volumes inspected, which were forwarded first to the governors, then to Peking. Apparently in at least some of the book bureaus, draft statements of the contents and significance (t'i-yao) of books were prepared and sent to governors and to the capital. The emperor was suspicious of the book bureaus, or at least of the potential for disruption and disorganization created when official functions were placed in semi-official hands (a suspicion which subsequent events were to prove justified) and urged governors to watch their subordinates carefully.

The book bureaus constituted bridges built between the central government and book owners, bridges built on the foundation of the examination system and its personnel, rather than the slightly less savory foundation of magistrates' yamens and underlings. Since local educational officials could be assumed to have fairly consistent standards in the matter of book selection, and could be assumed to share with the candidates they supervised a certain pride in their region's ability to contribute to national intellectual life, there was reason to believe that the book bureaus would perform with reasonable efficiency. Book bureaus thus illustrated the potential for gentry-state cooperation, a potential which could be realized when both scholars and the state recognized a common interest.

It was a potential which had its limits, however. In the first place, the book bureaus were temporary, task-oriented groups which disbanded when they were no longer needed. They neither could, nor were meant to, serve as standing ideological watch-dogs. Moreover, as attractive as the picture of the gentry and the state cooperating in pursuit of jointly-held aims was, the provincial book bureaus were, for several reasons, not up to the task for which they were created. Valuable books in China were very unevenly distributed. Some 4831 books submitted by provincial governors were eventually included in the *Ssu-k'u ch'üan-shu*. Of these four thousand, or 83%, came from the provinces of Chekiang (which submitted 1639 titles, 34%), Kiangsu (861 titles, 17.8%), Kiangsi (455 titles, 9.4%), Anhwei (327 titles, 6.8%), and the governor generalship of Liang-chiang (718 titles, 15%). Clearly, however successful the state was in creating an empire-wide network of book bureaus, the work of providing and evaluating books for the Ssu-k'u project would fall mainly on the Chiang-nan gentry who owned them. The northern China provinces made modest contributions to the effort — Shantung provided 211 titles; Honan, 67 titles; Chihli, 119 titles; Hupei, 60 titles; Shensi, 79 titles; and Shansi, 66 titles. The contributions of the other Chinese provinces varied considerably — Fukien contributed 160 titles; Hunan provided 33 titles; Kwangtung, 4 titles; while Yunnan, Szechwan, and Kwangsi provided none at all.[66]

The second problem with the book bureaus was that while the local educational officials who staffed them were suitable representatives of the court to most gentry in China, they were hardly the appropriate people to approach the very wealthy collectors who owned the finest editions. Private collectors significant enough to merit mention by name in the imperial catalog contributed 3426 titles to the Ssu-k'u effort, or 32% of the final collection. The care and feeding of these wealthy book owners was clearly essential to the Ssu-k'u effort. When, for instance, the governors of Chekiang and Kiangsu and the governor-general of Liang-chiang proposed that book bureau personnel be sent to the homes of salt merchants in their jurisdictions, the emperor responded that it would be better if someone "personally known to the book owners" like the merchant Chiang Kuang-ta, working under the guidance of salt commissioner Li Chih-ying, were sent. The emperor did not note the fact in his edict, but Li Chih-ying had had the recent experience of negotiating some rather large loans from the salt merchants to the imperial household, where he served as a bondservant.[67] As a token of imperial gratitude for their contribution to the Ssu-k'u project, the four largest private contributors of books — the Hangchow merchants Fan Mou-chu (1721–1780), Pao Shih-kung and Wang Ch'i-shu

(1728–1799), and the Yangchow salt merchant Ma Yü—each of whom contributed over five hundred titles, were rewarded with copies of the *T'u-shu chi-ch'eng*. Fourteen more book owners, who contributed between one hundred and five hundred titles, received copies of the *P'ei-wen yun-fu*. These rewards were probably largely symbolic for the wealthy men who received them although, to be sure, the symbolic value of a gift of books from the emperor was not to be underestimated. The emperor's gifts did have a measurable market value, however. In a contemporary but probably unrelated memorial, imperial household officials discussed what price copies of the *T'u-shu chi-ch'eng* would fetch on the open market. It was decided that the price of 12.46 taels set the previous month was not inappropriate, although only 44 of the original 896 copies offered for sale several months before had been sold.[68]

A final problem of the borrowing and copying system evolved by the emperor and provincial governors was that some book owners were evidently unwilling to loan books to the government. The reasons for this were hardly matters to be discussed in official or private sources; but it may well have been that book merchants or those who invested in books, realizing that reprinting under government auspices would lower the value of their holdings, sought compensation. To the extent that publishing was an industry in eighteenth-century China, the Ssu-k'u project put the government in competition with private entrepreneurs. The burden of paying for books was placed on the governors, and the emperor rejected one governor's perhaps somewhat wistful suggestion that books purchased by governors become part of provincial libraries. Instead, all books purchased for the project became part of the imperial library. The need for close accounting meant that governors had to keep track of numbers of books bought and copies borrowed. The Chekiang governor reported that in his province, 1875 titles had been bought and 2609 borrowed; in Kiangsu, 1071 were bought and 632 borrowed; in Shantung, 137 titles were bought, 155 copied, and 72 borrowed; in Honan, 91 titles were bought, 22 copied and none borrowed; and in Shansi, 57 books were bought and 31 copied. In 1777, the salt commissioner Yin Cho respectfully conveyed the request of salt merchants that the 932 titles which they had provided to the Ssu-k'u Commission be regarded as a gift to the imperial library.[69]

Once they arrived in Peking, the texts to be collated for the *Ssu-k'u ch'üan-shu* were assembled and cataloged in the Imperial Printing

Press and Bindery at the Wu-ying Throne Hall.[70] Then, if there were more than one version of the text, or if the text was to be recovered in whole or in part from the *Yung-lo ta-tien*, the task was assigned to a reviser. At least the most senior or famous of the revisers were allowed to work in areas of their specialty. Once the initial collation was complete, the reviser returned the manuscript to the Wu-ying Throne Hall, where it was reassigned to a copyist. When reproduced, the text was checked by two assistant collators working separately. Scholars earlier this century, examining books in the imperial household library, found inserted in them what would be called in modern bureaucratic parlance "routing slips" left over from the Ssu-k'u project. The form was as below, with blanks filled in by the appropriate administrative official:

> *Original of Volume (Blank) Pages (Blank)*
> Dispatched from the Wu-ying Throne Hall on
> (Date-blank) to (Location-blank) for collation.
> On (Date-blank) transferred to copyist (Name-blank).
> On (Date-blank) manuscript received.
> On (Date-blank) checked and returned. Received by rechecking
>   center on (Date-blank).
> Rechecking complete on (Date-blank).
> This volume contains (Number-blank) chapters.
> Including previous chapters (Number-blank) characters.[71]

After rechecking, the manuscript was given to a senior reader for final inspection before being submitted to the emperor. In spite of all these efforts, Grand Councillor Yü Min-chung complained of sloppy manuscripts, marred by yellow (perhaps tea?) stains, and had to admonish the collators not to use red ink in their work.[72] Both the emperor and Chief Councillor Yü made corrections from time to time in the manuscripts submitted to them. Finally, copies of the approved texts were made and placed in libraries especially built to house them at the four major imperial residences: the Forbidden City and the Summer Palace in Peking, the imperial summer retreat at Jehol, and the Ch'ing ancestral home in Manchuria outside the city known today as Shenyang. When these sets were completed, copies were made for libraries in the lower Yangtze valley cities of Hangchow, Yangchow, and Chinkiang.

By all accounts, a convivial and stimulating atmosphere prevailed among the collators, and they were given a fairly free hand in their work. A rare account of one day in the life of a Ssu-k'u collator was preserved in the autobiography of Weng Fang-kang, a reviser for books submitted from the provinces:

> Early each morning, I went to the Academy. A kitchen was established there to provide meals for us as we worked. After lunch I returned home. Then, at the Pao-shan-t'ing I discussed the books that I had collated in the morning with Ch'eng Chin-fang, Yao Nai, and Jen Ta-ch'un. Each of us contributed what we knew, and we compiled a list of the books that would have to be consulted. In the afternoon, I would hasten to the Liu-li-ch'ang book markets to look for them. At that time, the Kiangsu and Chekiang book merchants all competed with each other to acquire books that could be used for research. They were all assembled in the Five Willows Lodge of the Hall of Literary Refinement. Each day I found books that could be used and returned home with a cart full of them . . . If the price was not too high, I rented them. From the volumes that I could not afford to rent, I copied out the necessary portions, or hired someone else to do so.[73]

The Liu-li-ch'ang, a district located to the west of the central gate into the imperial city of Peking famous for its book markets, grew up in the early Ch'ien-lung years, and became particularly prosperous during the Ssu-k'u project.[74]

Chu Yun, as was his wont, contributed to the ambiance by giving parties. In the spring of 1773, Yao Nai, Weng Fang-kang, Lu Hsi-hsiung, and Chi Hsiao-lan all went on a picnic organized by Chu some ten *li* outside of the capital.[75] Each wrote a poem, and Chu provided a preface for the collection. Chang Hsueh-ch'eng commented that Chu's enthusiasm for parties during his Ssu-k'u years "was perhaps too great to permit his devoting a great deal [of time and energy] to his work."[76]

Some revisers, apparently, took the opportunity of working on seldom-seen texts to make copies for themselves or for sale. Chou Yung-nien, a particularly enthusiastic bibliophile from Shantung, hired a staff of ten copyists to assist him.[77] At least two of the texts recovered by Shao Chin-han from the *Yung-lo ta-tien* were subsequently privately published,[78] and others circulated among his friends in manuscript copies.[79] There appear to have been no legal or ethical objections to his practice; however, the collators' habit of physically taking manuscripts away from the work site complicated the government's task of keeping watch over all the manuscripts being collated, as the curious case of Huang Shou-ling demonstrates.

Huang was a Hanlin bachelor of the class of 1772, appointed to

serve as Ssu-k'u reviser. Sometime in the summer of 1774, he carried
one volume of the *Yung-lo ta-tien* out of the Hanlin Academy and lost
it. When the emperor heard of the matter, he was furious:

> The *Yung-lo ta-tien* [manuscript] is unique, revisers cannot be allowed to
> carry it off. Each day food is provided at the Academy and every colla-
> tor can eat his fill. If they work on public business, collating books at
> the Academy for a full day, the work schedule can be met. There is no
> need for them to burn the midnight oil. [Furthermore, we have] estab-
> lished proctors at the Academy. If the culprit informed the proctor of his
> intentions, then the two of them must be in this together.[80]

The capital police were ordered to investigate the matter thoroughly.
Several days later, the book turned up mysteriously under a bridge
in the imperial city, but the emperor was not mollified. He
commented:

> This book has been missing for a long time. It must be that the man
> who stole it took it around to book merchants and used paper shops and
> tried to sell it. But the merchants know that the *Yung-lo ta-tien* is govern-
> ment property and cannot be privately bought and sold. So the cul-
> prit . . . not daring to retain it, placed the book under a bridge in the
> middle of the night.[81]

The police were ordered to investigate further, but to no avail. The
incident had serious repercussions for those involved. A grand councillor, the head of the Imperial Printing Press and Bindery, and the
head of the capital police submitted a joint memorial of apology; the
salary of Huang Shou-ling, who had himself brought the matter to
the attention of Chi Hsiao-lan but evidently would not explain in full
what had happened, was suspended for three years. The salaries of
two proctors were suspended for six months. As a result of the case,
regulations on carrying of books from the collation site were tight-
ened. Chou Yung-nien, finding that he could no longer take out
books from the Academy, sadly discharged his staff of copyists.[82]

   The case is of interest for several reasons, aside from the view it
provides of the emperor-as-detective. First, it demonstrates the
emperor's genuine concern for the preservation of books and the
lively, if perhaps illicit, interest of collators and book merchants in
the process. Second, the daily work schedule envisioned by the
emperor contrasts sharply with the somewhat more relaxed schedule
described in Weng Fang-kang's *Autobiography:* the issue of how hard
the scholars were in fact working became a live one later in the

project. Finally, the case triggered an investigation of book collection and storage procedures, which provided some of the most important sources on compilation methods.

. . . To Forgive, Divine: The Errors List and Ch'ing Administration

While the most prestigious tasks of the Ssu-k'u project were entrusted to men of considerable scholarly reputation like Tai Chen, Shao Chin-han and Chou Yung-nien, the bulk of the work of collation was delegated to younger men appointed to the commission at the capital. These younger officials did not necessarily share their seniors' motivation or interests; their performance reflected their own attitudes and concerns and the sense of purpose and standards communicated to them by senior bureaucrats. Their work was far from flawless. According to many assessments, books that could have been recovered from the *Yung-lo ta-tien* were overlooked; collation was hasty and sometimes negligent, and proofreading perfunctory. In 1828 Ch'ien I-chi, an editor of the government-sponsored *I-t'ung chih* (Universal gazetteer), proposed that so many *Yung-lo ta-tien* books had been overlooked in the Ssu-k'u project that the Ming encyclopedia needed to be reedited. The Tao-kuang Emperor approved of the idea but, tied down by domestic rebellion and turmoil, had neither the personnel nor the resources to undertake the project.[83] Nor were the Ssu-k'u collators negligent only with respect to the *Yung-lo ta-tien*. Often, comparisons of different editions of a work, which could have resulted in a better text, simply went undone. As a result, prefaces were omitted, characters mistaken, maps misdrawn and mislabeled, and the appearance of original texts changed drastically. Such collation, the modern scholar Sun Chieh-ti has observed, "may have amused those engaged in it," but was of little value.[84] Contemporaries continually commented on poor proofreading in Ssu-k'u editions. Once, in no less exalted a text than a poem by the Ch'ien-lung Emperor's grandfather, a copyist's error turned a pear blossom into a plum blossom.[85] The last straw came for the emperor when he was proudly inspecting his new library in Jehol and found one volume entirely blank![86]

These assessments of the Ssu-k'u manuscript contained elements of truth, to be sure, but they also reflected their authors' perspective on the Ch'ien-lung government and its stewardship of Chinese affairs (see below). Perhaps the most interesting issue which Ssu-k'u errors posed in their own day was not one of assessment but of administration. What were the reasons for the mistakes, and how did Ch'ing officials perceive and respond to them? Yü Min-chung, at least, seems to have been aware of the quality of collators' work, but seemed unable to do anything about it. A "Record of Work Accomplished" (*ch'eng-kung-ts'e*) was kept in which the accomplishments and failings of each collator were recorded. Inspecting it, Yü remarked that "Except for Chou from Shantung [Chou Yung-nien], very few are really diligent."[87] On another occasion, Yü wrote: "The books you submitted for imperial review were returned in one day, but the emperor pointed out two errors. Reading through the books, these were very easy to see. If we are reprimanded like this again, [our colors] will indeed be faded."[88]

There were probably several reasons for the slipshod editing. One was surely the speed at which collators were expected to work. A primary concern of the emperor's, repeated over and over in his edicts and poetry, was that the project be completed quickly.[89] As the books piled up for collation, this became more difficult. "Previously," Yü wrote nervously to Lu Hsi-hsiung one summer, "the rule was that each reviser should read five pages per day. But so many books have been collected that each reviser must now read over thirteen hundred volumes (*ts'e*). If they read 160–170 volumes per month, that would be sufficient. If they could even read one hundred volumes per month, we could calculate a schedule." Assuming that even a small Chinese "volume" would have contained at least twenty folio pages, this meant each reviser would have had to read between 2,000 and 3,200 folio pages per month!

Caught between an impatient emperor and overburdened collators, senior officials could only temporize and mediate. Yü continued:

> The schedule that you and Chi Hsiao-lan worked out is of no value. Yet the emperor still inquires [whether it is being met]. How shall I answer him? I am really frightened. You must discuss this matter with the other collators and set a new schedule. Then write me a letter. You can figure from the number of volumes remaining. [Maybe we] can compensate for having deceived the emperor before.[90]

In the face of such demands for haste, the editors could hardly have expected unerring accuracy. Once, when the *Liang-ch'ao kang-mu pei-yao* (The essentials of the Kuang-tsung and Ning-tsung periods:

an outline history),[91] a history of the Sung dynasty written in the *kang-mu* form, was submitted for imperial review, Yü Min-chung commented that it was "all right if there [were] mistakes in the *mu* sections so long as the *kang* were correct." The *kang* were the headings under which events were described, the *mu* were the actual historical narrations.[92]

Another factor probably affecting collators' performance was the tediousness of the work and its often insignificant result. Even a man like Yao Nai, who shared the Chu Yun circle's enthusiasm for scholarly pursuits but not their commitment to evidential research, found that "although the Sung and Yuan commentaries are manifold, often only one or two lines are worth preserving in each."[93] References to books in the *Yung-lo ta-tien* were often so fragmentary or to texts of such little value that nothing could be done with them. Yü Min-chung remarked that the *Yung-lo* books were like "chicken ribs," numerous but not meaty; "however," he asserted, "we must continue with them. It will not do to turn back in mid-task."[94] The Sung dynasty *wen-chi* were particularly troublesome.

The environment of the Ssu-k'u project must also have affected the collators. In the competitive atmosphere of the Hanlin Academy during late Ch'ien-lung times, as we have noted, a crucial element in career success was the attention of senior officials. If one worked at the *Ssu-k'u ch'üan-shu* Commission, where success was measured in terms of the number of titles recovered and pages collated, the premium was on producing as much work as possible, regardless of its quality or significance. Regrettable as this situation was from the point of view of imperial or scholarly interests, it was an almost inevitable consequence of the bureaucratic environment of the project. Eulogizing Chou Yung-nien, Chang Hsueh-ch'eng charged that in recovering books from the *Yung-lo,*

> the officials at the Commission picked only the easiest books, then claimed there were no more to be found. Chou (Yung-nien) alone argued with them, claiming that there were many more texts that could be reconstructed. His colleagues ignored him . . . and left the task to Chou alone. Chou took no leisure; summer and winter, through the winds and the rains [he searched], reading over nine thousand volumes and eighteen thousand chapters . . . As a result, the *wen-chi* of the Sung brothers Kung-shih and Kung-fei and ten other books were recovered.[95]

Even allowing for some dramatic exaggeration on Chang's part, the differences between scholars and bureaucrats must be admitted.

Disciplining the collators was a complicated problem. On the one hand, the dynasty's prestige was at stake: errors in the manuscripts could undermine the effort to benefit scholars of "ten thousand generations." On the other hand, collators' errors were hardly matters of life or death, and there were many possible explanations, among them pressure from an impatient emperor, the nature of the work, and even the character of the Chinese language. This conflict, between dynastic prestige and bureaucratic practicality, was one characteristic of the era. It was never effectively resolved; as late as 1791, editors were ordered to travel around the empire at their own expense correcting errors in various versions of the manuscript. But the government's attempts to solve the problem of errors in the Ssu-k'u manuscripts suggest much about the quality and politics of leadership in the later Ch'ien-lung era.

On 13 November 1773, the emperor, finding several mistakes in the Ssu-k'u manuscripts submitted to him, ordered the Commission to deliberate and devise a method for preventing error. Two weeks later, Prince Yung-jung memorialized in response on behalf of the Commission.[96] His memorial argued that the problem was two-fold: the work of copyists was not being read thoroughly enough, and the successes and failures of the collators were not being monitored carefully enough. Each day, he explained, the thirty-two Ssu-k'u collators produced over four hundred thousand words, while the twelve *Ssu-k'u hui-yao* collators produced two hundred thousand words. The prince carefully avoided saying that this pace, probably maintained to please the emperor, was too fast, or that the collators were negligent. Instead, he declared that the chief collator established at the Wu-ying Throne Hall to prepare manuscripts for imperial review simply could not keep up with the flow of work. He recommended the designation of twenty-two proofreaders (*fu-chiao-kuan*) to recheck all manuscripts after they were submitted to the Wu-ying Throne Hall and before they were given to the emperor.

Prince Yung-jung also recommended that a system of credits and demerits be established. Each collator or copyist was to receive one credit for every correction he made in a manuscript. If the proofreader found an error, he would receive a credit and the collator and copyist responsible for the manuscript would each receive a demerit. If an editor discovered an error, copyist, collator and proofreader would all receive demerits. If the emperor found an error, all three would receive demerits and the senior editor would be turned over to the Board of Punishments for administrative discipline. At the end of five years, the collators and copyists were to be recommended for appointment to office on the basis of the amount of work they had accomplished and the number of credits and demerits they had accumulated. Ironically, those who had accumulated the most demerits were to be retained at the Commission for an extra two years before being appointed to office. One modern student of the project has argued that these sanctions were "very light,"[97] but in fact they seem to have been well suited to the framework of incentives and pressures within which collators worked.

Despite these measures, the emperor continued to find errors in manuscripts submitted to him. His response was shaped by two tendencies which, particularly in the later years of his reign, characterized his political behavior: to entrust his affairs to a relatively few men, and to hold them personally responsible for the successes and failures of his policies.[98]

The problem, the emperor reasoned in an edict of March 1774, must have lain with the men appointed as directors-general (tsung-tsai). "How can they cast aside their duties so lightly?" he exclaimed. But not all the directors were equally to blame:

> Among them (there are some) like my sixth son Yung-jung, Shu-ho-te, and Fu-Lung-an, who although they were appointed director were not expected to read manuscripts. . . . As for Ying-lien he is in charge of banner affairs and the imperial household and is too busy to read manuscripts. Chin Chien is also in charge of other matters. Yü Min-chung should be reading manuscripts but is probably busy with his secretarial duties and the managing of the Grand Council.[99]

The other directors, however, men like Ts'ao Hsiu-hsien, Wang Chi-hua, Ts'ai Hsin, Chang Jo-kuei and Li Yu-tang, had been negligent and should be turned over to the Board of Personnel for administrative punishment.

The list of directors reads like a "who's who" at the Ch'ien-lung court. Even those who were to be punished were men of considerable authority: Ts'ao Hsiu-hsien was president of the Board of Works, Wang Chi-hua was president of the Board of Personnel, and Ts'ai Hsin was successively president of the Board of Works, Rites and Personnel during the early years of the Ssu-k'u project.[100] In effect, the emperor was charging most of the senior officials of his government with responsibility for the failures of the Ssu-k'u project. He did not, he assured the directors, expect them to read every word of every manuscript submitted; but they should have read enough of the text to keep the collators on their toes and set a personal example. The same group was reprimanded again in the fall of 1774: "It has been five months since I left for Jehol. How is it that the directors have still not learned to collate?" In the winter of 1779, a new group of senior readers (*tsung-yueh-kuan*) was appointed to review Ssu-k'u manuscripts for error.[101]

The pressure on senior officials must have been enormous, as were, no doubt, the temptations to avoid reprimand by concealing evidence of failures from imperial view. In one of Yü's letters quoted above, the councillor confessed to Lu Hsi-hsiung that he was "really nervous" about the emperor's discovering that the work schedule was not being met, and urged Lu to devise a new schedule so that Yü could "make up for having deceived the emperor previously."[102] In another letter, Yü discussed how the errors in the *I lin* (Anthology of ideas), a T'ang dynasty anthology reprinted in the Wu-ying Hall Collectanea, could be corrected, and advised Lu to go ahead with the necessary procedures without imperial approval. "I will advise the emperor of the problem when the time is right."[103] In this situation of imperial impatience and bureaucratic misrepresentation, it would have been easy for a new man, or a man from a new faction, to earn imperial esteem by pointing out the deceptions and errors of his predecessor.

Such a man was rising in the Ch'ien-lung court of the late 1770s. Ho-shen probably first attracted imperial attention when he was appointed to the imperial bodyguard in 1772.[104] During the next few years, he rose rapidly through the inner court bureaucracy, being appointed in 1778 to the lucrative post of superintendent of the customs and octroi at the Ch'ung-wen gate in Peking. Ho-shen's first important political commission came in the winter of 1780 when he was sent to investigate charges of corruption against the governor-general of Yunnan, Li Shih-yao. This he did most thoroughly, with the result that Li and many of his subordinates were dismissed and banished.[105] The emperor was evidently delighted with the young bannerman's reports; even before he returned to Peking, Ho-shen was appointed president of the Board of Finance, and shortly thereafter he was betrothed to the emperor's youngest daughter, Princess Ho-hsiao.

Yü Min-chung died in late January of 1780, leaving vacant posts as chief grand councillor, literary secretary to the emperor, editor of the *Jih-hsia chiu-wen k'ao* (A history of the Peking region), the *Ming shih,* the *Liao-Chin-Yuan shih* (History of the late Liao, Chin and Yüan dynasties), and the *Man-chou yuan-liu k'ao* (A study of the origins and development of the Manchus), and director-general of the *Ssu-k'u ch'üan-shu* project, posts from which he was said to control the destiny of all the scholars in the empire.[106] Within two years, Ho-shen occupied all of these posts, except the editorship of the *Ming shih,* a text largely complete by the late 1770s. He was already editor of the *Man-chou yuan-liu k'ao* and the *Liao-Chin-Yuan shih* at the time of Yü's death, and was appointed editor of the *Jih-hsia chiu-wen k'ao* in 1781. A-kuei (1717–1797), another Manchu, was formally designated chief grand councillor, but as early as the spring of 1781, Ho-shen functioned as chief in his absence. On 4 November 1781, the former bodyguard was appointed director-general of the *Ssu-k'u ch'üan-shu* project.[107]

Ho-shen's first official act as director-general, performed the day after his appointment, was to indict Ts'ao Wen-chih (1735–1798), a senior Chinese statesman who was coeditor of the *Liao-Chin-Yuan shih,* for failing to report spaces left blank by the collator of *Ch'ing-hsia chi* (Collected works of the scholar of Ch'ing-hsia mountain) by the early Ming official Sung Lien (1310–1381). The spaces had been left blank, it seemed, in order to eliminate anti-Manchu references.[108]

Although this was by no means the first expression of concern about anti-Manchu references in literature, the fact that newly ascendant Manchu leaders used the issue against their Chinese counterparts had ominous implications. Throughout the Ch'ien-lung reign, there was a fairly regular alternation of the Manchu and Chinese leaders at the court. First the Manchu O-erh-t'ai, then the Chinese Chang T'ing-yü, then the Manchu Fu-heng, then the Chinese Liu Tung-hsun and Yü Min-chung, then the Manchus A-kuei and Ho-shen served as chief grand councillors. Associated with several of these leaders were significant groups of bureaucrats, although the Ch'ien-lung emperor was very sensitive to any public mention of them. There is evidence that at least in the cases of the transitions from O-erh-t'ai to Chang T'ing-yü, and Yü Min-chung to Ho-shen the partisans were also to use the fact of ethnic differences to their own factional ends.[109]

More relevant to the collation of *Ssu-k'u ch'üan-shu,* however, was the fact that Ho-shen, apparently from his very first day at the Commission, took upon himself the task of pointing out errors in the manuscript to the emperor. This role received imperial sanction early in

Ho-shen's tenure. In accordance with Prince Yung-jung's memorial of November 1772 (summarized above), reports of the demerits earned by each collator were submitted to the emperor quarterly.[110] The credits, if they were tabulated, were not reported to the emperor, hence the name for the documents, "errors lists" (*chi-kuo-chi*). The first errors list for 1781 was prefaced by an imperial order commanding Ho-shen and A-kuei to take charge of reading books and reporting errors.[111] The command was reprinted in every errors list after 1781, but was not mentioned in any lists before that date. In view of the number of duties that devolved on Ho-shen after 1780, it was unlikely that he personally read every Ssu-k'u text; nonetheless, he bore final responsibility and must have set the tone and guidelines for the task.

The watchful new administrators found many more mistakes than had their predecessors. During the last two years of Yü Min-chung's life, 221 and 259 demerits were reported. In 1781, 5,006 demerits were reported; in 1782, 7,072 were reported and in 1782, 12,033 were issued. The reports were not made after 1784.[112] Possibly, the errors in these reports were exaggerated, or at least counted retroactively, for at least one of the editors was charged with demerits posthumously.[113]

The consequences of new editorial vigilance for Ssu-k'u editors were most vividly demonstrated in 1787. At that time, five copies of the manuscript library had been completed. Leafing through his new library at random, the emperor found many mistakes and omissions, and so appointed a Committee headed by Ho-shen to inspect and correct the volumes.[114] After the new Committee had been at work for about a month, it became clear that much remained to be done. In late July former chief editor Chi Hsiao-lan, whose guilt in the matter had been termed "unspeakable" by both the emperor and Ho-shen, submitted a memorial of apology and offered to travel to Manchuria with assistants at his own expense to make the necessary corrections.[115] His offer was accepted, and in November he arrived at Jehol.[116] Working conditions were not what they had been in Peking. The group was located outside the main gate of the Wen-lu Pavilion where the manuscripts were stored. Chill November winds blew through the work rooms, and one collator was said to have perished from the cold. Possibly adding insult to injury, several Manchu soldiers from the garrison were assigned to watch the group as they worked and "to help them as necessary." After about two months Chi memorialized claiming that the work, which had taken years to accomplish in Peking, could be reviewed and corrected in about three months. "In the capital, each of us had private affairs which unavoidably extended the work," he explained. He therefore requested permission to return home within a month. "Why not stay a little longer?" the emperor remarked acidly on the memorial.[117]

Not surprisingly, many Ssu-k'u personnel were opposed to Ho-shen during his time in office, and were active in his demise. Ch'en Ch'ang-chi, a Yung-lo reviser, submitted a memorial probably in the late 1780s urging that the education commissioner in Chihli be charged with investigating corruption at the court. Yin Chuang-t'u, a senior reader at the Commission, barely escaped with his life when he suggested that "corruption was general throughout the empire and treasury deficits existed everywhere," with the apparent implication that Ho-shen was at fault. Another Yung-lo reviser, Mo Chan-lu, was sent to investigate the property accumulated by Ho-shen and his servants after his demise. Finally, one of the most explicit and telling indictments of Ho-shen, the famous letter to Prince Ch'eng, was written in 1800 by Hung Liang-chi. Hung was a very close friend of Chu Yun, and was also the associate and biographer of many Ssu-k'u collators, and probably served on the Commission in some capacity himself, though he is not listed on the personnel list of 1782.[118]

Certainly, Ho-shen's actions as director-general of the Ssu-k'u Commission were not the only causes for opposition to him. However, his role at the Commission was probably typical of the kind of action which brought about not only his own rise and demise, but those of many senior officials of the era. The *Ssu-k'u ch'üan-shu* was only one of several grand campaigns which the Ch'ien-lung emperor imposed on his bureaucracy and army in the last years of his reign. These projects would have taxed the resources of Ch'ing government even if it had not already been strained by a doubling population and concomitant social and economic tensions. Liu Tung-hsun's opposition to the grander proposals of Chu Yun's memorial reflected his recognition of the toll of time and energy these projects would exact. Caught between an impatient emperor and an overextended bureaucracy, senior officials faced an almost insoluble dilemma: either they could admit their inability to carry out the emperor's will, with possibly disastrous consequences for themselves, or they could try to conceal their failures with the attendant risk of revelation and denunciation. The errors in the Ssu-k'u manuscript, unfortunately, could neither be revealed nor long concealed. Much of Ho-shen's early career was spent in investigating and reporting on the deceptions and failures of his predecessors in Yunnan, at the Board of Finance, and at the Ssu-k'u Commission. After his own fall, many of his deceptions, particularly in the campaign against the White Lotus uprising, were denounced by his opponents. While much discussion of Ho-shen has focused on his greed and other personal characteristics, little attention has been given to the political and institutional environment in which he rose and fell. To say that Ho-shen profited from the existing situation of a corrupt and overextended bureaucracy, or that the powers he exercised over personnel and the deceptions he practiced on the emperor had their precedents is not to deny that his corruption and greed were more flagrant or his behavior more boorish than his predecessors. It is simply to suggest that the excesses of his tenure were weaknesses of an era as well as of an individual.

The Product

The products of the Ssu-k'u compilation project were: (1) Seven manuscript copies, each 36,500 *chüan* long, of the *Ssu-k'u ch'üan-shu*, the books selected by the Ssu-k'u Commission for inclusion in the imperial library.[119] (2) A list of the titles and authors of the books in this collection, with somewhat abbreviated reviews, entitled the *Ssu-k'u ch'üan-shu chien-ming mu-lu* (A shortened catalog of the *Ssu-k'u ch'üan-shu*).[120] (3) A shorter version of the *Ssu-k'u ch'üan-shu* compilation, 11,170 *chüan* long, containing only the most important works in the collection, entitled the *Ssu-k'u ch'üan-shu hui yao* (The essentials of the *Ssu-k'u chüan-shu*).[121] (4) A series of 134 titles, known as the *Wu-ying-tien chen-chu pan ts'ung-shu* (Collectanea printed from assembled pearls in the Wu-ying throne hall), which reprinted in moveable type the rarest and most valuable works in the Ssu-k'u.[122] (5) An annotated catalog, 4,490 pages long in its modern reprint, entitled *Ssu-k'u ch'üan-shu tsung-mu t'i-yao* (Annotated general catalog of the *Ssu-k'u ch'üan-shu*).[123]

Perhaps the most balanced evaluation of these products was by Yü Chia-hsi (ca. 1890–1960), a bibliographer who devoted his life to the study of the *Ssu-k'u ch'üan-shu* catalog:

> To say that there has been no work like this since Liu Hsiang's *Pieh-lu* would not be excessive. [Its benefits] were felt throughout the empire,

and soaked deeply [into the roots of Chinese scholarship]. Since the Tao-kuang and Chia-ch'ing periods, there have not been any scholars who have not partaken of its riches, or used it as a compass needle to find their way. Its accomplishments were great and its utility vast.

Of course, in a work which attempts to preserve all the accumulated insights of the past, some problems are unavoidable. Furthermore, this was a work of official scholarship, a product of many hands, accomplished within a limited period of time by men who had to meet bureaucratic standards. Within ten years, seven copies of the *Ssu-k'u ch'üan-shu* and two copies of the *Ssu-k'u ch'üan-shu hui-yao* were finished. Inevitably, copyist's errors were many and genuine insights few. Moreover, in these years, ten new works of imperial scholarship were commissioned, and several hundred books were reconstructed from the *Yung-lo ta-tien*. Between listing the books, editing and correcting them, banning some and making sure that the taboos were observed in others, when was there time for genuine research? Furthermore, there was the difficulty of obtaining reference works. If one borrowed them from the imperial library, [one risked] the penalty for losing them. If one relied on private holdings, then the reference collections were incomplete.[124]

A more thorough evaluation than Yü's would probably be beyond the capacities of any modern scholar. For this reason, Yü's suggestion that the faults of the *Ssu-k'u ch'üan-shu* were essentially the faults of a system is particularly valuable. Sponsorship of the project was hardly haphazard. Collators were selected, supervised, compensated and coordinated not by accident, but by a carefully formulated design that expressed the priorities and attitudes of emperor and bureaucrats toward scholars and scholarship. The accomplishments and failures of the Ssu-k'u Commission were not so much those of individuals as those of a government.

Conversely, many assessments of the Ssu-k'u Commission's products in the past two centuries have represented statements about the government which ordered the compilation as well as catalogs of lacunae and accomplishments. The earliest scholars who assessed the Ssu-k'u, Wang T'ai-yueh (1722–1785), Juan Yuan (1764–1849) and Shao I-chen (1810–1861), were men who shared the basic assumptions of Ssu-k'u editors. Their writings, really more supplements than critiques, tended to note the occasional omission or misprint rather than to call into question fundamental assumptions.[125] Miao Ch'üan-sun (1844–1919) was a bit bolder, or at least more rhetorical, in his criticism, observing that "many books were included which ought to have been omitted, and many were omitted that ought to have been included," with the result that the library had its strong and weak points.[126]

Miao's observations were too vague to constitute a critique, but many of his contemporaries, less constrained by loyalty to the Ch'ing and more inflamed by anti-Manchu passions, went further. Yü Yueh (1821–1907) noted that although the Ssu-k'u had its strengths, it was written to control thought rather than contribute to scholarship. The problem, as Yü saw it, was that no one could refute the Ssu-k'u as, for instance, Ch'en Yueh-wen (*chin-shih,* 1553) had refuted Yang Shen's writings on bibliography. If intellectuals' main function was to articulate irrefutable government positions, Yü asked, how could they ever serve or be appreciated by society? Yü's more radical disciple Chang Ping-lin (1869–1936) took up this theme in his "Lament for Burned Books," characterizing Chi Hsiao-lan and his assistants as henchmen of a foreign government bent on heavy-handed ideological manipulation.[127]

Post-1911 scholarship on the *Ssu-k'u ch'üan-shu* has been somewhat less shrill, but at least part of the burden of most Republican period writings on the text has been to explore ways in which the Ch'ing government sought to control thought by suppressing foreign doctrines and making textual emendations. The only English language work dealing with the Ssu-k'u project, L. Carrington Goodrich's *The Literary Inquisition of Ch'ien-lung,* argues that the book collection was "inseparably linked" with an effort of thought control and suppression.[128] Similarly patriotic concerns have fueled arguments that the anti-intellectualism of Ch'ing rulers resulted in a fairly cavalier attitude toward the texts of the *Ssu-k'u ch'üan-shu,* and carelessness in the matter of misprints and omissions. Another theme of twentieth century scholarship reflected in revisions of the Ssu-k'u was the introduction of European scholarly standards and attitudes into China. A 1929 study of the Ssu-k'u is prefaced with the observation that "as scholarship has become more complex, the need for more precise reference aids becomes more pressing, and the techniques for compiling them more vital."[129] Indeed, the several indices to the Ssu-k'u probably reflect this point of view. Twentieth-central nationalism, and internationalism, have also affected work on Ch'ien-lung's book collection. In 1917, the Japanese government turned over its portion of the Boxer Indemnity to the Oriental Cultural Enterprise Committee (Tung-fang wen-hua shih-yeh wei-yuan-hui), which in turn granted the request of a number of Chinese scholars and politicians that a revision and continuation of the Ssu-k'u be undertaken. The project was accomplished largely by Chinese scholars and continued until 1925, although many of the Chinese participants in the project fled south after the Japanese invasion. The product of this effort was stored in the Humanistic Research Institute of Kyoto University until it was published in thirteen volumes by the Commercial Press, Taipei in 1971.[130] More recently, the Ssu-k'u has been seen by at least some Chinese scholars as an object they can regard, or at least reprint, with pride. In 1982, the Commercial Press (Taipei) undertook to reprint the entire *Ssu-k'u ch'üan-shu* as a means "of ensuring that the treasures of Chinese thought and culture therein be preserved from extinction," and offered the product for sale to the public at an approximate cost of $27,000 U.S.[131]

No intellectual vision can stand forever. At the broadest level, the revisions of the *Ssu-k'u ch'üan-shu* proposed and accomplished in the last two centuries have demonstrated how China's turbulent modern history has influenced its intellectuals' view of their role in society and their scholarly heritage. At a more concrete level, the criticisms have suggested just how closely the Ssu-k'u reflected the series of intellectual and institutional compromises in the 1770s which produced it. Two of the most frequent criticisms, that the criteria for deciding which books were to be included in it were unclear, overly political and inconsistent, and that many texts in it were edited to suit the political or bureaucratic convenience of the rulers, have been particularly telling in this regard.

How complete should a "complete library" be? According to the *Annotated Catalog* at least 10,869 works were examined in the course of the collating and selection process; 3,697 of these were deemed worthy of inclusion in the *Ssu-k'u ch'üan-shu* and 134 were reprinted in the *Collectanea*. The remaining 7,038 were listed as "extant" in the *Catalog* and given somewhat abbreviated reviews. The Ch'ien-lung listing was by far the largest in Chinese history, and perhaps the largest in the world of its day.

The scope of the project was, in great measure, a consequence of imperial ambitions to supersede all previous book collection efforts. The emperor touched upon the importance of unprecedented scope in his edict of 3 March 1773 on the *Yung-lo ta-tien* but evidently discussed the matter more frankly later with his chief councillor, Yü Min-chung.[132] In a letter written during the summer of 1773, Yü reported to Lu Hsi-hsiung:

> At today's morning audience, the emperor inquired which dynasty's book collection project had collected the most books, and which was the most beneficial. Could you look into this matter carefully and report back to me by the seventeenth? His Majesty also asked whether or not the

Yung-lo ta-tien project is recorded in the *Ming-shih* (Official history of the
Ming dynasty). Please investigate and submit a memorial.[133]

On the 19th of August, Yü chided Lu: "You still have not responded
to my request for a report on which dynasty collected the most
books. I think we had better use *chüan* to count. The last report I received on the books examined by the book bureau and forwarded
from the provinces used titles (as a standard). Do you think we can
gather several tens of thousands of *chüan*?"[134] Numbers of books
collected became a standard of the success of the project. Yü once
counselled the editors: "It will not be a problem if there aren't many
collected writings of Ming authors so long as we list ten thousand
titles in the *Annotated Catalog*."[135] Expressions like "ten thousand"
(*wan*) were usually used figuratively in Chinese writing to mean "a
great many," however, it was probably not coincidental that the *Annotated Catalog* contained just over ten thousand titles.

The subject matter of the books included in the *Ssu-k'u ch'üan-shu*
also reflected imperial purposes. In his first pronouncement on book
collecting, the emperor sought writings which "clarified the study of
government" or "concerned man's essential nature."[136] In view of the
traditional concern of Confucian thought with man and how he was
to be governed, this was a mandate broad enough to include most
of China's literary heritage, but it was not unlimited. The crucial cri-
terion was utility. The "Principles" (*fan-li*) of the *Annotated Catalog*
asserted bluntly that the "learning of the sages emphasized clarifying
the essence in order to achieve usefulness. All writings in which realities cannot be seen clearly are wasted words."[137] An emperor cultivated learning in his realm not for its own sake, but for the insight
it gave him into government and the way of the sages.

Perhaps most controversial were the Commission's decisions about
which books were worthy of being printed in the *Collectanea*, which
were worthy of being copied into the imperial library, and which
only deserved treatment in the *Annotated Catalog* and which were to be
ignored altogether. As Yao Ming-ta has pointed out, the division of
books into categories by quality was unprecedented in imperial book
collecting projects, which had previously simply copied all extant
books. The Ch'ien-lung practice of dividing books into classes may
have reflected the emperor's desire to pass judgment as well as to
collect books, but it probably also reflected problems posed by the
enormous number of books in circulation in the late eighteenth

century.[138] It would have been a much more complicated process for the Ch'ien-lung government to have attempted to print or copy all the works in circulation.

Nonetheless, the assignment of a given text to one of these four categories was, as Yü Min-chung remarked in 1774, a matter over which scholars could easily have differences of opinion. The most severely restricted category was the first; only 134 titles, or 1.25% of the books inspected in the project were printed in the *Collectanea*. Another 3,697 books, or approximately 33.2% of the books inspected in the project were copied into the imperial manuscript library (see Table 3-2). Two criteria were mentioned by Yü for books which were to be included in the *Collectanea:* a book must have been either of "real benefit to human understanding" or else "very seldom seen." The publication of books which were not extant before the Ssu-k'u project, that is, those recovered from the *Yung-lo ta-tien*, would "increase their circulation and so be a real benefit to scholars."[139]

In a subsequent letter, Yü advised the editors that in assigning books to all categories, they could be "lenient with books written before [the end of the] Yuan dynasty and more severe with those written after the [beginning of the] Ming." This emphasis on pre-Ming texts was in effect an emphasis on Sung and Yuan texts, since very few pre-Sung texts circulated in any form. It probably was motivated by both scholarly and political considerations. The scholarly community had long urged that the project focus on the books of these two periods; and although Chu Yun's proposal that the project formally do so was rejected, his hopes were in fact realized. The denigration of Ming scholarship evident throughout the *Ssu-k'u ch'üan-shu* almost certainly also evinced a Ch'ing government desire to play up its own contributions to Chinese scholarship at the expense of the dynasty it had conquered.

Aside from these specific directions, Yü counselled the editors to use their own judgment and standards of praise and blame in selecting books for the compilation. If a reviewer found much to praise in a book and nothing to blame, it should be printed in the *Collectanea*. If he found reason both to praise and blame the book, it could be copied into the *Ssu-k'u ch'üan-shu*. Those books which contained little that was praiseworthy and much that was contemptible would be listed by title only in the *Annotated Catalog*. Evaluations of books, like Confucius' praising and blaming of historical figures in the *Ch'un-ch'iu*, were based on a reviser's assessment of their moral worth. Only

Figure 2. Number of Books Printed, Copied, and Listed in the
Ssu-k'u ch'üan-shu According to Subject

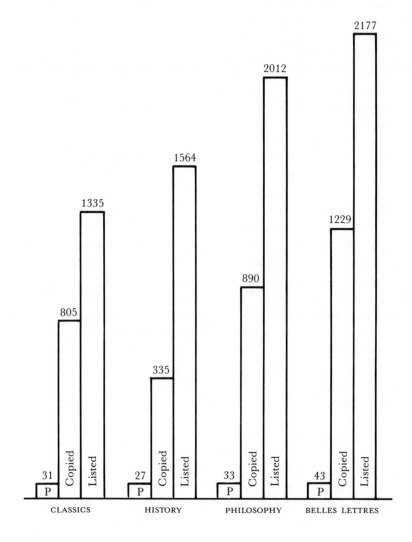

Legend
P (Printed): Printed in the *Collectanea Printed from Assembled Pearls in the Wu Ying Throne Hall*
Copied: Copied in the *Ssu-k'u ch'üan-shu*
Listed: Listed in the *Annotated Catalog*

if the compilation was based on such standards, Yü asserted, would it enable later scholars to "straighten out the quills and transmit the truth."[140] Another of Yü's letters made it clear that the classics and commentaries on them were particularly valuable; in recovering books from the *Yung-lo ta-tien* "if only one or two tenths of a text could be reconstructed, it was worth the effort."[141]

Several measures attest that Yü's dicta were closely followed. As befitted collections which sought to recapture the way of the sages, both the *Ssu-k'u ch'üan-shu* and the *Collectanea* included greater percentages of books from the classics division than from the other divisions, history, philosophy or belles lettres. Works recovered from the *Yung-lo ta-tien* were particularly favored by the editors: 21.02% of them were printed (as opposed to 5% of texts collected from all other sources). The emphasis on pre-Ming works was most evident in the *Collectanea* (see fig. 2), which published twenty-five books written before the Sung, eighty-five from the Sung and only ten from the Ming. The *Ssu-k'u ch'üan-shu* also showed this bias, however. While twenty-six *chüan* of the *Annotated Catalog* listings of books copied into the Ssu-k'u were devoted exclusively to Sung works, only ten *chüan* were devoted to Ming works. Conversely, in the section of the catalog which listed books reviewed but not copied into the *Ssu-k'u ch'üan-shu,* twenty-six *chüan* were devoted to Ming works, while only three were devoted to Sung works. Also, as Kuo Po-kung has noted, works by disciples of the Ming philosopher Wang Yang-ming (1472–1529) appear to have been consistently ignored: books by Wang's disciples Li Yung (1627–1705) and Huang Tsung-hsi (1610–1695) were only listed as extant in the catalog, and Wang's own works were listed under "belles lettres" rather than "classics."[142] The evident slighting of Wang Yang-ming may have reflected the editor's view that these works were not really useful, but it also reflected the anti-Ming bias of the whole collection.

The small size and apparent inconsistencies of the sections on Buddhist and Taoist texts have also been much criticized. The emperor's instructions were that the editors should include only such works as were necessary for reference. The Buddhist section of the catalog included only thirteen works, all written before the Yuan dynasty, while the Taoist section included forty-four books. Some of the editors' decisions were quite confusing. They chose, for instance, to include a Sung Buddhist work entitled *Kao-seng-chüan* (Biographies of eminent monks),[143] and ignored a T'ang text of the same title.

There was even some confusion among the editors themselves, for several works that Yü Min-chung specifically requested the editors to include do not appear in the catalog as we have it today.

Those who lamented the small size of the Buddhist and Taoist listings in the *Ssu-k'u ch'üan-shu* have often ignored the fact that not only the editorial principles which governed the selection of Buddhist and Taoist texts, but often the specific titles to be included were governed by historical precedent. Virtually all of the books listed in the Buddhism section of the Ssu-k'u had been included in earlier book collections, and the selection could perhaps be regarded as the consensus of generations of bibliographers on "What a well-educated emperor should know about Buddhism."[144]

A comparison of the Ssu-k'u listings with those of the seventeenth-century Buddhist bibliography *Yueh-ts'ang chih-ching* (A bibliographic guide to the Tripitaka) is suggestive.[145] The Yueh-ts'ang, compiled by Chih Hsu (fl. ca. 1650), was divided into four sections, the first three corresponding to the three traditional divisions of the Buddhist tripitaka. All of the Ssu-k'u listings came from the fourth division of the bibliography, that of miscellaneous writings about Buddhism. Within this category, the Ssu-k'u editors chose only works by Chinese authors, and did not choose works by authors clearly affiliated with any of the various schools of Buddhism. The Ssu-k'u selections included the largest extant works in each of the Yueh-ts'ang subcategories for "miscellaneous essays," "biographies of monks," and "presenting the faith." The only individual school to be represented in the Ssu-k'u collection was the Zen school, and the listings included four works by Sung Zen masters. What the Ssu-k'u editors, and the earlier imperial bibliographers whom they cited as precedent, appear to have been aiming for was a broad, secular view of Chinese Buddhism uninfluenced by sectarian concerns.

The Buddhist books so selected illustrated the cultural heritage which Chinese Buddhism and Confucianism held in common, a heritage which it was the purpose of a "complete collection" to highlight. Similarities in the historical development of Confucianism and Buddhism demonstrated the interrelatedness of the two doctrines. The editors of the imperial catalog were particularly fond of drawing a parallel between the development of Zen and of Neo-Confucianism. The basic theme of Buddhism, the catalog noted in one review, was the notion of karma, just as the basic theme of Confucianism was the interpretation (*hsun-ku*) of the six classics. Both Zen and the

Neo-Confucian study of *i* and *li* were departures from these basic themes, and both had emerged during the Sung. The argument here was not the usual one, that Zen had influenced the development of Neo-Confucianism, but that the two doctrines were parallel departures from original themes inspired by the same patterns of Sung thought.[146] In another review, the catalog lamented the unfortunate tendency of bodies of doctrine to spawn scholarly controversy. Before the T'ang, the editors wrote, Confucianism and Buddhism had contended primarily with each other. After the end of T'ang, however, factionalism increased daily and Confucians came to oppose other Confucians, while Buddhists fought with each other. Here the implication was not that Confucianism and Buddhism should go back to fighting with one another, but that the tendency to contend was a universal one, of which a universal ruler should be aware.[147]

Nowhere in the catalog did editors argue that an emperor should know something about Buddhism because his subjects believed in it; to have made such an argument would have perhaps conceded too much of the emperor's authority as an arbiter of absolute truth. Rather, the position of the imperial catalog was that there were many lessons that a good Confucian should learn from Buddhist books since, although Buddhism was originally a foreign doctrine, centuries of development within China had rendered it a part of the Chinese intellectual tradition. It was because Buddhism had been influenced by Chinese tradition, even influencing it in certain cases, that its books were useful to a Chinese ruler.

The issues involved in determining the utility of Taoist texts were somewhat different. Where Buddhism, for all its complexity, had an accepted body of texts to which editors could refer, the task before Ssu-k'u editors in the case of Taoism was to sort out what the original meaning of Taoism was, and to label and classify subsequent accretions. Issues of provenance, textual integrity and philosophical purity dominated the Ssu-k'u reviews of Taoist books. The Ssu-k'u editors' attitude toward these issues may well have been expressed in their comment in one review that Taoist books were useful in certain respects, but the Taoists themselves "were not very orderly."[148]

The themes of the Taoist section were set forth in its headnote. "In recent times, writing on ghosts and the supernatural have been classified as Taoist. And, as one might expect, Taoists have written about such matters, as for instance in the *Shen-hsien chüan* (Biographies of the spirits and immortals)," the editors wrote. "But at base, Taoism is

concerned with purity and self-discipline, succeeding through pa-
tience and self-restraint, controlling hardness with softness and
advancing through retreat." Much of what had traditionally been
classified as Taoist needed, therefore, to be reclassified; the *Shen-tzu*
and *Han-fei-tzu* belonged to the school of military strategy, and the
various writings on ghosts and spirits should be classified together
and apart from more philosophical Taoist writings. Most writings on
these subjects, the catalog observed, were later additions to the Taoist
canon. Despite the obvious need for some reorganization, Taoists
had been unable to clean their own house, and therefore there
seemed to the editors little point in doing so in the imperial catalog.
"But by examining the writings traditionally classified as Taoist, the
reasons for the evolution of the school can be established."[149]

The impulse to label and classify (perhaps even pigeon-hole)
ancient writings was, as noted above, at the heart of imperial compilation projects, and in the case of Taoist texts this seems to have been
particularly important. The basic organizational principle in the
imperial catalog was that the originator of a school or tradition
would be listed at the beginning of the section on that school, followed by various commentaries listed in the order in which they were
written. This means that in most sections, including the Buddhist
chapter, books were listed in chronological order. In the Taoist section, however, the editors separated what they considered to be different strains of Taoism, each deriving from an ancient text. Under
each text, several commentaries were listed. Thus the *Yin-fu-ching* was
listed first, followed by three commentaries; then the *Tao-te-ching*, followed by nine commentaries; then the *Lieh-tzu*, the *Chuang-tzu* and so
forth. The order in which the various strains of Taoism were considered in the Ssu-k'u catalog generally paralleled the order in which
they appeared in the *Tao-tsang* and in previous imperial bibliographies.[150] But the annotations of the Ch'ien-lung editors made explicit
a point which was only implicit, if present at all, in previous bibliographical treatments: that Taoism was an extremely diverse reflection of many trends within Chinese society rather than a consistent
philosophical whole. Appropriately, the first text which the editors
felt dealt fairly with Taoism as a whole was the *Yun-chi ch'i-ch'ien* (A
guide to Taoist writings), a compilation prepared under the patronage of a Sung emperor.[151]

The catalog was most suspicious of texts which confused various
strains of Taoism, or which confused Taoism with other strains of

Chinese thought. Like Chu Hsi, the Ssu-k'u editors rejected the claim of Su Ch'e (1039–1112) that Taoism and Buddhism derived ultimately from the same source.[152] They also rejected vehemently the interesting, and somewhat more likely claim of Hsu Ta-ch'un (1693–1771) that Taoism antedated Confucianism, and that Taoist texts more closely represented the ideas of the Yellow Emperor than the Six Classics. Lao-tzu, the catalog asserted, had lived in a time of chaos and had evolved his philosophy of purity in response to troubled times. It had not been Lao-tzu's intention that his work should ever be taken as the foundation for the government of a unified empire. Certainly, the sages of antiquity had meant the Six Classics, and not the *Tao-te-ching,* to be the guides for their followers.[153]

Ssu-k'u editors were also leary of those who practiced magic in the name of Taoism. But if magic were to be practiced, the editors seemed to argue, it ought to be written about under a Taoist label. For this reason, the editors were careful to point out, they classified books about divination which were based on the *I-ching* under Taoism. Previous catalogs had classified such works as commentaries on the *I-ching,* or under *wu-hsing.* But this seemed to the Ssu-k'u editors to confuse the issue, with the particularly pernicious effect of suggesting that Confucians had once approved of this sort of thing.[154]

Thus the goal of the Taoist section of the catalog was to establish clearly, and in some cases to distinguish between, the various philosophical impulses which were called Taoist. Books which furthered this goal were useful; others which confused Taoism with other schools of thought or which unduly stressed techniques or practices were not. As in the case of Buddhism, no mention was made of popular beliefs.

The line between correcting miswritten characters and actually editing a text to change its content was a fine one, and the editors of the Ssu-k'u undoubtedly engaged in both. The statutory record makes it clear that the emperor intended his editors to make changes in certain texts. He wrote of some collections of memorials by late Ming officials:

> Although there are many perverse words and sentences among the memorials, this cannot be held against their authors, for they were true to their own duties. Simply a few considered emendations; and the original works may stand for all time, to reveal to everyone how the Ming lost and how our own dynasty came to power, and to inform my posterity how their ancestors suffered to achieve empire.[155]

Yü Min-chung's letters suggest that the editors went along with such practices, although they occasionally did so with reluctance. In one case, he ordered editors to cut from a Sung collection a brief essay entitled "Chu-tzu chu-i" (Several proposals of Master Chu) containing several suggestions the Sung scholar Chu Hsi (1127–1200) made for opposing the Chin, or Jurchen invaders.[156] In another letter he remarked that one text was too commonly used to be edited.[157] To Yü's credit, he did instruct editors that deletions were to be clearly noted both in the *Annotated Catalog* and in the texts themselves, although this practice was not always followed.[158]

Editorial deletions were not confined to privately printed texts. In 1783, the emperor observed that several references to the Chin dynasty in the imperially commissioned *T'ung-chien kang-mu hsu-pien* (A continuation of the Comprehensive Mirror in outline form) were not in accord with the rules of transcription that he had recently prescribed.[159] He therefore ordered all provincial editors to collect and emend all printed editions based on officially prepared depository copies. The process went on for at least two years, with provincial governors required to report at the end of each year the number of copies they had changed.[160]

While scholars have been aware for some years that the Ssu-k'u editors made changes in texts, few have made the detailed comparisons of pre- and post-Ssu-k'u texts necessary to assess the significance of editorial deletions. Preliminary evidence indicates, however, that most of the deletions involved either the elimination of anti-foreign references, stories and legends circulating about the Manchu conquest or early complaints about the character of Manchu administration. For many years, for instance, it was assumed that Ch'ien-lung editors made significant revisions in the text of Ku Yen-wu's *Jih-chih-lu*. The assumption was based partly on the facts of Ku's life. A towering figure in early Ch'ing intellectual history, Ku was regarded as one of the most brilliant and creative minds of his era by contemporaries, and revered as a founder of modern scholarship by eighteenth-century intellectuals. The problem was that, although Ku's place in the shadowy world of seventeenth-century loyalist politics had yet to be carefully established, there were unmistakable signs of anti-Manchu activism in his biography. Was he also too seditious to be endorsed? Such, at least, was the suggestion of Chang Ping-lin who accused Chi Hsiao-lan and his "henchmen" of systematically emasculating Ku's work of its social and political content in the course of their editorial labors.[161]

Initially, Chang's suggestion seemed plausible, since the version of *Jih-chih-lu* in the *Ssu-k'u ch'üan-shu* differed from those circulating in the early twentieth century. However, closer examination of the catalog belied the notion of any systematic attempt to distort Ku's work. Some twenty-three of his works, almost his entire corpus, were noted in the *Annotated Catalog;* fifteen were copied into the imperial library and eight were given brief notice. Only seven other authors, one of them Chu Hsi, had as many books reviewed. Moreover, of the twenty-three works included in the Ssu-k'u, only two had previously been a part of the imperial library, and only one was described as an edition in common circulation.[162] The remainder of the works were drawn from private collections, in particular, those of the Liang-huai salt merchants. One consequence of the project, therefore, was to place editions of Ku's work, which had previously been found only in the hands of the wealthy few, into the public domain. How could this consequence be reconciled with the view that editors were censoring Ku's work?

The problem was resolved in the early thirties of this century when studies of the provenance of the *Jih-chih-lu* revealed that the unexpurgated version of the work available today was unavailable in the eighteenth century. The only version of the work which Ssu-k'u editors could cite was an expurgated one prepared by Ku's own students. shortly after his death.[163] What had seemed to Chang Ping-lin a crime of conscience proved to be a matter of accidents of historical provenance.

The complex interaction between scholars and courtiers in the editing and emending process was illustrated in the case of the *Chiu Wu-tai-shih* (Old history of the Five Dynasties). This work, a history of the fifty-three years immediately preceding the Sung dynasty, was commissioned by the first Sung emperor in 973 and presented to the court by prime minister Hsueh Chü-cheng in the following year. It drew on the official archives of the period to present a factual narrative. At least fifty years later, the great Sung prose master Ou-yang Hsiu compiled the *Hsin Wu-tai-shih* (New history of the Five Dynasties). Although Ou-yang's history was written in a crisp and attractive prose style and embodied the finest traditions of Confucian "praise and blame" historiography, it was not as reliable as the earlier

work. Both books were quoted during the Sung dynasty, but Ou-
yang Hsiu's proved to be the more durable. Hsueh's history was
copied in part into the *Yung-lo ta-tien,* but had been lost by the eighteenth century.[164]

In 1773, Shao Chin-han, one of the few Yung-lo revisers appointed to the Ssu-k'u Commission on the basis of his reputation as
a historian rather than his official position, set about reconstructing
the text of the *Chiu Wu-tai-shih.* He recovered large portions of it from
the Yung-lo encyclopedia, and supplemented them where necessary
with material drawn from relevant Sung dynasty sources. His reconstructed text was, therefore, as long as Hsueh's original, and probably contained much the same information, but it was not Hsueh's
exact text. To indicate where he had supplemented or changed the
original, Shao wrote notes on small, yellow slips of paper and inserted them into the manuscript, and prepared a long, bibliographical appendix.[165]

When the work was copied into the *Ssu-k'u ch'üan-shu* and printed
in the *Collectanea* it was decided, evidently after some debate, not to
include the bibliographical appendix, to print all the collation notes
at the end of the text, and to make quite a few changes of wording
in Shao's original. The product, which was neither primary fish nor
secondary fowl, must have horrified Shao who, as an enthusiastic
k'ao-cheng scholar, had often lambasted editors past and present who
made changes in texts without showing their sources. It certainly
offended many of Shao's friends.[166]

Time and circumstance have, however, mitigated the seriousness
of the Ssu-k'u editors' crime against the canons of evidence. Several
of Shao's friends evidently made manuscript copies of his work for
their own libraries, including the notes and appendix. One of these
was published in 1921, and another surfaced five years later; both
were used as bases for the *Ssu-pu pei-yao* and Commercial Press (Peking) editions of the text. In 1937, Ch'en Yuan published a short
study entitled *Chiu Wu-tai shih chi-pen fa-fu* (A reconstruction of the
working version of the old history of the Five Dynasties) in which he
detailed some 194 changes in wording made by the Ssu-k'u editors
in Shao's text. Most of these involved changing the traditional,
pejorative terms for nomadic peoples, *hu, lu, chieh,* and *i* into the
formal tribal name, *Ch'i-t'an,* for the people in question. References
to certain non-Chinese institutions, such as tribal chieftains, and
other mentions of bandits and traitors were also changed.[167]

The *Ssu-k'u ch'üan-shu* was characteristic of its era. Underlying the achievements and defining the tenor of late eighteenth-century China were certain patterns and potentials for cooperation and conflict between various elements of government and society. A project like the *Ssu-k'u ch'üan-shu* achieved its ends only to the extent that the elements of state and society with interests relevant to it were able to acknowledge a common purpose and cooperate in a common endeavor. The emperor, the elite "top-drawer" segments of the capital bureaucracy, and certain elements of the scholarly community were the crucial groups for the *Ssu-k'u ch'üan-shu* project. In many respects, their cooperation was successful. The most extensive catalog of Chinese learning ever compiled, useful today in spite of its biases, was produced and much more important work of bibliographic preservation and recovery was accomplished. In other respects, of course, the Ssu-k'u editors failed. But in a sense, the day for assessment of their products has passed: the text of the *Yung-lo ta-tien* is now destroyed, many of the editions with which Ssu-k'u editors worked are no longer extant, the Chinese quest for encyclopedism takes different forms today than in the past. The project is most relevant today in that it suggests how areas of common interest between emperor, bureaucracy and scholars made possible the successes of the project, points of disagreement presaged the failures.

Uniting emperor, bureaucrats, and scholars was a belief in the importance of the written word as the ultimate source of intellectual and political authority. The tremendous commitment of time, talent, space, resources and energy made by all three to the collation of ancient texts testified to the importance of this belief to them. Yet word and deed were differently related for each. For the emperor, words sanctioned deeds. Not only was the *Ssu-k'u ch'üan-shu* to be a collection of classical precedents for action, but the emperor's sponsorship of the largest book collection project in history, his ability to attract the most talented men of the era to his court and supersede all predecessors in literary reputation and munificence was meant to demonstrate his right to be leader of the literate community of the most civilized empire in history. His concern to achieve this demonstration during his own lifetime probably accounted for his constant demands for haste, and in part for his impatience with manuscript errors.

For the bureaucrat, the word, or at least the ability to manipulate language and write history, made possible deeds within the political arena. Literary ability opened doors to positions of power and authority at all levels of Chinese government, and under a leader like the Ch'ien-lung Emperor, this was particularly important. For a man who sought to wield influence and authority in the counsels of government, however, a position as collator of Ssu-k'u, or even editor, was only a stepping stone, not an end in itself; and the slow, painstaking work required to produce a perfect manuscript was not the sort of effort that earned recognition, commendation, and promotion. The institutional environment of the project thus inevitably influenced their attitudes toward it. Even the reporting of errors apparently served political as well as intellectual purposes. For the scholar, words explained deeds. Only when texts were carefully transmitted and meticulously edited could ideas and events past and present be understood. These differences in the perspectives of emperor, bureaucrats and scholars were, of course, present throughout Chinese history; they constituted more or less an obstacle to the achievement of a given goal to the extent that the importance of the tasks undertaken or the puissance of the authority involved were recognized.

In the late eighteenth century, a façade of harmony could be maintained over the Chinese government's activity; while differences between emperor, bureaucrats, and scholars were perhaps as explicit as they had ever been, they could be reconciled. The difficulties of cooperation and the realities of division would, of course, become more apparent in the nineteenth century. Indeed, the history of changing assessments of the Ssu-k'u project suggested some of the form this division took. When confronted with the unprecedented challenges of the nineteenth century, the emperor (or those who exercised power in his name), bureaucrats, and scholars, would come to stand for fundamentally different visions of social order and the actions necessary to defend it. But while nineteenth-century challenges magnified divisions within the Chinese elite, such divisions were not created in the nineteenth century. The *Ssu-k'u ch'üan-shu*, a grand 36,500 *chüan* summation of China's cultural heritage, illustrated both the potential for harmony and the realities of division and self-interest in the modern Chinese state.

5

Reviewing the Reviewers:

Scholarly Partisanship
and the *Annotated Catalog*

One of the most serious charges leveled by twentieth-century critics at the Ch'ien-lung Emperor and his editors at the Ssu-k'u Commission was that they sought to impose an intellectual straitjacket on the scholarly world, and to destroy through partial or total censorship works opposed to their "party line." In fact the *Annotated Catalog,* a work which expressed if any did the imperial view of scholarship and literature, was a complex tapestry, woven of many strands of imperial interest and scholarly conviction. Just as collation of texts for the *Ssu-k'u ch'üan-shu* was successful only to the extent that persons of diverse interests could be coordinated in a common endeavor, so the comprehensive scope and encyclopedic detail of the catalog could be accomplished only if many scholarly specialists could be incorporated into the political hierarchy.

The incorporation of scholars into government had implications for intellectual as well as political history. The question of how scholarly insight becomes received wisdom, an interesting one in any community, was particularly intriguing in imperial China. In theory, the emperor was a sage with broad understanding of and powers over the life of the intellect. In fact, new ideas and paradigms were developed as often outside the government as within it. How could the various demands of political authority and intellectual creativity be reconciled? How did sage and scholars interact to structure intellectual discourse? How were new insights communicated to the emperor and incorporated into the orthodox canon? How and by whom were the imperial powers in the intellectual world exercised and to what ends?

It was characteristic of the complexity of the eighteenth century that conflicts over these issues were not only, or primarily, between the emperor and a monolithic academic community. Within the scholarly community, different intellectual visions competed for imperial endorsement. Chu Yun and his colleagues had proposed undertaking several projects in conjunction with the imperial book collection effort because of their conviction that truth was contained in texts. In a broad sense, of course, the whole Ssu-k'u project represented an endorsement of that view. But the descriptive notes (*t'i-yao*) which were prepared for the *Annotated Catalog* offered opportunities for much more specific endorsements of individual views, and it was at this level that the competition for imperial endorsement was most keen. The success of the scholars of Han learning, and the frustrations of their opponents, were an important aspect of the Ssu-k'u project.

Despite the attempts of several historians to establish a single author or a small group of authors for the descriptive notes, it is now apparent that they were the work of many hands.[1] In the first edict concerning the Ssu-k'u project, the emperor ordered that a brief statement of contents and value be inserted in each work submitted to the imperial library. In fact, when Juan Yuan was serving as education commissioner in Chekiang in the 1790s, he found some volumes intended for submission to the imperial library which contained such statements.[2] Not all of the books for the Ssu-k'u came from outside the capital, however, and it seems likely that some of the revisers on the Ssu-k'u staff divided among themselves the tasks of reviewing books which came from within the imperial collection, or perhaps of rewriting the descriptive notes for books inadequately dealt with by provincial editors.

Three collections of draft reviews by revisers in the capital were published in the nineteenth century. One, a set of thirty-seven reviews by Shao Chin-han collected by Ma Yung-hsi, was included in the *Shao-hsing hsien-cheng i-shu* (Papers of the former masters of Shao-hsing) which appeared in 1883.[3] These reviews were mostly of books in the history division; among them were drafts for the reviews of all the twenty-two standard histories. But also among the titles reviewed were two books from the classics division and four from belles lettres. Draft reviews by Yao Nai were collected and edited by his student

Mao Yü-sheng (1791–1841) in 1841, and published in Mao's *Hsi-pao shu-lu san-chung* (Three unpublished works by Yao Nai).[4] Of the eighty-six reviews in this collection, eight were of books in classics, fourteen from history, twenty-three from philosophy, and twenty-eight from belles lettres; thirteen of the reviews were of books not listed in the *Annotated Catalog*. The drafts in these two collections clearly formed the basis for the reviews that appeared in the *Annotated Catalog*, but they underwent significant revision before publication. A third collection, said to include over a thousand draft reviews by Weng Fang-kang, was extant as late as 1875 but does not seem to exist today.[5]

The task of editing these drafts apparently fell to Chi Hsiao-lan. Liu Ch'üan-chih, one of the assistant editors for the *Annotated Catalog* wrote that "The emperor ordered my teacher Chi to oversee the writing of the *Annotated Catalog* so that all the reviews would be edited by one hand." Yu Min-chung wrote in a letter to Lu Hsi-hsiung: "I know the difficulty of the reviews. Unless it has passed through your hands or Chi Hsiao-lan's, it is not worthy of being called a draft. Although there are many among the editors who fancy themselves exalted, I have never dared trust them."[6]

The process was clearly open to many kinds of influence, and the extant draft reviews demonstrate in fairly graphic detail how scholarly partisans sought to use these opportunities. In fact, the authors of two of the published collections of drafts stood on opposite ends of the eighteenth-century scholarly spectrum. Shao Chin-han was a distinguished scholar of Han learning, adept at philology and intolerant of carelessness among his colleagues and contemporaries. Yao, on the other hand, had his doubts about the trend toward textualism, and articulated in his reviews a rather intuitive scholarly stance more in accord with the views of the great Sung scholar Chu Hsi. In terms of influence on the *Annotated Catalog*, Shao was clearly the more successful of the two reviewers; Yao's frustrations ultimately led him to resign from the court and return to his native T'ung-ch'eng, where he formed a rallying point for those dissatisfied with the political and intellectual trends of the times.

Shao's successes illustrated how insight could become received wisdom in imperial China, but they also illuminated the state of relations between the government and the Han learning movement in the eighteenth century. Ideas are not, of course, merely the handmaidens of politics, but in the case of the Han learning movement, the relation of political power and intellectual concerns seems especially worthy of investigation. Through much of the twentieth century, scholars have argued that the Han learning was essentially a movement of scholarly opposition to Manchu political repression and intellectual persecution. The editorial changes in Shao Chin-han's reviews suggest that this conclusion must be approached with some caution.[7] By the middle of the eighteenth century, a compromise appears to have been reached between scholars and the Ch'ing state. The changes made by Ssu-k'u editors in Shao's reviews suggests the nature of this compromise, and further demonstrates which of the contentions of the movement had come to be acceptable at court.

Conversely, the treatment of Yao Nai's reviews suggests that there were at least some levels at the Ch'ing court on which one could question old formulations of orthodoxy and evolve new ones, and some room for debate about which texts most effectively articulated ancient truths. They illustrate, in short, how the "politics of orthodoxy" could work in the late imperial Chinese court.

The Ssu-k'u ch'üan-shu *and Han Learning:* The Draft Reviews *of Shao Chin-han*

Shao Chin-han was born in 1743 into a prominent family of Yü-yao district, Chekiang. Little is known of his father, the only member of the family in five generations without an official degree, except that he enjoyed book collecting and on occasion returned home with his bibliographic treasures only to have his wife sell them to meet household expenses.[8] Chin-han was raised in the home of his grandfather Shao Hsiang-jung (1674–1757), a director of schools in a district farther east along the Chekiang coast, who was the major influence in Chin-han's early intellectual life and who passed on to him the scholarly traditions of the family and their native district.[9] These centered around the teachings of Shao Chin-han's great uncle, Shao T'ing-ts'ai (1648–1711). T'ing-ts'ai was an admirer of Huang Tsung-hsi and, through his own father and grandfather, a second or third-generation disciple of Wang Yang-ming.[10] At the end of his life, T'ing-ts'ai taught what he had learned to his younger half-brother Hsiang-jung,[11] and it was this tradition which Shao Hsiang-jung passed to Shao Chin-han. Although later associations stimulated other interests, Shao Chin-han remained faithful to his grandfather's teachings throughout his life, a fidelity noted by eighteenth-century and modern biographers alike.

During his examination career, Shao met other figures influential in his intellectual development. At the tender age of sixteen, he received his *hsiu-ts'ai* degree. At twenty-two he earned his *chü-jen,* and with it the lifelong respect and friendship of his examiner Ch'ien Ta-hsin. Ch'ien wrote of Shao's *chü-jen* examination:

> His five essays on government policy exceeded all others in breadth and penetration. I remarked that they could only have been written by a mature scholar. When Shao came to call, I was astonished to find he was a young man. But his learning was inexhaustible.[12]

In pursuit of his *chin-shih* degree, Shao made three trips to Peking, together with his friend Chang Hsi-nien.[13] On the strength either of his family ties or his contacts with Ch'ien Ta-hsin, Shao was able to meet on these trips most of the intellectual luminaries of the day. As a poverty-stricken fellow-provincial, Wang Hui-tzu noted, "From the time I met Shao Chin-han, I began to know all the great scholars of the day."[14] In 1767 Shao arranged through friends to read and copy Chi Hsiao-lan's manuscript of the famous study *Ku-wen Shang-shu k'ao* (A study of the old text version of the *Book of Documents*), one of the most famous works of eighteenth-century textual investigation.[15] On Shao's third trip to Peking in 1771, he received his *chin-shih* degree.

There is some question, however, of how Shao regarded the degree he had earned. Although he passed the Board of Personnel metropolitan examinations first on the list, a distinction that normally would have earned him a seat in the prestigious Hanlin Academy, Shao did not enter the Academy but instead embarked on a journey through southern China. Shao's modern biographer Huang Yun-mei, referring to a poem by one of Shao's associates, claims that Shao deliberately refused an appointment and chose travel instead.[16] Ch'ien Ta-hsin offered two explanations. In his formal biography of Shao, Ch'ien noted that although Shao passed first in the metropolitan examinations, he was ranked in the second class on the subsequent Palace examinations, too low to be assured of a seat at the Academy.[17] In a less formal writing about Shao's father, Ch'ien suggested that Shao had, in fact, chosen not to serve in government, as had several of his ancestors and early Han learning scholars.[18] Shao's travels to the south were not without effect, however, as they soon brought him to the yamen of Education Commissioner Chu Yun at

T'ai-p'ing, where he met Hung Liang-chi, Tai Chen, and Chang Hsueh-ch'eng, and probably participated in the discussions that led to Chu's memorial on the Ssu-k'u project.[19] In 1773, Shao followed Chu and Tai to the capital for appointment on the Ssu-k'u Commission.

Shao's service at the Ssu-k'u Commission was brief and was interrupted by an extended trip to Chekiang to mourn the death of his mother in 1776–1777.[20] It was long enough, however, for him to draft many reviews for the *Annotated Catalog*, recover several important works from the *Yung-lo ta-tien*, and receive the acclaim of the intellectual community. Shao and his colleagues Tai Chen, Chou Yungnien, Yü Chi and Yang Ch'ang-lin were known as the "five lords of evidence" of the Ssu-k'u Commission.[21]

After his work at the Ssu-k'u Commission, Shao served in several official sinecures, including compiler at the Hanlin Academy, senior supervisor at the Supervisorate of Imperial Instruction, proctor at the State Historiographical Commission, and chief editor of the Diaries of Action and Repose.[22] In these capacities, he worked on a number of imperial scholarly projects and wrote many of the drafts for the official biographies of the *Ch'ing-shih*, several times earning special imperial praise. He also completed his own commentary on the *Erh-ya*, began a history of the Southern Sung dynasty, and advised Governor-general Pi Yuan about a continuation of the *T'ung-chien kang-mu* (Comprehensive mirror in outline form) that Pi was contemplating.[23] Although Shao was in poor health for most of his life, he apparently died quite abruptly when a physician administered the wrong medicine to him on 19 July 1796.[24]

While the followers of Han learning shared enough assumptions to constitute a genuine movement, there were also differences among them, differences noted both by their contemporaries and modern historians. Shao's historical and philological work illustrated both the common assumptions of the movement and the points at issue within it.[25] Shao's allegiance to the teachings of his grandfather, and indirectly to the teachings of Wang Yang-ming and Huang Tsung-hsi, placed him in an intellectual lineage which Chang Hsueh-ch'eng labelled the "eastern Chekiang tradition."[26] The characteristics of this rather loosely constituted tradition were an interest in modern history, particularly that of the late Ming and early Ch'ing, and what Paul Demieville has termed an inclination toward "synthetic intuition" in historical interpretation.[27] The first characteristic probably derived ultimately from Huang's political loyalties. The eastern Che-kiang region resisted the Ch'ing conquest for a particularly long time and Huang, who participated in this resistance, took it as his charge to preserve its legends.[28]

The second characteristic was a compound of historiographical and epistemological assumptions. As a follower of Wang Yang-ming, Huang Tsung-hsi believed in a unity of knowledge and action, of learning and ethical cultivation. This belief was expressed in his motto: "Without broad reading, one cannot trace the changing of principle; without moral reflection, all learning is vulgar," which has often been contrasted with the motto of Huang's contemporary Ku Yen-wu, "in learning, breadth; in behavior, shame," the latter being said to express a greater separation of the worlds of learning and ethics.[29] One implication of Huang's belief for historical writing was that in order to provide a reader with sufficient material for moral reflection and intuitive perception of the ethical lessons that history was meant to teach, one should write what would be called in modern times "monographic studies" of entire eras or lives.[30] This historiographical stance can be contrasted with that of the "western Chekiang tradition" of which, ironically, Shao's friend Ch'ien Ta-hsin was the leader, which believed in slowly compiling apparently unrelated building blocks of historical evidence and recording them in "notation books" such as Ch'ien's own *Shih-chia-chai yang-hsin-lu* (Record of cultivating new knowledge in the Shih-chia Study) or *Tu-shu tsa-chih* (Miscellaneous reading notes) of Wang Nien-sun (1744–1832).[31]

In many respects, Shao was faithful to his grandfather's teachings. All of his historical studies were of entire eras, and he expressed in his writings a consistent concern for the moral role of the historian. In one of his Ssu-k'u reviews, for instance, he praised the Kung-yang commentary over the Tso commentary on the ancient *Spring and Autumn Annals* because the former supposedly distinguished between "accomplishments" and "inborn qualities," and dwelled on accomplishments.[32] Whether or not this was in fact the difference between the two schools,[33] Shao's remark was interesting in its implication that the historian's role was to judge what an historical actor made of himself, not simply to record his inborn nature. Shao also preserved the eastern Chekiang tradition of knowledge of late Ming history. Governor Pi-yuan was said to have remarked to Shao after one conversation:

You of eastern Chekiang are familiar with the writings of Liu Tsung-chou and Huang Tsung-hsi, the stories of factionalism and misgovernment at the end of the Ming, and the resistance of the Princes of T'ang and Lu. [This history] is passed on by word of mouth or in handwritten manuscripts and is outside of the official record. When you die, there will be no one to verify the events south of the river (i.e., in Eastern Chekiang).[34]

Shao, however, had many other abilities besides his historical knowledge, as was demonstrated in his commentary on the *Erh-ya*. While philology was very popular in the eighteenth century, it was not particularly associated with the school of Huang Tsung-hsi; if anything, it was associated with the followers of Ku Yen-wu.[35] The *Erh-ya* was a Han dynasty thesaurus which classified words and expressions in eighteen categories (mountains, birds, tools, and the like) according to their use in various classical texts.[36] Most commentators before Shao, as he noted in his preface, had pointed out only the allusions to the *Shih-ching* (Book of poetry), leaving unidentified the references to other texts. By assembling these references one could not only clarify the meanings of words in the *Erh-ya*, Shao claimed, but correct errors of transmission in the other texts.[37] The problem with such work was not so much assembling the texts to which *Erh-ya* could have been alluding, but deciding which allusions the ancient dictionary was actually making. Preparing a commentary, Shao said, was a labor of love: "Riding in my cart from north to south, I have never ceased to apply myself to it."[38] The work took him at least eight years, but the results were impressive. Huang Yun-mei claims that when Shao finished, several of his contemporaries, Wang Chung, Wang Nien-sun and Tuan Yü-ts'ai, put aside similar manuscripts on which they were working.[39] Hung Liang-chi composed a poem about the text, remarking: "On reading your book, all my doubts, as by the light of dawn, were swept out," and comparing Shao's knowledge and diligence with those of his examiner Ch'ien Ta-hsin.[40]

Shao thus typified many strands of Han learning. From his grandfather, he had learned the methodology and beliefs of Huang Tsung-hsi and his followers. Through his associates, and particularly Ch'ien Ta-hsin, he came into contact with other strands of the Han learning movement and developed other interests, including philology and classical studies. Through Chu Yun, Shao met several important eighteenth-century thinkers—Tai Chen, Chang Hsueh-ch'eng and Hung Liang-chi—who had not attained official prominence. All this prepared him personally and intellectually for his work at the Ssu-k'u Commission. One of the most interesting and intellectually revealing tasks he performed there was the preparation of draft reviews for many important history texts, including the standard dynastic histories. A comparison of Shao's wordings with the final versions that appeared in the *Annotated Catalog* will suggest some of the areas of conflict and agreement between the Ch'ien-lung court and the Han learning movement.

Shao Chin-han's modern biographer Huang Yun-mei has accused the Ssu-k'u editors of a consistent attempt to deflect the thrust of Shao's draft reviews. They did this, according to Huang, by incorporating Shao's discussions (*i-lun*) and excising his conclusions (*pien-cheng*), thus converting carefully argued essays into antiquarian treatises. A close examination of the drafts and final versions, however, suggests something other than systematic emasculation. Four reviews will be examined in detail below. In the first two, the editors adopted most of Shao's draft, changing only slightly his argument. In the latter two, the editors did significantly change Shao's draft, but the import of their change was to modify the style rather than the substance of the pieces, and the two versions are more suggestive of the differences between a scholar seeking imperial approval and an editor carrying out imperial will than of a systematic attempt to impose intellectual orthodoxy.

Shih-chi chi-chieh. The *Shih-chi chi-chieh* by P'ei Yin (ca. fifth century) was one of three early commentaries on the *Shih-chi* (Historical records), the other two being by Chang Shou-chieh (fl. 736) and Ssu-ma Chen (ca. eighth-century A.D.).[41] Evidently, it was Shao's favorite of the three; he remarked in his review of the *Shih-chi* that "With Ssu-ma Ch'ien's ability to explain what has survived of the classical canon, and P'ei Yin's ability to preserve the dispersed theories of the ancient Confucians, one who wanted to understand the meaning of the classics would certainly be able to attain his goal."[42]

Although the commentaries had been written centuries apart, it was the practice to print the three of them in one volume, a practice followed by the Ming Imperial College when it produced its authoritative edition of the *Shih-chi* and commentaries. The only edition in which the *Shih-chi chi-chieh* stood alone was that printed by Mao Chin, a bibliophile of the late Ming. The Ssu-k'u project had uncovered a copy of Mao's text (or a text based closely upon it).[43] The bulk of both Shao's draft and the final *Annotated Catalog* review was devoted to a comparison of the text of the Mao edition with the text of *Shih-chi chi-chieh* that could be reconstructed from editions like that of the Ming Imperial College in which the P'ei commentary was printed together with the Chang and Ssu-ma commentaries.

The point of both reviews was that many errors which had crept into the editions in which the three commentaries were printed together had been corrected in the Mao edition. This point was buttressed with numerous examples culled from a close comparative reading of the various versions. For instance, in the Ming Imperial College edition after the sentences "There were eight talented men of the Kao-yang family" and "There were eight talented men in the Kao-hsing family" the Mao text noted "see the *Tso-chuan* for biographies," while the Ming Imperial College edition made no notation. Or after the sentence "Ch'in Shih-huang paid no attention to the carts but valued the horses . . .", the Mao text quoted Hsu Kuang as saying "The original of this remark did not contain the words 'but valued'," while the other editions contained no quotation. As both the draft review and the final version noted, editions which imitated the Ming Imperial College edition were even less accurate.[44]

The two reviews were each about eight hundred characters long and differed in only ninety characters, of which forty constituted a list of the official positions held by P'ei Yin. All of Shao Chin-han's drafts lacked such a list, evidently a nicety of imperial compilation projects of which Shao was unaware or could afford to ignore. In the remaining fifty-two characters, just over five percent of the review, the editors made a very subtle change in the nature of the argument, a change which illustrated the different concerns of author and editor.

In the draft, Shao praised P'ei Yin's scholarship lavishly. From it, he claimed, "One could see [Han scholars'] tendency to draw on all the classics of the hundred schools, editing superfluous language and retaining only the essentials, keeping their words restrained (*yueh*) and their meaning broad (*po*)."[45] In sharp contrast to such meticulousness, later scholars had indiscriminately combined P'ei's text with those of other commentators and introduced errors into it. In short, where early scholars were precise, later scholars were sloppy; the Ming Imperial College edition was only one example of such carelessness albeit one of the worst.[46]

The Ssu-k'u editors changed Shao's praise of P'ei to a terse, four character notation that early scholars "drew on a wide variety of sources." The point of the final *Annotated Catalog* review was to contrast the accuracy of the Mao Chin edition uncovered by the Ssu-k'u project with the inaccuracy of the Ming Imperial College edition, and thereby implicitly to praise the achievements of Ch'ing imperial scholarship at the expense of Ming government-sponsored scholarship. Since all the texts commonly circulating derived ultimately from the Ming edition, the editors remarked in their final sentence, the edition reprinted in the *Ssu-k'u ch'üan-shu* was superior to all other texts.[47]

The changes the editors made in Shao's draft can be variously interpreted, but at base the differences between the two reviews were of form rather than substance. Both agreed that the Mao text was superior to all others, and both proved this point at what seems to the modern reader soporific length. All the Ssu-k'u editors did was to substitute praise of Ch'ing official scholarship (in its recovery of the Mao text) for Shao's praise of P'ei Yin. It is true that one of Shao's points, about the importance of reading ancient commentaries on ancient texts, was suppressed in the substitution. But the point was probably not lost on the intelligent reader, for it was implicit in much of the argument. And, the editors' change no doubt made the end product far more attractive to the Ch'ing rulers.

Sung-shih. Although the *Sung-shih* was the longest of the dynastic histories Shao reviewed, and has been praised by one modern critic as representing the "apex" of the "basic principles of Chinese official historiography,"[48] it was one of the least satisfactory for Shao. He seems to have had three objections to the work. Shao agreed with Chang Hsueh-ch'eng that although the *Sung-shih* contained a staggering amount of information, it was "inferior in literary quality, conveyed none of the meaning of historical events and exhibited no sense of selection in the presentation of data."[49] Second, much of the information in the history was in error and it made many contradictory statements.[50] Finally, it dealt primarily with the Northern Sung, ignoring what were for an intellectual historian compelling issues in Southern Sung history.[51] Shao's opinions on the *Sung-shih* were shared by many eighteenth-century thinkers, including Ch'ien Ta-hsin, Ch'üan Tsu-wang and Wan Ssu-t'ung.[52] Probably as early as 1774, Shao began work on or at least planned his own revision of the

Sung-shih, to be entitled *Sung-chih* (A treatise on Sung). The revision was evidently never finished and only its table of contents is extant today, but it was widely known among Shao's contemporaries.[53]

Shao made no secret of his scorn for the *Sung-shih* in his draft review. It began: "Previous critics have all remarked that the *Sung-shih* needed to be edited and sorted, but they have given few examples." This Shao proceeded to do, by citing and examining in detail a number of errors and inconsistencies pointed out by previous commentators K'o Wei-ch'i (1497–1574) and Shen Shih-po.[54]

Shao then turned to the philosophy of history in the book. Like his predecessors in eastern Chekiang, Shao was concerned with the moral significance as well as the factual accuracy of history. The *Sung-shih* contained much historical detail, but little interpretation; many had argued, in fact, that the historical information in the book made up for its lack of analysis. But for Shao, the book was unsatisfactory on both counts:

> Those who have criticized the *Sung-shih* have often observed that the biographies in it contain only the names of [their subject's] ancestors, and none of the realities of [their subject's lives] and in this they resembled grave inscriptions. It is as if the editors copied directly from government personnel records, making no emmendations or corrections. Those who like such history will perhaps argue that since the information in them is directly from long-preserved government records, it must be reliable. But when you compare them with other sources, the biographies in *Sung-shih* too often prove unreliable.[55]

Finally, Shao observed that the history was shaped by the political biases of Sung scholars who "liked to talk of the Northern Sung" but wrote little of the Southern Sung, and as a result dealt inadequately with the later period. The *Annotated Catalog* editors included all of Shao's criticisms in their final review.[56]

When they came to the question of why the *Sung-shih* was inadequate, however, both Shao and the Ssu-k'u editors had to tread lightly. Shao wrote: "When they compiled the *Sung-shih* the majority of editors simply took the official history as a draft. There was no time for further research."[57] In fact, Shao's observation was probably correct; as Yang Lien-sheng has remarked, the primary goal of the compilers of the Liao, Chin and Sung histories "seems to have been a quick finish," a goal which was achieved through "effective, if sometimes high-handed supervision and editing."[58] The problem was that the *Sung-shih,* like the *Liao-shih* and *Chin-shih,* was compiled during the

Yuan dynasty; the pattern of relations between warlike Mongol rulers, sinified Mongol bureaucrats and Chinese scholars was complex, perhaps resembling the pattern at the Ch'ien-lung Emperor's own court. Obviously, the implication that non-Chinese control of Chinese scholarly efforts was pernicious had to be avoided, particularly when the foreign rulers involved were the Mongol Yuan, to whose historical rights Ch'ing rulers were particularly sensitive.[59] Even the notion that scholarship sponsored by rulers, be they Chinese or foreign, could have harmful priorities and deadlines would have had to be approached with caution in the *Ssu-k'u ch'üan-shu*. Evidently Shao's passing reference—"There was no leisure for further research"—was too much for the Ssu-k'u editors, for it was cut from the final version.

Thus, while the Ssu-k'u editors reprinted all of Shao's criticisms of the *Sung-shih,* they generalized from them only cautiously. Their caution was evident in the introduction and conclusion, forty and eighty characters long respectively, which they appended to Shao's review. Drawing, perhaps, on other sources about the compilation of the *Sung-shih,* the editors began with an official biography of the Mongol compiler T'o-t'o (1313–1355). They then remarked that since the *Sung-shih* was the history of an entire era comprising over five hundred chapters, it must have been very difficult to proofread and correct. The final review then made the rather astounding comment that since the *Sung-shih* was written under the influence of the Sung learning, it was probably inaccurate on all matters except the Sung learning. The *Catalog*'s polemical tone here probably preserved Shao Chin-han's intentions in the review, but it may have served another purpose as well for, as Liu Han-p'ing has observed, attacks on the Sung learning in the *Catalog* were often associated with condemnations of Sung learning scholars' tendency to form factions.[60] At any rate, the editors were careful to provide some intellectual rationale for continued reading of the *Sung-shih:*

> From the time of K'o Wei-ch'i[61] on, many have edited the *Sung-shih.* But with the passing of the years, the sources have been lost and dispersed. One may take the original as a draft and supplement it in small respects, but it will never be superseded. Therefore, those who research the events of the two Sung dynasties must base themselves on the original text. It cannot be disregarded.[62]

In view of the Manchus' sensitivity to criticism of the Yuan, the Ssu-k'u editors were perhaps bold to include as many of Shao's criticisms of the *Sung-shih* as they did. That they did so illustrated the point that textual research *per se* was not perceived as a threat by the Ch'ing court. In editing Shao's drafts, the Ssu-k'u officials seemed to have little hesitation about adopting his major conclusions, method-ology or evidence, occasionally adding a proof or polemical sally of their own. In fact, much of what is useful in the *Annotated Catalog* today was probably the work of men like Shao. The changes that were made in Shao's reviews of the *Shih-chi chi-chieh* and the *Sung-shih* were significant in that they illustrated the differences between a scholar anxious to convey his vision to the world, and editors who would be held accountable for any political implications, but they were at base tangential to the issues at stake in the reviews. Were there matters more pressing to Shao or the editors?

Han-shu. The problem of forgeries in Chinese scholarship was dis-cussed in the review of the *Han-shu.* The first two-thirds of both the draft and final *Han-shu* reviews considered the differences between the "ancient" and "modern" versions of the work. The two versions were so labelled by Liu Chih-lin (476(?)–547) who, according to his biography in the *Liang-shu* (Records of the Liang dynasty), was sum-moned by Crown Prince Chao-ming (501–533) to compare two ver-sions of the *Han-shu* which the Prince held in his library. The version called by Liu "modern" had actually been in circulation since the Han dynasty and was in Liang times the standard edition. The "ancient" version had been discovered and presented to the prince by a subordinate. The implications of Liu's use of the labels were clearly that the "ancient" version was more authentic than the "modern."[63]

With this conclusion both Shao and the Ssu-k'u editors took issue. They offered six arguments. First, the dates of the memorial present-ing the *Han-shu* to the Han court contained in the "ancient" but not in the "modern" version did not correspond to the dates of comple-tion given in the biography of the author, Pan Ku, in the *Hou-Han-shu* (History of the Later Han dynasty). Second, the "ancient edition" placed an explanatory note (*hsu*) at the beginning of the work when it was the custom of Han dynasty authors to place such notes at the ends of their books. Third, the "ancient" contained a biography of Pan Piao, which would have been inappropriate in a history of the Former Han since Piao received his official degree and accepted office in the Later Han. Fourth, the "ancient" version was divided into 38 *chüan,* while Pan expressly described his history as having been divided into 106 *p'ien,* which was the form of the "modern" version. Fifth, the organization of the biographies in the "ancient" version differed from that in the "modern," and appeared to be the work of someone either unfamiliar with Han dynasty inheritance practices or with Pan Ku's basic aims. Finally, Prince Chao-ming himself quoted from the "modern" version in his famous literary collection, *Wen-hsuan* (Selections of literary writings). It was the conclusion of both Shao and the Ssu-k'u editors that the "ancient" work was in fact a Liang dynasty forgery.[64]

The argument was an elaborate one. It was also in a sense unnecessary since many of Liu Chih-lin's contentions had been refuted during the T'ang. But it provided Shao an opportunity for an extended and scathing attack on eleven centuries of Chinese scholarship. Most later commentators and collators, Shao wrote:

> liked to discuss the paper in Sung editions, or examine the antiquity or printing techniques, paying no attention to the meaning of words in the texts. They were all like this. The Han dynasty was close to antiquity and this book is orderly and thorough. Therefore, Kan Pao and Cheng K'ang-cheng (Cheng Hsuan) quoted it in their commentaries on the classics, while the Han classicists Fu Ch'ien and Wei Chao all wrote commentaries on it.[65] It most certainly can be used to supplement the classics. Yet for generations people wrote phonological and philological commentaries, never even suspecting that there could be a forgery. Their successors followed them in a common and thoughtless band, saying that (Pan) Ku could not write well and one could not draw implications from his work.[66] During the T'ang, Yen Shih-ku was called the "loyal minister of Pan Ku." Alas, all he did was to assemble old commentaries and weigh them.[67] He did not assemble other source materials in order to clarify the implications [of the text]. Those who followed him did not exceed his scope, so it would be appropriate to call them "loyal ministers" too.[68]

Forgeries, of course, aroused the particular ire of Han learning scholars. But the *Han-shu* review was not the only one in which Shao Chin-han intemperately attacked previous generations of scholars. In his review of the *Hou-Han-shu,* for instance, Shao wrote:

> [Intellectuals of the Six Dynasties] who discussed history did not consider the truth or falsehood of events, but blithely spoke of "praise and blame." They didn't give a thought to the literary merits of their own writings, but argued whether one or two words or expressions [in an ancient work] raised or lowered [a subject's stature]. Over and over they evaded each other's arguments, in and out [they bobbed] all without evidence.[69]

In a discussion of Ou-yang Hsiu's *Hsin Wu-tai-shih* (New history of the Five Dynasties), Shao remarked "What was most regrettable about Ou-yang Hsiu was his inadequate research."

Such comments were edited from the draft reviews before they were published in the *Annotated Catalog*. In the *Han-shu* review the editors replaced Shao's tirade with a calm, scholarly, survey of one of the controversies surrounding the *Han-shu*, whether Pan Ku had been paid to write it.[70] They also praised Yen Shih-ku's commentary with the observation that "just because a few words are out of order in it, there is no need to reject the whole commentary."[71] The reviews of the *Hou-Han-shu* and *Hsin Wu-tai-shih* took a different form entirely, relating the textual histories of the works rather than evaluating their worth as histories.[72]

Shao's carefully-mustered textual evidence, which was meant to provide support for his indictment of post-Han scholarship, was thus converted by the Ssu-k'u editors into the raw data for relatively dry surveys of textual history, and his major conclusions excised. But the reasons for the changes in Shao's reviews must be carefully assessed. Shao's views posed no threat to the Ch'ing court; if they had, why would the editors have included his evidence and suppressed only his conclusions? On the other hand, the comments did pose a threat to the intellectual consistency of the *Annotated Catalog* as a whole. One could hardly dismiss texts in one review, while carefully assessing their merits in another. Furthermore, Shao's statements were essentially partisan pronouncements. The *Ssu-k'u ch'üan-shu* was meant to be a source book for ten thousand generations of Chinese scholars, not a platform for scholarly controversy.

Ming-shih. Without corroborating evidence, it is difficult to prove that Shao Chin-han meant to be damning by giving only faint praise in his review of the *Ming-shih*. However, in his family and native district, Shao had learned more about Ming history than most eighteenth-century scholars would ever know. Yet his 575 character draft review of the *Ming-shih* was one of the shortest and least substantial that he wrote. On a number of subjects, particularly the legitimacy of certain late Ming pretenders, Shao may well have discreetly chosen to remain silent. Whatever the reasons for Shao's silence on these issues, the draft and final versions of the *Ming-shih* reviews had a distinctly different flavor. The final version depicted a government in its rightful role as teacher of moral lessons through history. Shao's draft, on the other hand, compared the *Ming-shih* with the writings of private historians, and stressed government editors' access to primary source materials on political and military events.

The draft and final versions of the *Ming-shih* review began very differently. Shao first observed that a number of Ming intellectuals had written histories of one reign or one event based on local traditions, gravestone inscriptions, and epitaphs. He cited as examples the *Wu-hsueh-p'ien* (Compilation of my studies) by Cheng Hsiao, the *Ming-shu* (Records of the Ming) by Teng Hsi-yuan, and the *Hsien-chang lu* (Chronology of statutes) by Hsueh Ying-ch'i.[73] All of these works suffered, Shao argued, because their authors did not have access to government archives and had written only partial histories. Even the methodical study by Wang Shih-chen, *Shih-ch'eng k'ao-wu* (An examination of errors in the historical records)[74] had not been able to correct all the mistakes. However, the compilers of the official history led by Chang T'ing-yü, working from a draft by Wang Hung-hsu and having access to the official records, had been able to compare the various private histories and compile a record that was both fair, comprehensive, and accurate.[75] Interestingly, none of the books Shao mentioned were copied into the *Ssu-k'u ch'üan-shu,* although all received brief notices in the *Annotated Catalog.*

The final version of the *Ming-shih* review made no mention of these earlier histories. It was more economical, a concern for all editors but particularly so for editors of a two hundred *chüan* imperial encyclopedia. But it was also more acceptable politically, for instead of discussing private historians' contributions to the *Ming-shih* it stressed the imperial role in the compilation process, quoting at length passages from the memorial presenting the *Ming-shih* to the Ch'ing court describing how the compilers had for over fifty years enjoyed the patronage and teaching of the K'ang-hsi, Yung-cheng and Ch'ien-lung Emperors.[76]

Both versions of the *Ming-shih* review then noted several innovations its compilers had made in traditional Chinese historiographical practice. The first of these was the inclusion of charts in the treatise on astronomy. Shao traced the idea to the works of Hsu Kuang-ch'i (1562–1633), and ultimately, to the influence of Jesuit astronomers who resided at court in the late Ming and early Ch'ing periods.[77] The *Annotated Catalog* simply ignored the contributions of private astronomers and wrote that "astronomy was created by numbers; numbers also created calculations. . . . Today the [methods of] calculations are much more complex than in ancient times. Therefore, if there were no charts, the points could not be made clear."[78] Perhaps the *Annotated Catalog* editors' reluctance to acknowledge the contributions of the Jesuits to Chinese astronomy reflected the restriction on Jesuit residence in China first issued late in the K'ang-hsi reign and reinforced during the Yung-cheng and early Ch'ien-lung reigns. Following closely after passages in which the emperors' guidance and patronage of the *Ming-shih* were praised at length, the comments in the *Annotated Catalog* had the appearance of attributing innovations to imperial wisdom.

Both reviews then discussed the *Ming-shih* compilers' practice of listing in the "treatise on arts and literature" only books written by Ming authors, and ignoring works by earlier authors. The final version of the review traced this practice to suggestions made by Liu Chih-chi (661–721) of the T'ang. Not incompatible, but somewhat more convincing was Shao Chin-han's explanation that because the records of the Ming imperial library had been lost, the *Ming-shih* compilers had nothing to consult in writing the treatise.[79] Finally both reviews noted that the creation of categories for biographies of eunuchs, and hereditary local commanders, and the addition of a chart of the presidents of the six boards and the Censorate reflected the importance of these groups during Ming times.

Although both reviews ended with praise of the Ch'ing emperors' role in compiling the *Ming-shih,* the *Annotated Catalog* editors added a final, fulsome comment on a subject that Shao may well have chosen to ignore, namely, the reigns of Ming pretenders. Shao's intellectual forebear, Huang Tsung-hsi, had served one of the pretenders, Prince Fu, until 1649; and Shao was said to know more about these reigns than any late eighteenth-century author.[80] The pretenders were also an especial concern of the Ch'ien-lung Emperor, who issued an edict on the subject on 17 December 1775 decreeing that Prince Fu's reign name could only be used in reference to events before 1644, and those of Princes T'ang and Kuei could not be used at all.[81] The *Annotated Catalog* carefully recorded the emperor's views on this subject, and asserted with the simpering sycophancy characteristic of imperial historiography that they demonstrated "the will of the sage to be both correct and fair", and further that "never since there have been historical source materials has there been a case of such insight."[82]

In the concluding passage, as throughout the *Annotated Catalog* review, the editors stressed the right of an emperor to establish the truth of history and determine the moral lessons to be drawn from it. Shao Chin-han's writing, at least on the *Ming-shih*, suggested that he believed just the opposite, that emperors and official compilers were bound by the same canons of evidence and proof as their subjects. The issues of whether private historians should participate in imperial compilations, and whether private and public scholarship were in fact compatible, were among the major questions dividing the Ch'ing court from the Han learning movement in the seventeenth century. By the eighteenth century, of course, Han scholars had overcome their ambivalence about government projects. But echoes of the earlier view, in which the private historian's role was valued at least as much as the official compilers', survive in Shao's review of the *Ming-shih*. These echoes could not, of course, be allowed into the *Ssu-k'u ch'üan-shu*.

But perhaps the more important point was that in spite of such divergence of opinion, Shao was allowed to write for the imperial catalog and to draft reviews of texts no less important than the standard histories of the Ch'ien-lung Emperor's imperial predecessors. It was one thing for the government to wish to avoid endorsing embarassing statements about the importance of private historians, and quite another for it to have embarked on a systematic campaign to distort the significance of one of the major intellectual movements of its day.

Many have discerned a pernicious intent in the Ch'ien-lung court's treatment of the Han learning movement; an intent, that is, to shift the focus of the movement from content to methodology, rob the Han scholars' pronouncements of any philosophical or political thrust, and convert them into antiquarian discussions with no relevance to eighteenth-century concerns. The editors' motives must be carefully assessed, particularly in light of the fact that they adopted so much of what Shao wrote. As recent critiques of modern linguistic philosophy have shown, method and content cannot be so easily separated. Embodied in any intellectual methodology, particularly one so sophisticated as textual investigation, are presumptions about the nature of the problems to be solved. The notion that one could, by examining the authenticity and integrity of ancient texts, reestablish the wisdom of the sages and see more clearly present errors, was just such a presumption, and it unquestionably lay behind much of the reasoning of the *Annotated Catalog*. Although it did not reprint the most provocative conclusions of Han learning scholars like Shao, the

Annotated Catalog reflected many of the basic assumptions of Han learning. In fact, the close identification between the Han learning movement and the *Ssu-k'u ch'üan-shu* produced a major reaction among thinkers opposed to Han learning, as the following section will show.

The Ssu-k'u ch'üan-shu *and Sung Learning: The Draft Reviews of Yao Nai*

The terms "Han learning" and "Sung learning," commonly used by historians to describe the two schools of thought whose conflict dominated the intellectual life of late eighteenth and early nineteenth-century China, are in a sense deceptive. The Han learning scholars were by no means engaged in the same sort of philosophical endeavor as their first-century namesakes; nor were the Sung scholars animated by the same metaphysical concerns or anti-Buddhist sentiments as their predecessors. The issue in the eighteenth century was how to interpret China's intellectual heritage. The advocates of Han learning felt this could best be done through the interpretation of texts; the Sung learning was in part a reaction to the excesses of the Han learning, a reaction which lamented the overspecialization of textual studies and the resultant fragmentation of learning. To refer to these two intellectual orientations as schools may be to exaggerate their differences, for they were never really institutionalized, and shared many common assumptions. It may be more accurate to regard the Sung learning as a counteractive tendency which arose from the midst of the Han learning. A pivotal figure in this process was Yao Nai, a Ssu-k'u reviser, classicist, and litterateur from T'ung-ch'eng district of Anhwei province.

Yao's experiences at the Ssu-k'u Commission may well have been a precipitating factor in the formation of Sung learning, for it was only after his resignation from government in protest against the editorial policies of the Commission that Yao began to express his dissatisfaction with Han learning. In subsequent letters and prefaces, he laid the blame for the excesses of Han learning specifically on the Ssu-k'u editors and the small group of Peking literati who surrounded them and dominated intellectual life at the capital. Yao himself died in 1815, but the banner of Sung learning was enthusiastically taken up by his students. In 1825–26 one of Yao's most famous students, Fang Tung-shu, wrote a broadside against the Han

learning movement entitled *Han-hsueh shang-tui* (A polemic against
Han learning) which was published in 1831.[83] A decade later, a volume of Yao Nai's draft reviews for the *Annotated Catalog* were prepared
for publication by another student, Mao Yü-sheng.[84] Taken together,
these two documents sounded most of the themes of the Sung learning movement.

The Japanese intellectual historian Hamaguchi Fujio has recently
argued that Fang Tung-shu's criticism of Han learning in *Han-hsueh
shang-tui* was primarily "directed against the anti-social character of
the movement," and represented "a reawakening of the spirit of
applied statesmanship in Ch'ing thought." This reawakening,
Hamaguchi further contends, produced some remarkably far-
sighted suggestions of policy toward the opium trade at Canton in
the 1830s.[85] There was no indication of practical policy suggestions
in Yao Nai's draft reviews for the *Annotated Catalog*. The reviews did
demonstrate, however, that the followers of Sung learning drew different lessons from the classical canon about the nature of truth and
the role of the intellectual in verifying it than did the followers of
Han learning.[86] The nature of these different lessons, and their relationship with the Ssu-k'u project, are the subjects of the present
chapter.

Although Yao Nai's native T'ung-ch'eng district was a scant 225
miles away from the eastern Chekiang region that produced Shao
Chin-han, the two areas were worlds apart in intellectual traditions.
They had had very different experiences at the time of the Ming-
Ch'ing transition. The early years of Ch'ing rule were punctuated
with outbursts of Ming loyalism in eastern Chekiang, but a calm
acquiescence, in which the new rulers and old elites made common
cause in maintaining the status quo, prevailed in T'ung-ch'eng.[87]
Thus, at the very time when Shao Chin-han's grandfather was
learning the tales of heroic resistance from Shao T'ing-ts'ai, Yao Nai's
great-great grandfather was serving as president of the Board of
Punishments, charged with the task of passing sentence on captured
Ming loyalists.[88] The tradition of service at the Ch'ing court was
strong in T'ung-ch'eng, and one which Yao Nai was to abandon only
with the greatest reluctance.

Eastern Chekiang and T'ung-ch'eng also had different philosophical traditions. While the dynamic "unity of knowledge and action"
held sway in Chekiang, the stern moralism of Ch'eng-Chu Neo-
Confucianism dominated T'ung-ch'eng. Yao learned this tradition

within his family, one of the two most prominent lineages in the district, and from Fang Pao (1668–1749), scion of the other important lineage.[89] Despite his implication and imprisonment in the literary suppression case of Tai Ming-shih, Fang held his court office until the end of his life, and was influential in the Ch'ien-lung Emperor's project to print the Thirteen Classics. He was also a firm believer in the philosophy of Chu Hsi, and was said to have remarked in a letter to a friend that the death of the friend's eldest son was a punishment for the latter's attacks on Chu Hsi.[90] Perhaps the most famous characteristic of T'ung-ch'eng's intellectual traditions was an interest in writing, particularly of the *ku-wen* (ancient prose) style.[91] Yao Nai learned this technique both from Fang Pao and from other local masters and was the undisputed master of it in his day. Chou Yung-nien once proclaimed that "all the great writers [of ancient prose] were from T'ung-ch'eng."[92]

One rather curious incident from Yao's early life suggested that, despite his strong Sung Neo-Confucian background, Yao may not always have been as firmly opposed to Han learning as he would later appear. Sometime in the early 1750s Yao evidently wrote to Tai Chen asking if he could be Tai's student. Tai responded politely, thanking Yao for his praise, but noting that between two friends there could be no question of discipleship.[93] Chang Ping-lin has claimed that Yao's disappointment over the incident was the origin of his animus toward Han learning.[94] Yao, however, never mentioned it in his own writings nor do any of his condemnations of Han learning seem even remotely reminiscent of it. While factionalism and personal animosity did undoubtedly play important roles in the intellectual disputes of the era, it seems unlikely that such a major dispute would begin over such a small incident. Far more likely, a young and impressionable Yao was struck, as were so many of his contemporaries, by the force of Tai's work and declared an intellectual allegiance which he would later repudiate.

Yao's more immediate concern was his examination and official career. Following the best traditions of his district, he passed in the top third of his *chin-shih* class in the examinations of 1763, and was appointed to the Hanlin Academy, where he served from 1763 to 1766. Between 1766 and 1771, he was secretary in the Boards of War and Rites, and a department director in the Board of Punishments.[95] He evidently attracted Yü Min-chung's attention, for Yü asked that Yao become his protégé (*ch'u ch'i men*) and slated him for office as a censor.[96] In 1771, on the recommendation of Liu Tung-hsun and Chu Yun, Yao was appointed reviser for books submitted from the provinces at the Ssu-k'u Commission.[97]

After several years at the Commission, however, Yao found himself chafing at the intellectual biases of his colleagues and superiors. His son described the situation as follows:

> The compilers all competed with each other to make new finds, and despised the Confucian scholars of the Sung and Yuan. Considering their works empty commentaries and attacks, the compilers ridiculed and laughed at [Sung scholars] and exerted no effort to [compile or comment on them]. My father repeatedly debated with the various editors but, although it would not have been difficult, none made any effort to help him. When he was on the point of leaving, Weng Fang-kang wrote a note . . . [lamenting] that "all the editors want to read books that they have never seen before, no one wants to look at common books." At this time, Liang Shang-kuo, one of [my father's] friends, told him, "If you resign, I will be able to obtain an appointment." Yao thereupon resigned in Liang's favor.[98]

Such a course was not easy for Yao, as he confided to a certain Mr. Chang shortly after his resignation. The excitement of the Ssu-k'u project was so enormous that no one could ignore it, "even if one were deaf and blind, it would have reached the eyes and ears; even if one were halt and lame, it would have stirred the feet." Furthermore, Yao had his ancestors to think of: "My forebears all passed their [official] robes to one another, and walked in each others' footsteps serving at court, but now there are none from our family serving." Painful as the choice was, it was one Yao felt he had to make. There were some in the world, Yao observed, who were able to drink one hundred cups of wine and some who had no tolerance for wine at all. Why should it be different in the matter of capacities for state service, he asked.[99]

After his resignation Yao had a long teaching career at a number of important academies, including the Chung-shan shu-yuan in Chiang-ning, where he acquired a number of loyal students and articulate defenders. Among these were Ch'en Yung-kuang (*chin-shih* of 1801), Mao Yü-sheng, and Fang Tung-shu (1772–1851).[100] Yao wrote and dedicated to these and other friends an extraordinary number of prefaces and letters in the later years of his life, in which he articulated his views on scholarship and writing. These writings are the major source on Yao's thought today.[101]

Despite the vehemence of some of their later condemnations, Yao and his students did not object to all of Han learning. Yao praised, for instance, Hsieh Ch'i-k'un's *Hsiao-hsueh-k'ao*,[102] and a work on the *Shang-shu* entitled *Shang-shu pien-wei* (A discussion of forgeries in the *Book of Documents*).[103] Evidential research and study of Han dynasty texts had their place. All scholarship, Yao asserted, was composed of three elements, research and evaluation, principle, and literary practice. The best writing kept these three elements in balance. If principle were over-emphasized, writing became confused and disorderly; if research were over-emphasized, it became meaningless and trivial.[104] This idea was not new with Yao—it had been expressed previously by Tai Chen and Chang Hsueh-ch'eng among others[105]—but it was expressed so frequently by Yao and his followers that it came to serve as a slogan for their movement.

What Yao objected to were the excesses of the Han learning movement which, he felt, could lead in two harmful directions. First, Han learning had a tendency to fragment scholarship, focussing the scholar's attention on the trivial and arcane, rather than on the larger moral and political questions which had inspired China's early philosophers. Perhaps more important, Han learning had a tendency to sneer at the very people who seemed to Yao to come closest to embodying in their lives and writings the wisdom of the sages, men like the Sung philosophers Chu Hsi and the Ch'eng brothers.

Yao expressed his first objection most forcefully in a preface he wrote for a work by Ch'ien Tien (1744–1806).[106] The faults of Han learning, Yao argued, could be traced to the Han dynasty:

> When Confucius died, the great *tao* was diminished. When Han Confucians continued [the Master's work] after the Ch'in's burning of the books, they began by establishing specialties. Each scholar concentrated on one classic, and the techniques were passed on from teacher to disciple. Thereupon, resentments and jealousies arose, and [the scholars] could not fully understand each other. It was as if they had built walls and blocked up their doors and alleyways to the teachings of the Sages. Comprehensive knowledge was gradually lost. To penetrate the classics, to support them with proofs and clarifications, and to choose the most worthy theories, of these tasks they were unworthy. [They were only able to] mix up the situation with calculations, and confuse with strange, prejudiced, petty and trifling examples.[107]

Yao's view of the formation of separate traditions of learning during the Han contrasted sharply with those of Chi Hsiao-lan, Sun

Hsing-yen, and Chang Hsueh-ch'eng,[108] who saw this process as contributing to the sharpening of intellectual methodologies, and the more accurate transmission of sagely teachings. Yao saw the formation of schools as merely confusing the issue, and accused eighteenth-century scholars of the same sin:

> They concentrate on determining ancient names, systems of social and political organization, pronunciations and publications, and consider breadth to be the goal. They particularly value observing minutiae and accomplishing difficult [textual recoveries]. The worst of them even want to abandon Ch'eng and Chu, and respect only Han thinkers. This is like planting a branch and neglecting the root.[109]

Those who went so far as to abandon the teachings of the Sung masters earned Yao's special scorn, for they willingly cast aside the one group who seemed to have obtained some insight into human nature and morality:

> Since Han times, there have been many who have written about the classics. On some questions they agree, on other questions they disagree; there is no one road to be followed. But Ch'eng and Chu of the Sung really had many profound insights into the teachings of the ancients. Their expression of ordinary sentiments in language was indeed appropriate, unlike the Han scholars' clumsy words which did not accord with feelings. Furthermore, [the Sung scholars] were able to establish and cultivate virtue in their own lives, so that they really put into practice what they preached. For these reasons, they were admired by their successors.[110]

Although native place associations and personal friendships were obviously important in eighteenth-century scholarly disputes, there were also genuine philosophical differences between the Han learning and the Sung learning. As has been argued above, the Han learning represented a departure in Chinese intellectual history. To such a departure, there was inevitably a response. Yao Nai's letters and essays suggested the basic issues at stake between the two schools; his Ssu-k'u draft reviews clarified what these differences meant in practice.

Historians occasionally pose counter-factual questions to set broad patterns of events in relief, but seldom can such questions be answered. There is evidence, however, on the question of what the *Annotated Catalog* would have been like if it had been dominated by the Sung learning rather than the Han learning, for the *Hsi-pao-hsuan shu-lu* (Book notes by Yao Hsi-pao [Nai]) included only reviews which had been rejected by the Ssu-k'u editors. Prefaces made it clear that the work had been published in order to show Han learning scholars the errors of their ways. In this respect it differed from the collection of Shao Chin-han's reviews which was published by a member of the eastern Chekiang gentry who sought to celebrate the achievements of the district's native sons, and contained only drafts which had formed the basis for Ssu-k'u reviews. Given the nature of the collection of drafts by Yao Nai, one must assume that each of the reviews in it was, therefore, in some way unacceptable to the editors.[111] In the first two reviews discussed below, the conclusions of Yao and the Ssu-k'u editors were broadly similar but the methodology and style of argumentation were quite different. In the third and fourth reviews discussed, the conclusions of Yao and the Ssu-k'u editors differed substantially.

Ku-shih. Yao Nai's review of the *Ku-shih* in many respects parallelled Shao Chin-han's review of the *Shih-chi chi-chieh;* Shao's review expressed the views of Han learning on an early commentary on *Shih-chi*, while Yao's review expressed the views of Sung learning on a Sung commentary on the same text. Not only did the two reviews suggest the differences in the scholarly styles of Han learning and Sung learning, the two texts under review exemplified the intellectual tempers of the two eras which gave the Ch'ing movements their names. The *Ku-shih* (Ancient history) was a work by Su Ch'e (1039–1112), brother of the poet Su Tung-p'o, which purported to supplement and correct basic annals, biographies, and descriptions of hereditary and noble houses in Ssu-ma Ch'ien's *Shih-chi*.[112] Among the items included in the *Ku-shih* were the story of Lao-tzu's meeting with the Buddha, and an account of the Confucian disciple Tzu-ssu's teaching of Mencius. Although the language of the book was elegant, Naito Kōnan has pronounced it "practically useless" from the point of view of modern critical historical scholarship.[113]

The historical opinions in the book were, however, much admired by Neo-Confucians, particularly Chu Hsi, who wrote: "Of all recent histories, this work comes closest to expressing the truth." Chu particularly admired Su's statement that "the kings and emperors of old must have been good just as fire must be hot and water must be wet."[114] Living in an age much more sensitive to the canons of proof and evidence, neither Yao nor the Ssu-k'u editors could accept Chu Hsi's views on the *Ku-shih* at face value. Yao, however, did not address the question of the factual accuracy of the work in his review. Rather, he presented a series of brief and impressionistic characterizations of the two authors, Su Ch'e and Ssu-ma Ch'ien, which he composed into a type of critical fugue. The *Ku-shih* was written when Su Ch'e was advanced in both age and wisdom, Yao said, and the work had earned the respect of Sung scholars despite its weaknesses. On the other hand, Ssu-ma Ch'ien had done a magnificent job in compiling the *Shih-chi*, particularly in view of the dispersion of source materials which occurred with the burning of the books by Ch'in Shih-huang-ti. Although many corrections could be made in the *Shih-chi*, the work of those who corrected (lit. nurtured and decorated) earlier works could not be compared with the labors of the original authors. Still, Ssu-ma Ch'ien did have his faults, particularly his excessive liking for the novel and strange, which Su Ch'e had corrected according to Confucian principles. All in all, Su Ch'e ought to be viewed as Ssu-ma Ch'ien's "good friend across a thousand years."[115]

Instead of relying on brief biographical statements to prove their points, the Ssu-k'u editors used examples drawn from the *Ku-shih* and other texts on Han history. They began by citing a number of the more patently unbelievable stories in the book, such as that of Lao-tzu's meeting with the Buddha, and noting their source. Then, the editors quoted Chu Hsi's praise of the book, but also noted that Chu had criticized some of Su's assertions in his *Tsa-hsueh-pien* (Distinguishing true and false in miscellaneous teachings).[116] Like Yao, the Ssu-k'u editors commented that Su Ch'e's work presented no challenge to Ssu-ma Ch'ien, although they did so in slightly stronger terms than Yao:

> Ssu-ma Ch'ien is to history as Li Po and Tu Fu are to poetry, Wang Hsi-chih is to calligraphy, and Ku K'ai-chih is to painting. . . . Su Ch'e's effort to correct and annotate the work unavoidably involves some nonsense.

Nonetheless, the editors conceded that when Su drew upon reliable sources, his emendations to *Shih-chi* were sometimes worth consulting.[117]

One obvious difference between the two versions of the *Ku-shih* review was length; the final version was almost three and a half times as long as Yao's draft. But the difference in length reflected a more profound difference in styles of argumentation: where Yao dealt in images and impressions that could be captured in a metaphor or other rhetorical figure, the Ssu-k'u editors carefully noted where Su Ch'e had supplemented Ssu-ma Ch'ien, and where Su had taken issue with him. One can well see how Yao might have argued that the editors' review .diverted the reader's attention from the central issue, the worth of a Sung history text, with details of Han history, while the editors might have argued that Yao's draft was insubstantial and empty. Behind this methodological difference was a fundamental disagreement over how the documents of Chinese history should be approached. This disagreement, at once methodological and substantial, was also apparent in the two versions of a review of one of the most colorful Han dynasty texts, the *Shan-hai-ching* (Classic of mountains and seas).

Shan-hai-ching. The problem in reviewing the *Shan-hai-ching* was to decide what sort of interpretive paradigm should be used to measure the work. The Chinese have, at least in historical times, regarded geographical features as being endowed with supernatural powers, and believed it possible for man to chart the resonances between the natural and the supernatural orders with various sorts of geomancy. The *Shan-hai-ching,* a text of uncertain date and authorship, was a catalog of mountains, rivers and other topographical features in China and her supposed borderlands, and the legends and powers associated with them.[118] It has been interpreted as a heterodox text on magic and astronomy by Chu Hsi,[119] as a book of folktales by Yang Shen (1488–1559),[120] and as an ancient geography book by Hui Tung and his followers in the early Ch'ing.[121] Each of these interpretations implied a different set of standards and expectations.

Yao Nai's draft review of the *Shan-hai-ching* centered around Chu Hsi's assertion in his *Ch'u-tz'u pien-cheng* (Critical comments on the *Ch'u-tz'u*) that the *Shan-hai-ching* was written after the *Ch'u-tz'u* to explain and elaborate certain astonomical points in it.[122] In defense of this argument, Yao reviewed the evidence on the authorship of *Shan-hai-ching.* Dismissing claims that it was written by the legendary figures Po-i and Yü on the grounds that many of the place names in the book dated only from the Ch'in dynasty, Yao suggested instead that the book was the work of Wei and Chin dynasty authors who drew on popular legend to annotate the ancient poetry. Yao Nai agreed with Chu Hsi that the work was probably originally accompanied by charts, although he said that these charts were probably lost by Sung times, and rejected as fake a Ming dynasty edition of

Shan-hai-ching charts. Brief statements about the provenance of the text, and about certain deletions which Ssu-k'u editors had made in the text concluded Yao's review.[123]

The *Annotated Catalog* review of *Shan-hai-ching* contained much the same information as Yao's draft, but was organized around different questions. Like Yao, the Ssu-k'u editors examined the authorship of the *Shan-hai-ching* but they did so in order to establish the character and authenticity of the text, rather than to prove or disprove Chu Hsi's theories. They noted the views of the Sung philosopher, but as only one opinion on the question, and at that a rather late and implausible one. It was their conclusion that the resemblances between the *Shan-hai-ching* and the *Ch'u-tz'u* showed only that the two texts drew on the same fund of early legends, not that they were related as text and commentary. The editors agreed with Yao, however, that if any charts had originally accompanied the text, they had now been lost.[124]

Both Yao and the Ssu-k'u editors therefore viewed the *Shan-hai-ching* as a collection of legends; they differed only in their ideas on why the legends had been collected. But this difference had important implications for the question of how the text was to be viewed. Seeing *Shan-hai-ching* as an annotation of the *Ch'u-tz'u*, Yao regarded it as a heretical text, and described its contents as "extravagant and false theories." Seeing the work in another light, the Ssu-k'u editors placed it in a section of the catalog entitled "tales," remarking:

> In the discussion of topography in the book's preface, there is much discussion of ghosts and spirits, therefore it was included in the *Tao-tsang* (Taoist tripitaka). . . . But if you examine carefully its basic intent, the work does not belong to the Huang-Lao school at all. In its discussion of mountains and rivers it draws upon various sources, whatever happened to reach the eye and ear of the author. Although several groups regard it as the first geography book, this too is inappropriate. If you look at the work carefully, then [it is apparent that] it is the oldest collection of tales.[125]

Although the factual content of the two reviews was very similar, the implications drawn and the critical categories employed were quite different. Yao was reading the book, as had Chu Hsi, as an example of early anti-Confucian thought; the Ssu-k'u editors were looking at it as an example of early tales. To put the matter in more general terms, Yao was trying to evaluate the relationship of the *Shan-hai-ching* to a core of orthodox beliefs which had been founded in ancient times, and given forceful expression by eleventh-century Neo-Confucians. Although Yao Nai was willing to think critically on some issues, he was clearly not willing to do so when such would call Chu Hsi into question. Traditional beliefs were more important to him than the textual foundations on which they rested. By contrast, Han learning scholars were asking new questions about provenance and meaning, and evolving powerful new techniques for resolving many traditionally intractable questions in Chinese intellectual history. Inevitably, these new questions undermined the old orthodoxy in some respects, and led Han learning thinkers to new conclusions, as the examples of the next section will demonstrate.

Mo-tzu. The *Mo-tzu* was a much-maligned and little-studied text in imperial China and as a result the text had become corrupt by the eighteenth century. Only sixty-three of the original seventy-one chapters of the *Mo-tzu* were extant during the Ch'ing.[126] Furthermore, as both Yao and the Ssu-k'u editors pointed out, portions of the text — most probably the ones on military strategy, and the classical canon and its exposition — were not by Mo-tzu himself, but by disciples or forgers.[127]

For Yao, however, the major faults of the *Mo-tzu* were not the corruptions in the text, but its inelegant literary style, and the philosophical heresies it espoused. Mo-tzu, Yao observed, was not like Lao-tzu, Chuang-tzu, Han Fei-tzu and Sun-tzu, whose literary abilities were sufficient to express their ideas. On the contrary, Mo-tzu seemed deliberately to avoid elegance in writing, fearing, as one of his disciples remarked to the King of Ch'u, that people would dwell on his style while ignoring his main point.[128] To Yao, this fear was misplaced:

> If the writing is insufficient, then the truth will not be clear. Therefore, it is said that when words have no polish, they will not travel far. It is clearly impossible for a [a philosopher's] disciples to admire [his] language and ignore [his] theme. Why should one worry that elegance in writing [will distract the reader]?[129]

The *Mo-tzu's* uninspired style was, Yao argued, symptomatic of its philosophical errors:

> The truth of the sages must accord with human feelings. Therefore, the sages [preached] restraint of lusts and desires but did not prohibit [such natural sentiments]. But Mo-tzu would have men end up miserable

spendthrifts. Naturally, Mo-tzu's condemnation of music, and urging of miserliness in funeral rites is poorly written! How far his ideas were from the common mind! . . . It is not even worth arguing with such nonsense![130]

Starting from the same observations about the text, the Ssu-k'u editors reached entirely different conclusions. For them, the corruptions in the *Mo-tzu* text were only a symptom of the neglect which the *Mo-tzu* and texts like it had suffered over two centuries. They attributed the neglect of the *Mo-tzu* to Mencius' condemnations of Mo Ti and Yang Chu,[131] and argued that as a result of Mencius' attacks, no major philosopher in imperial times had styled himself a Mohist. This was unfortunate, in the editors' view, since in Chou times the *Mo-tzu* had been an influential text. Moreover, certain of the *Mo-tzu*'s doctrines were especially worthy of admiration, particularly its emphasis on temperance and the timely utilization of resources. For all these reasons, the editors concluded that the *Mo-tzu* deserved to be ranked among the major early Chinese philosophical works.[132]

The Ssu-k'u editors were not alone among eighteenth-century scholars in lamenting the baneful influence of Mencius' attacks on *Mo-tzu*. In 1783 several prominent classicists, working under the patronage of Pi Yuan, brought out the first critical edition of the *Mo-tzu* to appear in over a millennium. Piecing together fragments and restoring ancient readings, these scholars performed an act of rehabilitation which, as one student of the text has observed, could probably only have taken place after the book had been favorably reviewed in the imperial catalog.[133] Wang Chung (1745–1794), a close associate of many of the imperial catalog editors, wrote two prefaces for the new edition. In the first he noted that the *Mo-tzu*'s emphasis on frugality, which entailed opposition to some of the music and ceremonies of the Confucian school, was his solution to pressing economic problems of his day. He further argued that Mencius' condemnation of Mo-tzu should not be regarded as absolute truth, but as ancient polemics, evidence of the competition of early philosophical schools for adherents and attention. In his second preface, Wang endorsed Chuang-tzu's comment that Mo-tzu shared with the sage Yü an admirable spirit of dedication and self-sacrifice.[134]

These statements drew a haughty and vindictive rebuke from Weng Fang-kang (1733–1818). Weng was a close friend of Yao Nai and like Yao, was afraid that eighteenth-century textualists went too far in their contentions and undermined ancient teachings and the philosophical foundations of Confucianism. In fact, when Yao resigned from the Ssu-k'u Commission, Weng had written him a brief note of sympathy.[135] Commenting on Wang Chung's prefaces, which were in some respects milder than the Ssu-k'u review, Weng asserted that the *Mo-tzu* had been considered benighted, at the very least, by most bibliographers and philosophers in the past. Weng further proclaimed that those who "read the *Mo-tzu* and seek to clarify its text are definitely outside the way of the sages." Weng could tolerate the publication efforts sponsored by Pi Yuan, but Wang Chung's arguments were pure sophistry. "How can he say this!" Weng blustered. In a final condescending flourish, Grand Secretary Weng noted that although in other respects Wang Chung had demonstrated talent and a scholarly temper, in his work on the *Mo-tzu* he had obviously been led astray. The vituperation in Weng's remarks may well have been a result of the fact that he had long restrained himself. If Wang's preface earned such a rebuke, one can imagine the feelings of Yao and Weng when they saw such views openly and unimpeachably proclaimed in the imperial catalog.[136]

Chung-yung chih-lueh. The Ssu-k'u editors' treatment of Sung texts was, if anything, even more frustrating for the opponents of Han learning. The elevation of the *Chung-yung* (Doctrine of the mean) to a central position as one of the Four Books of Confucian philosophy was a central and characteristic achievement of Sung Neo-Confucianism. An essay of about 3,500 characters dating probably from the Warring States period or early Han dynasty, the *Chung-yung* has been termed "the most philosophical work in the whole body of Confucian literature."[137] It articulated a concept of truth which was at the heart of Sung Neo-Confucianism, a view, as Tu Wei-ming has paraphrased it, that "heaven-endowed human nature defined what the Way is, which in turn, characterizes what teaching ought to be."[138] It was to counter precisely such a view of truth, as part of man's heavenly endowment to be preserved through ethical self-cultivation, that the Han learning scholars advocated study of texts as the best means of pursuing truth.

Credit for the elevation of the *Chung-yung* to canonical status belongs principally to Chu Hsi, but actually his view of the work reflected those of earlier Sung thinkers. Sometime in the twelfth century, Shih Tun (*chin-shih* of 1145) collected the early Sung commentaries on the *Chung-yung* in a volume entitled *Chung-yung chi-lueh*

(Collected interpretations of the *Chung-yung*). Later in the century, Chu Hsi favored Shih Tun's work with a preface and published it in two separate editions.[139] But Shih's work was overshadowed and indeed ceased to circulate independently after the publication of Chu Hsi's own works, *Chung-yung chang-chu* (A textual exegesis of the *Chung-yung*) and *Ssu-shu huo-wen* (Questions and answers about the Four Books).[140] During the late Ming, a scholar from Shih Tun's native district obtained the printing blocks for the Shih Tun commentary and reprinted it.[141]

Yao, however, did not mention this later edition, and barely mentioned Shih Tun in his draft review. He concentrated instead on the connection between the Shih Tun commentary and Chu Hsi's later work. Yao noted that Chu wrote his first preface to the *Chung-yung chi-lueh* in 1183 when he was 54 *sui*. Six years later, after Shih Tun was dead and Chu's own commentary on the *Chung-yung* was complete, Chu reedited Shih's text and reprinted it with his old preface. Chu had also written an epitaph for Shih Tun. For Yao, then, the most important fact about the Shih commentary was its influence on Chu Hsi.[142]

The *Chung-yung chih-lueh* was the first text to be reviewed in the section of the *Annotated Catalog* which dealt solely with the *Chung-yung* and as such afforded the Ssu-k'u editors an opportunity to explore the question of the provenance of the *Chung-yung*, and the disputes the work had aroused among Sung scholars. Han dynasty authors, the editors observed, had not known what to make of the *Chung-yung*. Liu Hsiang had placed it in the general essays (*t'ung-lun*) section of his catalog, while others had seen it as part of the *Li-chi* (Book of rites). It was not until the Sung Neo-Confucians had rediscovered the text that it was seen as essential for an understanding of Confucian philosophy. After that rediscovery, however, the commentaries on the text had become more numerous every day. The editors further noted that Ssu-ma Ch'ien's attribution of the *Chung-yung* to Confucius' disciple Tzu-ssu, which Chu Hsi had accepted unquestioningly, was only a suggestion, and had been rejected by several Han dynasty scholars.[143]

Furthermore, as the *Annotated Catalog* observed elsewhere, the *Chung-yung* was not an easy text to read, and even the Sung thinkers had had their disagreements about its meaning.[144] Chu Hsi had established his own interpretations in the *Chung-yung huo-wen,* and in the course of so doing had refuted many earlier Sung thinkers. The value of Shih Tun's effort was that it preserved many of these early interpretations, so that one could see how and why Chu Hsi had reached his conclusions. In short, far from emphasizing the connection between the Shih commentary and Chu Hsi's work, the Ssu-k'u editors emphasized the differences between the two commentaries, and stressed the difficulties involved in interpreting the often ambiguous early text.[145]

If the suggestion that the *Mo-tzu* had some value was heretical to the opponents of Han learning, the notion that the *Chung-yung* was not necessarily sacrosanct must have seemed even more offensive to them. The reviews of the *Mo-tzu* and *Chung-yung* suggested how the different methods of Yao Nai and the Ssu-k'u editors could lead to fundamentally different conclusions. The Ssu-k'u editors approached the *Mo-tzu* and the *Chung-yung* as documents of intellectual history with certain values and certain weaknesses, values and weaknesses which could be determined from the histories of the texts themselves. Yao dealt in absolutes: the *Mo-tzu,* despite its date, was heretical; while the *Chung-yung,* despite its somewhat fuzzy origins, was an invaluable guide to the wisdom of the sages hallowed by both historical traditions and the endorsement of Chu Hsi.

The Origins and Significance of the Sung Learning: An Hypothesis

Modern intellectual historians have not been kind to Sung learning. In his biography of Fang Pao for *Eminent Chinese of the Ch'ing Period,* Fang Chao-ying writes that "bigotry was characteristic of the T'ung-ch'eng school, which limited itself to the study of Chu Hsi's commentaries and the prose writings of a few men, branding other types of literature as harmful to the mind."[146] Liang Ch'i-ch'ao characterized Sung learning in similar terms: ". . . if we judge this school by its writings, it was imitative, overly punctilious and devoid of substance; if we judge it by its teachings, it encouraged abstractness and stifled creativity and was therefore not beneficial to society. Moreover, it never occupied an important place in Ch'ing learned circles."[147] Certainly, such descriptions capture the spirit of the polemics which the followers of Han learning aimed at Yao Nai and his disciples in the early nineteenth century. They may also reflect the influence of the earliest students of Ch'ing intellectual history—men like Chang

Ping-lin (1868–1938) and Hu Shih—who admired and furthered Han learning.

It is the contention of this chapter, however, that the differences between Yao Nai's draft reviews and the final versions that appeared in the *Annotated Catalog* illustrated the intellectual issues at stake between the Han and Sung learning more clearly than did either the polemics which the two schools exchanged or subsequent historical treatments of Sung learning. The essential difference between the two schools was over the concept of truth and how it was to be verified. For both Han and Sung learning thinkers, the ancient sages were, of course, the ultimate sources of philosophical and moral wisdom. But for Han learning, that wisdom was to be found in and verified by texts, while for Sung learning truth was a matter of judgment and understanding, more often experienced than documented. As a result of these two fundamentally different visions of truth, the two schools asked different questions; what seemed trivial to one school seemed of central import to the other. Thus, there were few differences between the two schools on point of fact, but very substantial and important differences on questions of interpretive stance, differences which seem to have been almost irreconcilable. There seems to have been, in fact, very little interchange between the two groups, except on a polemical level, because they were primarily arguing on different planes.

If, as the evidence above suggests, Sung learning was basically a response by certain elements of the scholarly community to some of the new departures of the Han learning movement, why was such a response articulated only in the 1780s and 1790s?[148] The Han learning movement had been active since the middle of the seventeenth century, and by the mid-eighteenth century had already produced many fundamental challenges to traditional beliefs. The movement does appear to have reached a new level of maturity and intensity in the last years of the eighteenth century, passing from what Yü Ying-shih has called "intellectualism" to "textualism," and this may have been responsible in part for the Sung learning reaction.[149] But the significance of the Ssu-k'u project in this regard cannot be ignored, for the project changed the relationship of Han learning both with its opponents and with the government.

Despite the Ssu-k'u editors reluctance to print the more radical conclusions of thinkers like Shao Chin-han the *Ssu-k'u ch'üan-shu* was, in concept, methodology and argumentation, a document of the Han learning movement. Might not the Sung learning reaction have been triggered by the very position of political and intellectual dominance Han learning had attained in the project? At the very least, it can be argued that Sung learning began with Yao Nai, that Yao's experiences at the Ssu-k'u Commission were a turning point in his career and intellectual life, and that the editing of his draft reviews provided him with first-hand experience of Han learning which dominated political and intellectual life at the capital. The proposition that the spread of Sung learning was precipitated in part by the Ssu-k'u project is only a hypothesis, but it is one based on strong evidence from both Yao's life and the Ssu-k'u project itself.

Whether or not the above hypothesis can be sustained, the foregoing evidence suggests that the Ch'ing government indeed had a role to play in the process by which scholarly insight became received wisdom affecting established conceptions of the moral and social order in eighteenth-century China. But in playing that role, the government was guided by the opinions and perceptions of its most articulate subjects, serving more as an arbitrator than as an ideological policeman. As chapter 6 will show, it was not that the Ch'ien-lung government had abrogated its prerogative of censorship, or that the emperor and his ministers were insensitive to the political implications of ideas. But in their view the relationship between the court and the Chinese elite, or at least its more dynamic and articulate elements, was critical to the success of Ch'ing rule in China. To preserve that relationship, it was important that the court listen to the views of the Han learning movement. As the concerns of Chinese intellectuals changed, the orthodoxy of the Ch'ing court had to change with them. While it was probably a source of regret to some that Yao Nai left state service, it was probably a source of pleasure to more that Han learning, with its promise of realizing the age-old ideals of Chinese society, had found its place in Ch'ien-lung's reign of virtue.

6

Ch'ui-mao ch'iu-tz'u:
Blowing Back the Fur and Examining the Faults

Twentieth-century critics have often seen the *Ssu-k'u ch'üan-shu* project as a cover, even a pretext, for the massive campaign of censorship and suppression which raged through China in the 1770s and 1780s. In his study of the Ssu-k'u project, Kuo Po-kung asserts that the suppression of certain forms of literature was a primary purpose of the Ssu-k'u compilation.[1] The first four of the thirteen "real motives" for the Ssu-k'u project (as opposed to the real reasons given in edicts and rescripts) which Yang Chia-lo imputes to the Ch'ien-lung Emperor in *Ssu-k'u ch'üan-shu kai-shu* all involve destruction of various forms of history and literature.[2] Even L. Carrington Goodrich, whose 1935 work *The Literary Inquisition of Ch'ien-lung* introduced the campaign to Western readers and proved, in the face of some doubt, that it had actually occurred, declares that from the opening of the Ssu-k'u project, "a systematic search [for seditious books] on a huge scale was inseparably linked with it."[3] Implicit in these characterizations are two assumptions; first, that the emperor planned the campaign from the very beginning of the Ssu-k'u project, and second, that its massive scope was in accord with his wishes.

This picture of the relation between the Ssu-k'u project and the campaign against sedition is problematic in several respects. Why would an emperor, in the midst of a reign of unparalleled prosperity and power, undertake a campaign that was bound to disrupt the intellectual community, would require immense commitments of time and energy from provincial and local officials, and would create a potential for social disruption within the realm? If, on the other hand, it was not the emperor's intention to launch such a massive undertaking, how did it begin and grow? What was the role of the intellectual community, which had otherwise been so influential in shaping the course of the Ssu-k'u project, in what has been called the worst crime against Chinese literature since the burning of the books by the First Emperor of the Ch'in? To answer such questions, central to understanding Ch'ien-lung and his era, a reconsideration of the literary inquisition is necessary.

Fortunately, two sources, available since the works quoted above were published, make possible such a reconsideration. One is the collection *Ch'ing-tai wen-tzu-yü tang* (Archives of literary cases during the Ch'ing dynasty), edited by Ch'en Yuan and published serially between 1931 and 1935, which includes edicts, memorials, and legal depositions of the major cases of the inquisition. The second is the archives of the National Palace Museum, Taipei, opened to the public in the last twenty years, which contain local officials' reports to the central government on book collection procedures and results.

These new sources, as well as previously published materials, suggest that the censorship developed in three stages. The first, lasting from February 1772 to September 1774, saw most of the book collection for the Ssu-k'u project but very little banning. Only gradually was the policy of censorship and suppression formulated; not until 10 September 1774 did the emperor order a systematic campaign against seditious books. During the second phase, from September 1774 until December 1780, the campaign grew rapidly but largely for reasons outside the emperor's control. Partly because the nature of sedition was still ill-defined, and partly because the rewards for successful prosecution were great, the effort afforded opportunities for many to advance their interests with hastily formulated indictments and overly severe enforcements. As these opportunities and the punishments for dereliction of duty became known, the campaign acquired a momentum perhaps beyond the imaginings of its initiators. The central government stepped in during the third phase of the campaign, December 1780 to the end of the reign (1796), to systematize procedures for identifying and collecting treasonous books, and to reduce the potential for social disruption. A shift of leadership at the highest levels of the Ch'ing bureaucracy accompanied this change of policy. This chapter will examine the three phases of the campaign.

The Origins of the Campaign
(February, 1772–September, 1774)

Censorship was not new to the Ch'ien-lung Emperor in 1772; he had banned several works during his reign, as had all his Ch'ing predecessors. Political interference with the written word was, in fact, as old as the Chinese imperium itself. Had the emperor desired in 1772 to suppress certain works or categories of works, he need not have created an elaborate cover for doing so. But the imperial motive in the campaign of literary suppression of the 1770s went beyond this; the court sought not merely to suppress a few books, but to expunge an idiom of protest from the Chinese political lexicon. Just as in the compilation of the *Ssu-k'u ch'üan-shu,* what the emperor did not say about book suppression in his initial edict on the subject was as important as what he did say. The startling stridency of imperial language on censorship masked complex motives.

The first mention of censorship in connection with the Ssu-k'u project occurred in an edict issued in April of 1773. The debate within the Grand Council over Chu Yun's proposals of tasks to be undertaken in conjunction with the Ssu-k'u project had just been completed and the Ssu-k'u Commission established. The April edict began with a summary of that debate and the goals of the book collection project. The Emperor then noted that rather few provincial governors or private collectors had complied with his order that they submit rare and valuable books to Peking, and speculated:

> It must be that provincial governors, knowing that the books were not all from one hand, feared that there might be some expressions of a rebellious or seditious nature in them, whereupon they themselves would be held liable. . . . In their turn, the owners of libraries, noting the apprehension of the governors, were secretive and not forthcoming. . . .

Digressing briefly, the emperor offered a relatively tolerant view of how differing opinions on philosophical or historical issues could arise.

> When men of learning write books to set forth their theories, each pours forth his own beliefs. There are sometimes contradictions, and sometimes untruths. This is unavoidable. In fact, the contradictions and untruths are usually obvious, and there is no reason why books [containing them] should not be collected and stored together with [more trustworthy accounts]. And if the wording is sometimes offensive or

treasonous, . . . this is because these are the bigoted views of former men which have no contact with the present. Why should there be so much fear?[4]

Such an edict would surely not have been issued unless there had been some evidence of fear on the part of local officials and book owners. Yet, in principle, there was nothing new here: the views that the state had a role to play in intellectual life, and that when mis-understandings arose, it was the duty of a wise and benevolent emperor to correct them, were widely held. The fear book owners felt probably derived not so much from the abstract principle the emperor was expressing, as from the specific "bigoted views of former men" which all knew to exist. As the works of Hsieh Kuo-chen in China and Lynn Ann Struve in the United States have shown, a considerable literature describing the seventeenth-century transition from the Ming to the Ch'ing dynasties in China survived into the eighteenth and nineteenth centuries.[5] Most of this literature was written from the point of view of the Chinese, and was written a few years after the events it described. It chronicled in vivid lan-guage the not-always savory events of the day, and portrayed the emotions of the Chinese who saw the native Ming dynasty fall to the Manchu house. Moreover, eighteenth-century scholars occasionally used such sources; as recently as April of 1772, Ch'a Shih-kuei, a lit-eratus who held office as a district magistrate, was punished for com-piling a treatise on the late Ming period and the Ch'ing conquest based on private histories (*yeh-shih*) unfavorable to the Manchus.[6] All of the official documents about the book suppression campaign, and most of the subsequent analyses of its aims, suggested that the pri-mary concern of the campaign was the destruction of this anti-Manchu literature.[7]

In the edict of April of 1773, the emperor probably meant to allay officials' fears by acknowledging a potential that all knew to exist. By September of 1774, however, the balance between gathering the best books, and suppressing offensive ones had shifted. Anti-Manchu lit-erature was clearly on the emperor's mind, and the problem was to establish a procedure for finding and destroying it. The emperor set forth his concerns:

Now, over ten thousand books have been submitted by the several prov-inces, but none has been singled out as offensive. How is it possible that among such a quantity of literature bequeathed by former generations, not one should contain a trace of sedition? During the period at the end of the Ming, unauthorized histories were very numerous, and in them both defamatory and eulogistic comments were expressed, according to the authors' own prejudices. It stands to reason that among the rumors and gossip, there must have been some defamatory to our dynasty.

He further ordered provincial governors to publicize his orders promptly:

> Therefore, let an order be transmitted to the governors and governors-general to delegate trustworthy men to go to those households which have already supplied them with books, and explain my edict clearly. Let them be told that if they own books which they should not possess, they must hand them over with all speed and will not be held to account.
> . . . But if after this edict there are still those who secrete mischief-working books, then it must be that these books are being intentionally withheld. If such books should be discovered, their owner's crime be beyond forgiveness.

One should not underestimate the significance of this shift in tone. In both language and substance, this edict represented a departure from previous officially stated policy. The emperor's references to sedition were far more specific and threatening in 1774 than they had been in 1773; "the bigoted views of former men which have nothing to do with the present" became "rumors and gossip . . . defamatory to our dynasty." Where the emperor sought in 1773 to allay officials' fears, he seemed to want in 1774 to arouse them. Referring by name to ten governors, he asked:

> [These] are all officials . . . whose fathers too have served our dynasty. If they see books hostile to the dynasty, be they privately preserved anecdotes or collected works of poetry or prose, there is not one of them but ought to show hatred. How can they permit these things to be hidden and circulated illicitly to mislead future generations? We cannot understand what these officials did when they encountered these books in their investigations. We command them to memorialize faithfully in this regard.[8]

Punishments were authorized for the first time for those who withheld books.

On the other hand, one should not overestimate what was being attempted. Very little, if any, of the literature seized in the book suppression campaign of the 1770s incited readers to rebellion against the government as it existed in the eighteenth century. Unlike *The Rights of Man* against which a contemporary British government would soon direct its sedition statutes, or *The Revolutionary Army* which the Ch'ing government would proscribe in the twentieth century, the literature of the eighteenth-century "inquisition" in China did not propose new conceptions of government or oppose existing models of authority. It was seditious only in the sense that it echoed historical challenges to Ch'ing authority, and questioned Manchu morality in rather abstract terms. The material which the court sequestered in the 1770s and 1780s was treasonous, in that it revealed matters the court would probably have preferred kept secret, but the secrets so exposed could only be used as weapons in an ideological combat. There were, to be sure, truly seditious and treasonous materials around in the eighteenth century; but such works—collections of official memorials which did reveal state secrets, or the manifestos of White Lotus rebels—were involved in the suppression campaign only tangentially. From the point of view of content, the Ch'ien-lung Emperor and those around him seemed like peacetime generals who were determined to fight the last war; they focused not so much on eighteenth-century challenges to Ch'ing rule, as on seventeenth-century idioms of protest. It simply would not do for a secure, legitimate, and powerful Chinese ruler to tolerate racist slurs particularly when, as the emperor had every reason to expect, such works could be fairly easily eliminated. Just as the compilation of the *Ssu-k'u ch'üan-shu* was ultimately meant to demonstrate the legitimacy, in traditional Chinese terms, of Ch'ing rule, so the campaign of book suppression was meant to show that the regime had always held a justifiable control over the Chinese state.

At least six provincial governors responded to the emperor's order to memorialize faithfully on book collection procedures. Two governors merely reported that they had not seen any seditious books in the course of their investigations, and that new procedures for locating and destroying treason had been instituted.[9] Four governors submitted more informative reports, however. Governor P'ei Tsunghsi of Anhwei wrote that he had submitted a total of 516 books on six occasions for the *Ssu-k'u ch'üan-shu*, but had not found any instances of sedition. Since other figures show that Anhwei submitted a total of 516 books for the Ssu-k'u, it is clear that the process of assembling books for the compilation was complete in that province before the search for sedition began.[10] Similarly, Governor Yü Wen-i of Fukien reported that 203 volumes, virtually the entire contribution of his province to the compilation, had been submitted but no seditious materials found.[11] In Kwangtung the collection of books for the *Ssu-k'u ch'üan-shu* was complete in 1773, a year before the order to collect seditious books was issued; and in Kiangsi, half the provincial contribution to the compilation had been dispatched before the censorship began.[12] It is, of course, possible that the governors were not reporting truthfully on the extent of sedition in their provinces; but they could not be lying about the number of books forwarded to Peking. These figures show most clearly that the collection of books for the *Ssu-k'u ch'üan-shu* and for the literary inquisition were two separate endeavors, and that the policy of collecting books for censorship was a new one in 1774.

Why the emperor undertook this policy direction in 1774 was less clear. No single event in the political or intellectual history of the 1770s can be said to have triggered the censorship campaign; nor did it develop from the suggestions of officials in Peking or in the provinces.[13] The initiative for the campaign came from the emperor himself, or from his very closest advisors. It came, in short, from deep within the inner court, which is to the modern historian one of the most impenetrable areas of the Ch'ing government. The only clue to the emperor's motives was a shift in his literary and scholarly interests which took place in the 1770s, a shift reflected in his patronage of scholarly projects in the early 1770s.[14]

During his sixty-year reign, the Ch'ien-lung Emperor commissioned over ninety scholarly works. Fifteen dealt with Manchu language, Manchu history, or the history of the last years of the Ming dynasty. Eleven of these were commissioned between 1772 and 1781, and two which had been commissioned earlier were expanded and reissued in the late 1770s. The result of all this work was a new, officially certified record of the Manchu's rise to power and a standardized system for transcribing Manchu and Mongol names into Chinese. The censorship of the late 1770s seems to have been closely related to this series of publications.

The sequence and timing of the imperial commissions for works on Manchu history and language were suggestive. The first books were concerned with Manchu and Mongol language. In 1771, a new edition of the *Ch'ing-wen chien* (Glossary of Manchu language) was commissioned.[15] Shortly thereafter work was begun on the *Liao Chin Yüan shih kuo-yü chieh* (A Chinese gloss on the histories of the Liao, Chin and Yuan dynasties).[16] In the mid-1770s the imperial attention seems to have shifted from correcting extant historical texts to writing new histories; and several works tracing Manchu genealogy and recording the victory over the Ming were ordered.[17] The new definition of sedition formulated in the edict of September 1774 perhaps reflected this shift in imperial interests: in order for the new works on Manchu history to be believable, older ones had to be destroyed or discredited. The sequence of works on Manchu history reached a culmination in 1781 when, in a single year, three new studies of Manchu customs were ordered.[18] With the exception of a reediting of the *Liao Chin Yuan shih kuo-yü chieh* ordered in 1791, no further Manchu history books were commissioned after the early 1780s. As will be argued below, a significant change of procedure and general retreat from the goals of the censorship campaign also took place in 1781.

The emperor's view of the new record of Manchu history which he was commissioning was expressed in edicts ordering the works undertaken, and prefaces he wrote personally for the books produced. One theme of these writings was that Manchu language and history, though different from Chinese, deserved as much respect and attention from scholars and bureaucrats. He worried, for instance, that because the sound *t'u* in Manchu names (a common syllable) was sometimes represented in the *Ming-shih* with the Chinese character meaning "rabbit" rather than the character meaning "map" or "portrait," later scholars would assume that the *Ming-shih* editors were ridiculing Manchu culture in the manner of Confucian "praise and blame" historians. He therefore ordered a committee to correct and standardize transcriptions of Manchu and Mongol place names in the *Ming-shih*.[19] He remarked in another edict that the Chinese habit of classifying alien peoples according to the direction from which they approached China; as northern barbarians, southern barbarians, and so forth, obscured ethnic differences among various tribal groupings, and thus did an injustice to the Manchu heritage.[20] A clear retelling of Manchu and late Ming history would, the emperor remarked in his preface to the *K'ai-kuo fang-lueh,* show Manchu rule in China to have been "more glorious than the achievements of Han and Ming, not to mention T'ang and Sung."[21] Underlying each of these projects was an effort to rectify a perceived historiographical injustice done the Manchus by earlier historians. The search for anti-dynastic books which the emperor ordered in 1774 would have been a natural outgrowth of this effort.

While no one event or concern can be shown to have been

responsible for the Ch'ien-lung Emperor's assertion of the importance of Manchu language and history in the 1770s, several possibilities may be suggested. The emperor may have been troubled by the decline of Manchu military readiness which was particularly apparent in the second campaign against the Chin-ch'uan (1770–1776) and the suppression of the Wang Lun rebellion in 1774. The former was an extremely protracted and costly affair in which, on one occasion, a Manchu army under General Wen-fu was completely annihilated.[22] A similar rout in the Wang Lun campaign was avoided only when Manchu troops fled from the scene of battle, an event which elicited an edict from the emperor in October 1774 castigating the Manchus for forgetting their military heritage and neglecting the skills of archery and horsemanship associated with it. The Wang Lun Rebellion itself, and the unnerving ease with which White Lotus-inspired rebels were able to organize and attack a strategically critical area along the Grand Canal, may also have served to remind the Ch'ing court of its own vulnerability.[23]

The emperor's interests and fears may also have been aroused by a cache of documents which was apparently rediscovered in the course of the editorial work of the early 1770s. Both the *Ch'ing-wen-chien* and *K'ai-kuo fang-lueh* were based on papers in the Manchu language which have come to be known as the "old Manchu archives" (*chiu Man-wen tang*). These documents recorded in vivid and revealing detail the history of the Manchus from 1621 until 1633 and 1635 until 1636. In a recent doctoral dissertation, Gertraude Roth Li has compared the texts of the old Manchu archives with the *K'ai-kuo fang-lueh* produced by Ch'ien-lung's editors. She finds that the editors suppressed information on the social and economic condition of the early Manchus, the opposition of Chinese living in Manchuria to Manchu rule, and the opposition among Manchu princes to the rule of Hong Taiji (1627–1644). Some of the items suppressed, such as tales of Chinese poisoning the wells in Manchuria to eliminate their powerful but uncouth overlords, had ominous implications for Sino-Manchu relations. One can well see how the discovery of such material could have triggered an imperial desire to set the record straight, and to search for other writings of the same sort.[24]

Finally, it is possible that the emperor's interests in Manchu history and literature reflected the situation of factionalism at the Ch'ien-lung court. As has been suggested in chapter 3, there appear to have been bureaucratic factions at the court throughout the reign; in the early 1770s, a faction of Chinese scholars was led by chief grand councillor Yü Min-chung, while another group was coalescing around the Manchu bodyguard Ho-shen. The emperor may have wanted or needed to balance this patronage of Chinese scholars with comparable support for a collection of works on the Manchu heritage. Certainly, the ability to attract imperial patronage was an important measure of the strength of any faction at court. It cannot have been completely coincidental that most of the Manchu compilations were edited by Ho-shen and A-kuei, or that the series of publications culminated in 1781, just at the time when Ho-shen and A-kuei became chief grand councillors.

The historical circumstances surrounding the origins of censorship have been discussed at some length, in order to suggest that the collection of books for the banning and for *Ssu-k'u ch'üan-shu* were two separate albeit parallel endeavors. The *Ssu-k'u ch'üan-shu* was meant to be a monument to the success of the dynasty and the prosperity of the reign, and a resource for scholars and rulers of ten thousand generations. The censorship was undertaken to expunge from the historical record signs of early Sino-Manchu conflict and Chinese disrespect for Manchu custom, heritage and tradition. Both projects, of course, demonstrated the Ch'ien-lung Emperor's profound belief in the importance of the written word as a source of ideological justification, and his almost obsessive concern with his place in history. But the two efforts were not related as pretext and reality; they were two distinct outgrowths of the combination of power and vulnerability, pride and sensitivity that characterized the rule of Manchus, and particularly the Ch'ien-lung Emperor, in China.

The Growth of the Literary Inquisition (1776–1782)

The search for sedition initiated in 1774 became, over the next seven years, a campaign of major proportions. Thousands of suspect volumes were sent to the capital, provincial governors and their subordinates spent an increasing proportion of their time examining seditious books and processing sedition cases, families were turned against themselves, and a major potential for social disruption was created. There was no indication that any of these consequences was intended by the court in 1774. But once initiated, the campaign acquired a malignant momentum of its own. Censorship grew in scope and ferocity as new groups with differing interests were swept into the process, and as bureaucrats and the scholarly community realized that the procedures of search and destruction could be turned to their personal ends. The evolution of inquisition procedures and the growing response to them therefore reflected as much the character and interests of the literate community as they did imperial initiative.

The censorship of the 1770s and 1780s took place in three stages. First, book collectors were informed of the types of works sought and were instructed to submit any questionable works in their holdings to provincial authorities. Then, the submissions were evaluated and a preliminary determination of the character of works was made in the provincial capital. Finally, books judged to be censorable were forwarded to Peking for final evaluation; when it was decided that a particular book was to be destroyed, governors were ordered to search for the woodblocks used to print it and ship them to the capital as well. There was no single sedition statute; rather, as provincial governors and their subordinates encountered difficulties in carrying out their orders, they evolved *ad hoc* solutions, reporting them to the court and receiving imperial sanction as necessary.[25]

These procedures paralleled those used to assemble and edit the *Ssu-k'u ch'üan-shu* but, owing to the rather different character of the tasks involved in the two efforts, different actors became dominant and different interests emerged. As in so many cases in Ch'ing history, similarity of form masked important differences of function. Although it seems an exercise in an historian's *ex post facto* moral casuistry to differentiate too sharply among those who collaborated in the two efforts, some distinctions can be made. The intellectuals who dominated the Ssu-k'u project were men of influence and position, or scholars whose intellectual achievements gave them access to the world of influence. They collaborated willingly in an enterprise which they expected to benefit them individually and collectively. The book suppression campaign became essentially a police action, albeit one carried out in the context of a society willing to sacrifice a great deal for the maintenance of orthodoxy. Almost inevitably, such an action offered different opportunities to those who participated in it than the Ssu-k'u project, opportunities based more on the punishments involved in such an effort than the benefits accruing to those involved.

The task of informing book collectors of the imperial will was entrusted, at least in Hupei, Chekiang, and Anhwei provinces and the jurisdiction of the governor-general of Liang-chiang, and probably in other jurisdictions as well, to a group of functionaries known collectively as "local educational officials." These were, of course, the same individuals who had staffed the book bureaus responsible for gathering, evaluating and copying.[26] Since they were often natives of the districts in which they served, and responsible for the intellectual development and political activities of those who took the examinations in their districts, they were the natural officials for the court to rely upon in the search for anti-Manchu books. As the gatherers of materials for the *Ssu-k'u ch'üan-shu*, they had been responsible for gathering about half the books used in the compilation (see chapter 3). They were, however, rather less successful in the initial phases of the book suppression campaign, and the reasons for this suggested something of the nature of local educational officials as a group. The image of these officials, at least in court sources, was of failures, men who had grown old trying to succeed on examinations, and who were unable to secure any other positions. Judging from the number of times the adjectives "feeble and debilitated" (*lao-shuai*) were applied to them in imperial edicts, this image was common at the highest level of Ch'ing government.[27]

There is reason, however, to suspect this image. Although seldom the subject of essays and biographies — educational officials were most likely to turn up in historical sources as the preceptor of beloved memory of one accomplished scholar, like Shao Chin-han's grandfather Shao Hsiang-jung, or the sworn enemy of another — they were not insignificant figures. As events in the book suppression campaign were to show, educational officials, or at least expectant educational officials could be quite vigorous. Moreover, local educational officials had some power in Chinese society: they had authority over the licentiates of a district and controlled the lands whose proceeds provided the stipends for those enrolled in state academies. In a sense, they were poised between the world of central government appointees like Provincial Education Commissioner Chu Yun and local students, sensitive to the wishes of both groups but the captives of neither. The court image of local educational officials as old and tired perhaps did not reflect demographic realities so much as central government concerns that local school officials could not be readily controlled, a concern which was certainly in the Yung-cheng Emperor's mind when he established the post of provincial education commissioner to oversee them. In the case of a project like the Ssu-k'u effort, local educational officials performed with moderate success. The search for anti-Manchu literature, a far more delicate task likely to involve an official in what must have been one of the most distasteful cans of historical worms in eighteenth-century China, was quite another matter, and it seems likely that many local educational officials were successful in avoiding the onerous burden placed upon them.

Several other procedures of book collection at the local level were also indicative of the problems and priorities of these early years. In only two provinces, Kwangtung and Chekiang, were mutual guarantee (*pao-chia*) units utilized to canvass the literati. This was not surprising in view of the fact that the gentry, who were most likely to be book owners, frequently evaded the mutual guarantee system. Furthermore, the tasks required in the literary inquisition were not those to which the mutual guarantee system, essentially a police and surveillance mechanism, was most suited. The emperor's orders were not directed at the population at large, but at a small segment of the elite, and required judgment and discretion to enforce. They could hardly be entrusted to the village headmen.[28] In fact, in several provinces, the governors relied not on officials but on unofficial "trustworthy and capable" literati to perform the tasks of notification and collection. In the book suppression campaign, as in the *Ssu-k'u ch'üan-shu* project, it was important to inspect bookshops as well as private collections. This was particularly true in the prosperous provinces of Kiangsu, Kiangsi and the jurisdiction of Liang-chiang, where committees were formed to search for seditious books in bookshops. In fact, it was a consignment of purchased, rather than confiscated, books which triggered one of the major procedural innovations of the inquisition.[29]

In areas where book bureaus had existed prior to 1774, these organs probably continued to serve as censorship boards. The number of such groups grew as the campaign proceeded and governors discovered that areas which had not been able to contribute much to the Ssu-k'u effort, might have more to do in the book suppression campaign. By 1778, a bureau had been established in Hunan; in 1779, one was formed in Shantung, and four were established in Szechwan. By 1781, bureaus were functioning in Shantung and Chihli.[30] In the Ssu-k'u project, book bureaus had done reasonably well at the task of separating the bibliographical wheat from the chaff and forwarding the best books, or copies of them, to Peking.

They did not do so well in the book suppression campaign, for reasons that can best be described as structural. Few standards were laid down for these groups. The members of provincial bureaus had to extrapolate a definition of anti-Manchu literature from rare books for the Ssu-k'u project from edicts condemning individual works which came from Peking. In part, the absence of a formal definition of sedition reflected the fact that origins of the literary inquisition lay in a generalized imperial concern over anti-Manchu references rather than an attack on any specific genre or idiom of protest. The emperor could not easily codify his aims, because he did not know what sort of anti-Manchu literature existed, or where it was located. The vague definition of sedition may also have reflected the character of Chinese law, which tended to be either very general or very specific in its prohibitions and prescriptions. Certainly, the functioning of book bureaus in the early years of the inquisition demonstrated how dependent the Manchu court was on Chinese literati of the realm, relying on them even to define anti-Manchu literature.

The leeway that bureaus had in determining the nature of sedition posed problems both for the people who staffed the bureaus, and for modern historians of the campaign. The fact that local gentry, in effect, censored themselves must have opened up infinite possibilities for manipulation by local magnates. Such manipulation, as well as difficulties of coordination, meant that the standards of provincial bureaus could vary widely, accounting for what many have seen as wild inconsistencies in the standards of proscription (see below).

After preliminary decisions by the local bureaus, books judged to deserve censorship were sent to Peking for final evaluations and disposition. In the capital, they were stored in the offices of the Military Archives Commission, which was subordinate to the Grand Council, and to which only employees of the council and the Commission had access. In theory, the volumes were then inspected by grand councillors for the purpose of making a recommendation to the emperor. Several sources suggest that the grand councillors delegated their authority to Ssu-k'u editors Chi Hsiao-lan and Lu Hsi-hsiung.[31]

The first burnings seem to have occurred in 1778. In June of that year, the Grand Council reported to the emperor that so many condemned books and the woodblocks used for printing them had piled up in the Military Archives Commission that the councillors were afraid some volumes would be lost or misplaced. The imperial rescript, dated 11 June, ordered the councillors to "cast the books to the flames."[32] That imperial orders were carried out is indicated by a report of 12 November 1781, that 52,480 woodblocks for printing seditious books, weighing 36,530 catties, had been broken up for firewood. "Since," the memorialist continued, "firewood costs 2 taels, 7 cash per thousand catties, 98 taels, 6 cash has been saved on palace expenses since 1774."[33] It should be noted, however, that when the premises of the Grand Council were examined in the twentieth century, copies of books banned in the inquisition, and marked for burning, were found intact.[34]

The use of expectant educational officials as inquisitors. These early procedures obviously had many loopholes. Many volumes escaped the inquisition through educational officials' negligence, inconsistencies in the standards of provincial boards, and the like. The effort to close these loopholes, however, necessarily involved a much larger segment of the population in the campaign. As more people became involved, not only were more seditious books discovered, but more private aims and interests came to fuel the search. The growth of the campaign reflected the character of the literate community as much as imperial interests. This growth began in 1776, when the duties of publicizing proscriptions and collecting banned volumes were turned over to expectant educational officials.

The reform had a complex history. In 1776, Governor Hai-ch'eng of Kiangsi hit upon the happy idea of paying book collectors and book sellers double the price of all suspect volumes, then ordering his subordinates to search through the purchases, delivering the seditious volumes to him personally. As a result, Hai-ch'eng presented over eight thousand volumes for destruction in December, 1776. The enormous consignment of books, and Hai-ch'eng's accompanying memorial, attracted the emperor's attention, suggesting the existence of a hitherto unsuspected amount of anti-Manchu material. On 21 January 1777, the emperor ordered copies of Hai-ch'eng's memorial sent to the governors of Chekiang and Kiangsu:

> What [Hai-ch'eng] has done is very good. We find that Hai-ch'eng, in this search for books of former generations, has shown the greatest zeal and thoroughness. Therefore, from first to last, the collection he has made of prohibited books that ought to be burned exceeds those of Kiangsu and Chekiang. Now, the number of their literary productions, of private libraries and of bookshops is double that of other provinces.

These two provinces ought not collect fewer books than Kiangsi. . . . But the two provinces of Kiangsu and Chekiang, since the first few deliveries, have sent up no later consignments; and further, they have not truthfully informed me how they were proceeding in the search for and purchase of works, and whether or not they required more time for the satisfactory and thorough completion of the job.

Kao-chin and San-pao have been prosecuting this business for several years. Yang Kuei has also been in office for half a year. How can they treat this matter so superficially? This edict is issued as a severe reprimand to these officials.[35]

Yang-kuei, governor of Kiangsu, and San-pao of Chekiang both responded to this edict in February. San-pao professed himself "agitated, uneasy, and ashamed" on reading the imperial edict. He continued:

In Chekiang province, the book collectors are mostly concentrated in Hang-chou, Chia-hsing, Hu-chou, Ning-po, and Shao-hsing [districts]. Among the collectors in these areas, many are degree holders, and they are unwilling to buy or store books which are of a seditious or taboo nature. Booksellers seldom sell them. Furthermore, your edict has been widely proclaimed and its teaching made clear, so that all are aware of its admonitions. None would dare withhold sedition, thereby implicating himself. However, Chin-hua, Ch'ü-chou, Yen-chou and Ch'u-chou prefectures are rustic and far from the thoroughfares. Perhaps the heterodox works of the Ming dynasty which were once possessed by the ancestors [of families living in these regions] have now been passed to sons and grandsons who cannot read or write, and cannot inspect the books. For this reason, . . . I have frequently ordered my subordinates to search thoroughly so that no holding is neglected.[36]

Governor Yang-kuei wrote in the same vein:

The people of Kiangsu are book-loving and cultured. There are many who print histories, anthologies, literary collections, and collections of personal letters. But the type of work being sought does not circulate widely, and has probably been hidden away for a long time. Men today may not even know that it exists. Although we widely proclaim your edict and collect books, it is not possible to eliminate all such books at one time. . . .[37]

Under imperial pressure to bring seditious books to light, the two governors not only spoke with greater urgency, they directed their efforts at a new stratum of book collectors. They took aim, not at the big collectors who lived in cities and traded at the major book markets, but at those who lived "in rustic areas far from the major thoroughfares" and who might not even be able to read the books they were holding. In order to reach such collectors, a more active policy had to be pursued. Governor Yang-kuei outlined the essentials of such a policy when he spoke of the need to "have the rural book collectors inform each other [of the search]" and "compete with each other to hand in" offensive volumes.

It was San-pao, however, who first proposed a specific procedure, in a memorial of April, 1777:

> Your servant has observed that there are a great number of expectant educational officials at leisure in the province. I have selected the most able of them and dispatched them to their native districts, ordering them to reinforce in person the injunctions of your former edict, to search among their own friends and relatives from whom it should be easier to collect books, and to carry money for buying books and do all they can to purchase volumes. I can then judge their ability by the number of books they have submitted. When the time comes to submit names to the Board of Personnel to fill vacancies [among the educational officials], the candidates can be ranked according to the number of books they have submitted. . . . This procedure will cause little trouble and is bound to produce more books.[38]

In institutional terms, shifting the burden of the book search from educational officials in office to those awaiting appointment was a fairly minor change, but it produced a major shift in the character and motivation of the inquisitors. By the middle of the eighteenth century, a post as a local educational official was one of a very few positions in the government hierarchy to which a man who held a *chü-jen* degree but not *chin-shih* might reasonably aspire. Vacancies in the posts seldom arose, however. Although provincial governors and educational commissioners were required by law to examine an educational official's fitness for office once every six years, this task was often performed perfunctorily, with the result that educational officials tended to remain in office until they died or retired. Governor T'ien Wen-ching reported to the Yung-cheng Emperor in 1725 that some expectant educational officials had been on the waiting list for office for over fifty years. There was every reason to expect that the expectant educational officials would take advantage of the opportunity offered them in San-pao's procedure to increase their chances for appointment.[39]

The submissions of books to the capital over the next few years revealed vividly the efficacy of the procedure. Prior to April, 1777, only 315 books were submitted to Peking from Chekiang province. Between April 1777 when San-pao proposed the procedure and July

1779, 4,811 seditious volumes were forwarded to the capital.[40] In Kiangsu where the procedure was also utilized, the results were more striking. In 1776–1777, 644 volumes were sent to Peking, but in the ten months between March and October of 1777, over 10,640 volumes were forwarded.[41] San-pao was transferred to the governorship of Hupei in early 1778, and instituted the same procedures there as he had in Chekiang, with the result that, whereas 279 volumes had been submitted to the inquisition before March of 1778; 5,713 volumes were submitted between March of 1778 and December of 1779.[42] There could be little doubt about the reasons for this growth. Both Yang-kuei and his superior the Governor-general of Liang-chiang, Kao-chin, reported in separate memorials that the use of expectant educational officials as inquisitors was responsible for the new flow of books.[43]

It is possible that the expectant educational officials canvassed the countryside and, as San-pao had predicted, found seditious works sequestered in rural hiding places. It is also possible, and perhaps somewhat more likely, that the new inquisitors searched everywhere, in cities and in the countryside, and that their success was due more to the vigor they brought to the task than to the new sources they discovered. Whatever the source of the books, the new ferocity of the campaign must certainly be attributed to the energy and ambition of expectant educational officials.

Manufacturing anti-Manchu literature: The case of Wang Hsi-hou and its aftermath. The new flow of books changed the character of the censorship, for it made the emperor not only more aware of anti-Manchu literature, but more suspicious of those provincial governors who failed to find it. The new imperial attitudes, in turn, had an impact on provincial governors, who sought to allay the emperor's fears by every means possible. Since the definition of anti-Manchu literature was never clearly articulated by the court, it was possible for a governor, consciously or unconsciously, to "pad" his submissions to Peking with books whose seditiousness was questionable at best. This phenomenon was illustrated in the aftermath of one of the most famous cases in the day, that of Wang Hsi-hou, the hapless lexicographer.

Wang earned his *chü-jen* degree in 1750, but never qualified for any higher degree. By 1777, he had written ten books, including a volume of poetry, a local gazetteer, and his dictionary, *Tzu-kuan*

(Comprehensive dictionary). This last work was the cause of the trouble, being condemned on two counts: First, in defense of his own scholarly efforts, Wang criticized in his preface the organization of a dictionary commissioned by and named after the Ch'ien-lung Emperor's grandfather, the *K'ang-hsi tzu-tien* (Dictionary of the K'ang-hsi Emperor). Second, Wang wrote in full the temple names of the K'ang-hsi, Yung-cheng and Ch'ien-lung Emperors, which was considered treason. A third charge against Wang was that in his family genealogy, he traced his ancestry back to the mythical emperor Huang-ti.[44] The case was brought to the attention of the court by one Wang Lung-nan who, it developed, had been banished from Wang Hsi-hou's native Hsin-ch'ang county some years earlier for "fomenting litigation." When Lung-nan tried to return to the country in 1777, Wang Hsi-hou and others caught him and turned him in to the district magistrate; Lung-nan, in turn, accused Wang Hsi-hou of writing anti-Manchu literature. A copy of Wang Hsi-hou's dictionary was forwarded to Kiangsi Governor Hai-ch'eng who transmitted it to his book bureau for inspection. The personnel at the book bureau read it, marked the questionable passages, and returned it to Hai-ch'eng, who forwarded it to Peking with the recommendation that Wang Hsi-hou be deprived of his *chü-jen*.[45]

Wang was certainly not a Ming loyalist. But his book called into question the scholarly achievements of the Ch'ing in a way that was particularly offensive to the Ch'ien-lung Emperor. Since the beginning of his reign the emperor had emphasized, probably for reasons as much political as personal (see chapter 2), the contribution to Ch'ing intellectual life of K'ang-hsi era patronage. The *K'ang-hsi tzu-tien* was one of the shining examples of that patronage, a work which, in the words of the senior officials who reviewed the Wang case, was meant to "serve as a model for a thousand generations of scholars." For a private scholar to compare his own work to a book which was meant to be the last word in lexicography, let alone to do so favorably, was bound to be offensive to an emperor as exquisitely sensitive to the political implications of scholarship as Ch'ien-lung. Moreover, Wang had failed to accord to Ch'ing emperors the respect traditionally due Chinese rulers. And Wang had done this in spite of the fact that he held a *chü-jen* degree. Serious though his errors may have been, Wang Hsi-hou's crime hardly seemed to merit the punishments that were meted out to him and to his family. After due deliberation by the Board of Punishments, Wang himself was sentenced to death, a sentence which was carried out on 22 December 1777, and twenty-one members of his family were made slaves.[46] More was at stake here than lexicographic hubris; most probably, the emperor was using the Wang case to make a statement to the literary community about his determination to preserve his dynasty's reputation. The singling out of one offender, repugnant though it may seem today, was not an uncommon means of communicating, in the eighteenth century, to a large and diffuse community uncertain of imperial directions. The imperial statement in the Wang case had at least the dubious merit of reaching a large audience; an empire-wide search for Wang's works was launched and over two thousand copies of his books were found, including 125 copies of his dictionary, some from provinces as far away as Yunnan.[47] The fate of Wang Hsi-hou and his relatives well illustrates the dangers of publishing in an empire where the ruler had almost unlimited power even over the world of knowledge, particularly when the ruler happened to be so insulated from the realities of life in his empire as Ch'ien-lung was.

Wang Hsi-hou's tragedy was compounded by the fact that, judging from the subsequent edicts in the case, the imperial verdict was directed not so much at the literary community as at the bureaucracy. In his pronouncements on the case, the emperor directed his ire not only at Wang but at the governor of Kiangsi, Hai-ch'eng who had, in the view of the court, conducted a lax investigation and recommended an overly lenient punishment. How could the governor have entrusted a matter as serious as the dignity of the dynasty to his subordinates at the book bureau, the emperor asked; why had he not personally reviewed the book himself? Most serious of all, the emperor discovered Hai-ch'eng's name among the list of those who had contributed funds for publication of the volume, a fact which the governor never explained. The possibility of collusion between Ch'ing officials and the authors was too much for the emperor to tolerate. In an edict of December 1777, Hai-ch'eng was described as "completely blind to heavenly [imperial] virtue, and ignorant of the Greater Duty." The emperor continued, "I cannot but express myself through strong punishment. This man has been appointed as a high provincial official; he has received our grace without any sense of conscience at all." The case against Hai-ch'eng was turned over to the Board of Personnel, which recommended that he be dismissed from office and handed over to the Board of Punishments. There is no record of the deliberations of that Board, or of Hai-ch'eng's eventual fate, but in a rescript of January 1778, the emperor decreed that Hai-ch'eng's punishment should be lightened to imprisonment awaiting execution in the fall. Two other provincial officials were dismissed from their posts.[48]

The case was in its day and probably is today the most widely known of the literary inquisition cases. Two taboo character cases reported shortly after its conclusion suggested the impact the case had on the provincial bureaucracy. In May of 1778, the governor of Shansi reported the case of Wang Erh-yang, a *chü-jen,* originally from Liao-chou, who held office as a local educational official. A *sheng-yuan* of Shansi, Li Lun-yuan, asked Wang to write an epitaph for his father. Wang composed the epitaph using the expression *huang-k'ao* to refer to Li's father. Whereas the character *huang* meant 'emperor', the compound *huang-k'ao* had long been used in Chinese as an honorific expression for one's deceased father, and was defined as such in the classic Chinese treatise on rites, *Li-chi.* The governor judged, however, that Wang had committed sedition in using the character *huang* without elevating it as was required in all reference to the emperor. An investigation of the case was begun. Li's house was searched, and the governor requested that Wang's ancestral home in Liao-chou be searched as well.[49]

When the case came to the emperor's attention, however, he ordered that investigation cease, finding that the usage in question was not seditious. The emperor pointed out that the expression had been used throughout Chinese literature, as, for example, in *Li Sao,* and in Ou-yang Hsiu's essay in honor of his father, "Lung-ch'ien kang-piao." While there were instances in the history of Chinese literature of classical expressions being changed because one character in them was taboo, these cases involved characters in the personal names of emperors, not generic terms like *huang.*[50] Not only was *huang-k'ao* a perfectly valid classical locution, but the epitaph in which it was found had nothing to do with the dynasty or its ruler. The emperor declared that such a case must be distinguished from true sedition; "In our management of affairs, we have always tried to attain fairness. The case of Wang Hsi-hou was truly seditious, and we are unwilling to relent in the slightest. This case, however, is not rebellious, and we are unwilling that there should be no distinction made [between the cases]. The former case was punished severely. This matter need not be investigated further. Let this be widely proclaimed and known!"[51] The phrase "more like Wang Erh-yang than

Wang Hsi-hou" was used again and again by the emperor to mean
that a case need not be further prosecuted.

A second case involving the misuse of taboo characters followed
shortly. On 3 July 1778, the Governor of Hunan, Li Hu, reported
the sedition of a student at the Imperial Academy, Li Ta-pen. Governor Li's attitude toward the literary inquisition was probably shaped
by the recent history of the Hunan governorship. Li Hu had held the
post for twelve days when he reported the case of Li Ta-pen. In June,
a commoner had left an essay in front of the yamen of Li Hu's predecessor, Yen Hsi-shan. On reading it, Yen concluded not only that it
was seditious but that it had been presented to a previous Hunan
governor. Unfortunately, the man Yen sent to investigate the case
was a distant relative of the former governor. This aroused the emperor's suspicions: were Yen and the new man trying to cover up for
the former governor? Yen was ordered to turn over his duties to Li
Hu and proceed to the capital for interrogation and punishment.
Governor Li Hu was, understandably, a model of caution and vigor
in his investigations.[52]

The case against Li Ta-pen involved a volume of poetry entitled
Tzu-hsiao chi (Accumulated filial piety collection), containing poems
written by various relatives in honor of Li's mother's eightieth birthday. The main charge against the volume was that in it Li's mother
was compared to a number of famous Chinese ladies, including kins-
women of the legendary emperors Huang-ti and Shen-nung. This
comparison was not without its seditious overtones. If one's mother
was like the mother of an emperor, the implications for oneself were
obvious; imperial pretensions were not welcomed in Ch'ing China.
However, the governor cited no evidence that Li's intention in the
book was other than flattery. A second charge was that in one of the
poems, a character in the K'ang-hsi Emperor's temple name was
written in full. There were strong echoes here of the Wang Hsi-hou
case; Wang was charged both with writing the names of various
Ch'ing emperors in full and with tracing his geneology back to
Huang-ti. Despite these echoes, there was no evidence of any questioning of the Manchus' right to cultural and political leadership in
the volume by Li Ta-pen. While the Emperor found it exaggerated,
he did not find it treasonous. It was "more like Wang Erh-yang than
Wang Hsi-hou."[53]

Several factors could have accounted for the governors' actions in
these two cases. Perhaps, in the absence of precedent and standard,

the governors misunderstood the basic aims of the inquisition and generally misjudged the significance of the volumes they were submitting. It was also possible that the governors, under pressure from an impatient court, became distracted from the emperor's basic purposes and reported cases of marginal significance in order to maintain an image of vigor and effectiveness. The timing of the cases, and the nature of the accusations strongly suggested the latter possibility. Both cases involved the misuse of taboo characters, and followed closely one of the most famous taboo character cases in Chinese history. Furthermore, the governor who reported the case of Li Ta-pen had special reason for showing vigor and concern in his prosecution of cases. Whatever the reason for the phenomenon, the cases of Li Ta-pen and Wang Erh-yang were not isolated instances.[54] For the remainder of the inquisition, the emperor and his officials carefully sifted through the books submitted for proscription, weeding out many that they felt did not deserve official action.

Gentry accusations. The effort to collect all the seditious books in China would have made no headway at all if it had met with major resistance, active or passive, among the literati. On the other hand, literati participation could not be welcomed incautiously, for people could hand in books for a variety of reasons. Fear, as well as respect, could motivate an individual to hand in his own books; a desire for vengeance, as well as a sense of duty could lead one person to accuse another of possessing seditious books. As the number of proscriptions grew, and the penalties for holding seditious books became widely known, the number of gentry accusations grew as well. Separating true accusations from false, and useful information from that which merely involved the court in innumerable petty disputes unrelated to the inquisition was a task which came to preoccupy the emperor and bureaucrats alike in the later years of the campaign. When accusers had motives unrelated to the basic purpose of the inquisition, the fact was carefully recorded in memorials and edicts. A fairly clear picture of the significance and causes of the gentry accusations can, therefore, be reconstructed.

Approximately one-third of the cases documented in *Ch'ing-tai wen-tzu-yü tang* involved one member of the gentry accusing another. About sixty percent of these accusations led to convictions; while about forty percent were shown to be trivial or false. The accusations were a mirror of the tensions in eighteenth-century society. As land was a major form of wealth, it was not surprising that disputes over land ownership and use were commonly found to lie behind accusations. The majority of these were disputes among relatives, and many centered around commonly-held, or corporate land. As might be expected, all the cases of this form came from the southeastern provinces of Fukien and Kiangsi, where such landholding was common. As one of these cases will be discussed as a case study below, no further examples will be considered here.[55] Land ownership was not, however, the only cause of accusations, nor were all gentry lineages the highly articulated communities focussed exclusively on landholding described in recent scholarship on China.[56] One need only to examine the biographies of Chang Hsueh-ch'eng or Yuan Mei to realize how important unofficial posts were in the lives of eighteenth-century literati, and to understand how critical a network of personal contacts was in obtaining these posts. Predictably, lost employment and scholarly jealousies were also frequent causes of accusations, as was the bitterness of arranged marriages gone awry.[57] Gentry conflicts in the eighteenth century took place, of course, against the background of a very significant population change. From a level of about 150 million at the turn of the eighteenth century, the Chinese population increased to about 275 million by the time of the censorship campaign, and about 313 million by the end of the century.[58] Undoubtedly, the silent pressure of population growth complicated all disputes in the century. But while the new competitive atmosphere probably exacerbated the conflicts behind accusations, the sources of conflict were common ones in late imperial Chinese history. Ultimately, book owners' actions in the censorship campaign were more the manifestation of long-standing tensions in China than the creation of new ones.

An interesting case involving several sources of tension was reported in 1778. Implicated in it were *sheng-yuan* Ch'en Hsi-sheng and a commoner, Teng Hui, both of whom made their living by teaching and fortune telling. Before 1778 they seem to have been good friends — their houses were quite close and they worked and read the *I ching* together. Ch'en had promised his daughter, when yet a minor, in marriage to Teng's son; consequently, the girl was raised in Teng's household. But in 1778, a falling out occurred. It seemed that Ch'en lured some of Teng's archery students away and proceeded to teach them himself. Teng was angry and the two men quarrelled. Ch'en wrote a document condemning Teng and wanted to break off the marriage agreement between the two families. But a marriage promise, even one involving minors, was not to be broken off that easily. The groom's family claimed that they had already invested significant financial and emotional resources in raising the potential bride.[59] Teng wanted to accuse Ch'en at the district magistrate's office, but was restrained from doing so by his friends and relatives. Relations between the two families appear to have become so complex, and their quarrel so involved that the standard mechanisms of dispute resolution in Chinese society were inadequate to reconcile them. At this point, Ch'en realized that he had a copy of one of the books banned by the court and hit upon the idea of claiming, not only that Teng had sold it to him, but that Teng was in the business of selling proscribed books. As it happened, Teng did have in his possession some books that had been banned. As the magistrate investigated the case, the story narrated above emerged from villagers familiar with the situation and finally from the principals themselves. The magistrate decided to accept Teng's statement that he had not known the books were banned. Ch'en, on the other hand, was found guilty of making a false accusation, and it was recommended that he be beaten one hundred strokes and exiled to a distance of three thousand *li*.[60]

Inevitably, the court's attitude toward accusations was ambivalent. Accusers' information could be valuable but their motives were always questionable. Some of the largest cases of the inquisition, including those of Wang Hsi-hou and the Cho, Tai and Chu families, came to light through accusations. On the other hand, about forty percent of the accusations lodged proved false or trivial. In short, people should not be encouraged to make accusations but every accusation had to be carefully investigated. Such was the thrust of an important imperial statement on the problem of accusations in 1778:

> If, in the future, there are men who accuse others of treason, the governor involved should make a careful investigation of the truth or falsehood of the charge. If false, then the accuser should be dealt with as if he had committed the crime he accused [others of committing.] True accusations should be completely reported by memorial.[61]

Regardless of court policy, accusations were probably unavoidable. The emperor went to the heart of the matter when he observed that: "those who murmur against a family and plot to harm it need only submit one paper, and the accused family will be involved in a legal case, even though they are without guilt." The position of a member of the eighteenth-century Chinese gentry was determined by a delicate balance of a number of factors including his wealth, his position within a lineage, his circle of friends, his literary output, and his rank in the official hierarchy. With the inquisition, the court had, probably inadvertently, created a new way in which the social status of an individual, group, lineage or lineage branch could be temporarily or even permanently altered. In short, a new variable had been created in the network of interpersonal relations that was eighteenth-century Chinese gentry society.

Scholars have long been puzzled by the apparent quiescence of the intellectuals in the "literary inquisition." Actually, as the phenomenon of gentry accusations, the actions of provincial governors, and the role of expectant officials in the inquisition indicated, the intellectuals played an active role in the campaign. Literati in and out of office located, condemned, investigated and evaluated most of the books that were sent to Peking for proscription. In doing so, however, literati were not subservient to the emperor's will, nor did they resist it; they were simply pursuing their own interests. In the complex, multi-centered world of eighteenth-century China, interests could overlap without necessarily being identical or even congruent with each other. The campaign grew through the interaction of gentry, bureaucratic and imperial interests, shaped by all but dominated by none. In the following section, this interaction of interests will be examined in an individual case.

A Case Study: The Proceedings Against the Chu Family of Kiangsi

The proceedings against the Chu family of Te-hsing, Kiangsi have been selected for examination here not because they changed the character of the censorship campaign, but because they were typical of the investigations and litigations which took place all over China in the late 1770s. The Chu case took place in the fall of 1779, during the height of the censorship in Kiangsi, a province which had sent a disproportionate number of its books to the capital for banning. The Chus were a large, landholding lineage with a record of examination success stretching back to the seventeenth century. The Kiangsi bureaucracy, which had recently been shaken by the dismissals of Hai-ch'eng and several subordinates in the Wang Hsi-hou case, was certainly well aware of the dangers and importance of literary litigations.

A gazetteer from the Chus' native county published in 1872 describes the area, and perhaps also the family. Crossed by shallow streams and broken by gently rolling hills, Te-hsing was one of seven counties which comprised Jao-chou prefecture in the red earth district of Kiangsi; the northeastern border of the district formed the boundary between Kiangsi and Chekiang. The names of the offenders did not appear in the nineteenth-century gazetteer, but enough evidence remained to suggest that the Chu family lived in the southeastern section of the district, near modern Nuan-shui. There were two *chin-shih,* one *chü-jen* and six *sheng-yuan* surnamed Chu from this area among the successful candidates on the examinations during the Ch'ing. One of the *chin-shih* and one of the *sheng-yuan* were identified in censorship documents as being members of the offending Chu family. According to the gazetteer there was a hall, the *Pi-chien tz'u,* in the area of Nuan-shui for persons surnamed Chu, and censorship documents recorded that the Chu family lived clustered around their ancestral hall. It cannot be conclusively proven that the Chus of Nuan-shui were the Chus of the case at hand, but both the government documents and the local history suggested that the Chus were a large, well established, land-holding and segmented gentry lineage.[62]

Little is known of Chu T'ing-cheng, a *sheng-yuan* who died in 1750, except that at some point he wrote a volume entitled *Hsu san-tzu-ching* (Supplement to the *Three Character Classic*) which he used to teach his grandson Chu Chieh how to read.[63] It was this writing which eventually brought the family to grief. The *San-tzu-ching,* originally written by Wang Po-hou of the Sung Dynasty, has been used as an elementary reading primer in China for over a thousand years. Composed in short, easily memorized, three character sentences, it was divided into sections entitled "Man's Original Nature," "Duty to Parents," "General Knowledge," and "Advice as to Learning and Diligence."[64] The section on general knowledge contained a brief sketch of Chinese history from legendary times to the founding of the Sung dynasty, and it was this sketch which Chu T'ing-cheng evidently supplemented by extending it to the middle of the Ch'ing. Subsequent criticism alleged that Chu not only listed the dynasties and emperors since Sung, but made moralistic comments on them in the tradition of Chinese "praise and blame" historiography. It is not hard to guess who was being blamed. One line from the book recorded in the archives of the case was: "With hair loose and clothes folding to the left, even harder to bear is the fact that China is filled with monks." The first part of this line was an allusion to the *Analects* where Confucius remarked on how fortunate China's escape from barbarian control had been. The second half of the line referred to the fact that the Manchus shaved the front part of their heads, which made them look like monks. This practice, along with the custom of braiding the back part of the hair into a queue, was forced on the Chinese when the Manchus took power and had met with considerable resistance. Thus one theme of Chu T'ing-cheng's comments on Chinese history was very probably opposition to Manchu rule.[65]

The original copy of Chu T'ing-cheng's work was lost. Many years after Chu T'ing-cheng's death, when his grandson Chu Chieh, a *sheng-yuan* himself, came to teach his own son how to read, he had to write out a copy of the text from memory. In this copy, seized by local officials, the taboos on Ch'ing emperors' names were not properly observed. In one particularly blatant instance, after a mention of the famous early Ch'ing rebel Wu San-kuei (1612–1678), a reference to the Ch'ing government was not elevated the requisite number of spaces. In a subsequent interrogation, Chu Chieh claimed that in writing the text he had followed the model of the original *San-tzu-ching*, arranging the sentences parallel to each other for ease of reading and memorization and for this reason he had not observed the Ch'ing taboos properly. Another fault of Chu Chieh's text was that it confused the reign names of the rulers of the Yuan dynasty. Since the rulers of the Yuan were Mongols, and were the most recent non-Han rulers of China before the Ch'ing, the court was very sensitive to any infringement of their historical rights. It is difficult to tell whether these were inadvertent errors on the part of Chu Chieh or intentional acts of sedition; at the very least it was certain that the manuscript was not intended for publication.

Sometime in the mid-1770s, there seems to have been a dispute within the Chu family between Chu Chieh's brother, Chu Hui, and a distant relative of both, Chu P'ing-chang. In this squabble, Chu Hui accused Chu P'ing-chang of selling corporate land. The exact nature of the dispute was not specified, but Chu P'ing-chang's behavior was later officially characterized as "destroying the ancestral hall." Lineages in south China often used the proceeds of corporate lands to suport their ancestral hall, or pay educational expenses; but conflicts over the resources from such lands could easily arise.[66] The two litigants were at first reluctant to take the case to court. They asked another member of the Chu family who had just earned his *chin-shih* degree, Chu Huang-fan, to arbitrate. Huang-fan evidently refused. The case went to court and Chu P'ing-chang lost, with the result that he was stripped of his *sheng-yuan* degree.

While this litigation was going on, Chu P'ing-chang had occasion to visit his kinsman Chu Chieh. A copy of the *Hsu-san-tzu-ching* was lying open on Chu Chieh's desk which P'ing-chang picked up and took home with him. In October, 1779 after Chu P'ing-chang had lost his degree, he sought a means of taking revenge on Chu Hsi and his brother. He decided to hand in the volume he had taken from Chu Chieh's house, and accuse the latter of sedition. P'ing-chang also wanted revenge against Chu Huang-fan. As it happened, Chu Huang-fan had written a volume of poetry and dedicated it to Chu Hui. P'ing-chang also obtained a copy of this volume and submitted it to the magistrate.

Thus on 13 November 1779, the magistrate of Te-hsing county received copies of the *Hsu-san-tzu-ching* and an untitled volume of poetry by Chu Huang-fan. After examining them, he forwarded them to the governor of the province with a report on how and why he had received them.

The governor who received the volumes was Hao-shuo, whose appointment in December 1777 followed the disgrace and execution of Hai-ch'eng for his role in the Wang Hsi-hou case. The Chu case was the first that Hao-shuo prosecuted as a provincial governor. On receipt of the magistrate's report, Hao-shuo ordered that the magistrate of Te-hsing and the prefect of Jao-chou to interrogate thoroughly all the principals, and search the Chu households; he then forwarded the books to Peking. Several weeks later the report of the magistrate and prefect arrived, containing the outlines of the story above. But the governor was not satisfied. He had learned through his own sources that Chu T'ing-cheng had written another book, entitled *Tz'u-t'ang pi-chi* (Desultory jottings from the ancestral hall) which was not mentioned in the report of the magistrate and prefect. The governor did not reveal the sources of his information on the second volume but, castigating the magistrate and the prefect

Figure 3. Members of the Chu Family Involved in
         Litigation Proceedings in 1779

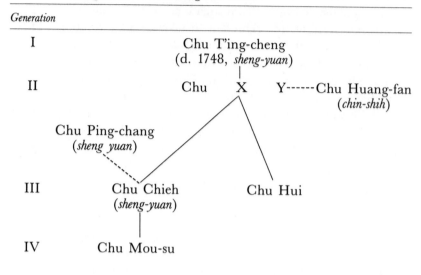

NOTE: Solid line indicates direct descendants; dotted lines indicate uncertain relationships. Nothing is known of the relationship between Chu Huang-fan and Chu
Hui, except that Huang-fan was of the same generation in the clan as Hui's
father. Chu P'ing-chang was a grade 5 mourning relative (in Maurice Fried-
man's parlance) of Chu Chieh. The implications of this latter relationship will
be discussed in the text.

for the "leaks" in their report, ordered a second investigation. This
time, the investigation was to be conducted by the censor of circuit
for northeastern Kiangsi, O-erh-teng-pu.

Two months later, O-erh-teng-pu reported back that he had found
it too late. Word had spread that the governor was looking for the
volume and there had been a scramble among the Chus to eradicate
certain names and phrases from the book. No less than seven
members of the family had made revisions. Each was interrogated
and searched. On March 19, the governor reported all the circumstances of the case to the court in a long memorial, and recommended punishments for those he found guilty.

The first criminal discussed in the memorial was Chu T'ing-cheng
himself, who had had the good fortune to die thirty years before his
case was discovered. Chu had committed two crimes—harboring
anti-Manchu sentiments and recording them in an historical text.

He was condemned as a "wild and reckless" man, "living in seclusion and deriding the dynasty." Even worse, he had received the bounty of the dynasty in the form of a *sheng-yuan* degree and still dared to speak against it. Chu was not only ungrateful for the benefits he had received but had in conduct and in writings scorned his benefactors. He was clearly a man of anti-social tendencies and immoral character. Or so he was pictured. There was no indication that his true feelings and intentions were investigated, nor can they be determined today.

Chu T'ing-cheng was also condemned for "praising and blaming former kings according to his own fancy" and "having the audacity to criticize previous emperors." The key phrase here was "praise and blame." Wm. Theodore de Bary has noted that while all men were free to study the past, "only a sage might dare actually to record moral judgments in his writing."[67] Chu T'ing-cheng was, therefore, condemned in terms thoroughly grounded in Chinese tradition, for usurping the prerogatives of a sage and expressing his opposition to the Manchu house. The governor recommended that, as punishment and as example to others, Chu's corpse be disinterred and beheaded. He also recommended that all the property Chu T'ing-cheng had bequeathed at his death be confiscated by the state, and that all of T'ing-cheng's descendants over 16 *sui* be executed.[68]

Punishments were also meted out to those who had resisted the Governor's will by hiding and altering the *Tz'u-t'ang pi-chi*. The individual who had altered the most was sentenced to a beating of one hundred strokes and exile to a distance of three thousand *li*. Another member of the family who had erased one name was deprived of his status as a licentiate. Others, who made only a few changes were sentenced to be beaten eight strokes, but were forgiven when they pled for mercy on New Year's Day 1780.

Only one member of the Chu family accused in the case escaped official condemnation. When the poetry of the *chin-shih* Chu Huang-fan was examined, no seditious wording was found. Huang-fan was not a descendant of Chu T'ing-cheng and thus bore no responsibility for the preservation of the *Hsu san-tzu-ching*. Officials accepted his word that he had never seen the volume. Probably Huang-fan's exalted rank helped to establish his credibility. In any case, neither his life nor his career was harmed; he went on to serve as magistrate in three counties before retiring to care for his aged mother and write his commentaries on the classics.[69]

The interests on the court, bureaucracy and gentry accusers were all served by the outcome of the Chu case. Although there were no imperial edicts in the case, several of the emperor's aims were obviously served. A book which, had it been published, would have been of considerable subversive potential was destroyed. The *San-tzu-ching* was more than a series of amusing jingles; it expressed the basic beliefs of Chinese high culture. Reading and memorizing it were the first steps in the socialization of the children of the Chinese elite; under no circumstances could a version of it which contained anti-Manchu language be allowed to circulate. Also, the practice of passing seditious volumes from generation to generation within a family was attacked forcefully, and it was made clear that any who engaged in the practice would lose not only their lives and writings, but also any property they might bequeath to their descendants.

The process of investigation in the Chu case illustrated the impact of the censorship on the provincial bureaucracy, and the crucial role of the bureaucracy in it. Not only did Hao-shuo send three sets of officials to investigate the Chus, he evidently sought information from private sources as well. In his final memorial, the governor recommended that the prefect and the magistrate, whose reports had not met his standards of completeness and accuracy, be dismissed from their posts. As a result of his diligence, an extra volume by Chu T'ing-cheng was discovered, several more people punished, and a minor panic created among the Chus. While the governor clearly felt a vigorous investigation of the case was necessary, it was not clear that the results of his search were worth the effort and disruption it entailed. Only one phrase, a standard classical allusion to the strenuous life of an emperor and his officials, was quoted from the volume, and it appears on none of the indices of works suppressed. It is an ironic comment on the difficulties of holding office during the Ch'ing that four years after his most thorough prosecution of the Chu case, Hao-shuo was summoned to Peking, cashiered and permitted to commit suicide for extorting money from his inferiors.[70]

The interests of the accuser in the case, Chu P'ing-chang, presented a slightly more complicated picture. Although P'ing-chang's quarrel was with Chu Hui, he accused Chu Chieh. While it was recommended that Chu Hui be executed for not restraining Chu Chieh, there were many other unfortunate consequences of the accusation. Why did P'ing-chang employ such a potentially disastrous means of revenge? He was not a sufficiently close relative of Chu

T'ing-cheng that he would have been held responsible had the book been discovered, nor did he give any moral or ideological reason for his act.

Anthropological literature on southeast China lineages provides some clue to the puzzle. In his early work on the subject, Maurice Freedman found that some segments or lineages seemed to have more power than others, and he hypothesized that the more powerful branches owned more land than the weaker ones.[71] Jack M. Potter, another student of southern China clans, tested this hypothesis and concluded that "the distribution of political power within Chinese lineages may be said to have been determined by ancestral property . . . wealth in the form of ancestral estates enabled wealthier branches to educate sons for the examinations, making it probable that wealthy branches would have more prestige and political connections."[72] There is evidence to suggest that the quarrel between Chu P'ing-chang and Chu Hui was not a quarrel between two individuals, but between two branches of the Chu lineage. The fight was over corporate land, the selling of which was regarded as "destruction of the ancestral hall." It was a fight of sufficient importance to go to the district magistrate's court. Chu P'ing-chang probably belonged to a different branch of the family than Chu Hui and Chu Chieh, for it was recorded that Chu P'ing-chang was a "coarse silk" mourning relative of Chu Chieh. This referred to an ancient Chinese system of classification of lineage members, wherein two people's relationship to one another was expressed in terms of the type of cloth they would have to wear at each other's funeral.[73] The terminology was not sufficiently precise that Chu P'ing-chang's relationship to Chu Hui and Chu Chieh can be pinpointed. However, Chu P'ing-chang could not have been a descendant of Chu T'ing-cheng, that is of the same lineage segment as Chu Hui and Chu Chieh, unless he were Chu T'ing-cheng's great-great-grandson. In view of the ages of the parties involved, this is unlikely.

If Chu P'ing-chang and the descendants of Chu T'ing-cheng were of different, feuding branches of the Chus, then the outcome of the case was clearly in P'ing-chang's interest. For, as a result of the finding of sedition, it was recommended that all of Chu T'ing-cheng's descendants over the age of 16 *sui* be executed, which would have eliminated the entire branch of the clan with which P'ing-chang was feuding. Whether or not Chu P'ing-chang anticipated this outcome when he lodged his accusation, he clearly had motives unrelated to the censorship and knew enough of government precedent and procedure to be certain that his accusation would receive a prompt and thorough investigation. Such widespread knowledge of censors' aims and procedures, and the ability of local gentry to manipulate them were clearly major factors in the growth of the censorship.

The Systematization of Censorship (1780–1782)

By 1780, the book banning had reached its zenith. All the provinces had instituted procedures for locating and collecting seditious books, and the consignments of books sent to the capital from the largest provinces averaged two thousand volumes annually, with some provinces sending as many as five thousand volumes per year. The potential for social disruption was becoming more and more evident, however. Litigations, like that of the Chu family, required more and more time to investigate and prosecute; book holders indicted each other regularly, with or without basis; and governors padded their submissions with books of marginal significance. In the face of mounting evidence of the disruptive side effects of the campaign, the Ch'ien-lung court seemed to shift its policies. For the first time, centralized censorship boards which could enforce consistent standards on various genres of literature were created; a master list of banned books and a statement of the criteria for determining sedition were promulgated at court and distributed to provincial officials; and governors were reprimanded, not for negligence, but for overzealousness in pursuing the goals of the campaign. The shift was a subtle one. The court neither retreated from its stated policy of finding and eliminating all anti-Manchu references in Chinese writings nor renounced any of the procedures of the early years of the campaign. The goals were rather to systematize and regularize the effort, and bring under central control forces previously left uncontrolled.

Contemporaneous with these developments was the rise of a new group of leaders in Peking. Many of this group, which coalesced under the leadership of Ho-shen, were ethnically Manchu or of imperial household origins. At least two of the changes in procedures during the early 1780s were directly associated with these new men: Ho-shen and others staffed the new censorship boards while Ying-lien, Ho-shen's uncle, prepared the banned book lists. The reprimands of governors were embodied in imperial edicts drafted by the Grand Council which was dominated by Ho-shen and his followers.

Ironically, it was Manchus who led the retreat from the excesses of the campaign against anti-Manchu literature. Considered more carefully, however, this development was not so paradoxical. The censorship put senior Chinese officials in a difficult position; they could not condemn its excesses without appearing to condone its objects. Only Manchus could confidently condemn the excesses of the campaign. Furthermore, in this respect as in many others, the early efforts of the new Manchu leadership appear to have constituted a genuine and needed reform of previous administrative procedures. The fact that these leaders' own activities became excessive and corrupt, and were so labelled by their successors, should not obscure the significance of their early actions.

The origin and development of censorship boards. Under the procedures of the early years, if a book contained even a few references offensive to the regime, local book collectors and provincial bureaus had no option but to submit it to Peking for destruction *in toto.* This meant that there was probably more destruction of literary texts than was necessary to achieve the court's goals and, with public knowledge of such destruction, a climate of fear and uncertainty was created which obfuscated the government's basic aims. These consequences could have been avoided if there had been a central censorship board with the power and confidence to censor selectively. Such a board was created for the censorship of drama in December 1780; shortly thereafter, similar boards were created for other genres. The central government's desire to minimize the potential for social disruption was stressed throughout the official documents creating these boards. That the first censorship board created had to do with drama may have reflected the calculation that, since the stories and language of plays were known to a larger segment of the population than other written texts, the dangers of public misunderstanding of drama censorship were greater than those of censorship of other sorts of texts.

On 6 December 1780, the emperor issued his first edict on drama:

> Formerly, we ordered the officials of each province to seek energetically writings containing seditious words and phrases and send them to the capital. The successive submissions of the various governors have been numerous indeed. It has occurred to us that playscripts cannot be without seditious passages. For instance, among those which depict events at the end of the Ming dynasty, there must be some references to the

present dynasty which ought to be investigated thoroughly. As for poetic drama which concerns the Southern Sung and the rise of the Chin, there must be many scripts in which the roles [presumably of the Jurchen rulers, who were ancestors of the Manchus] are portrayed so excessively that the truth is lost. These have been in circulation so long that the uneducated masses take them for the truth when in fact they bear no relation at all to the truth. They also must be investigated thoroughly.[74]

Instead of charging provincial officials with the task of evaluating seditious drama, the emperor entrusted it to Ch'üan-te, a Manchu of the Plain Yellow banner who had distinguished himself in the campaigns against Burma and the Chin-ch'uan rebels.[75] It was essential, the emperor admonished, that Ch'üan-te carry out his tasks without arousing any noise or commotion, or making any display of force. The emperor probably felt that the Manchu military officer could be relied on to carry out his orders quietly and effectively; but as a drama censor Ch'üan-te had one liability, namely, that he could not read Chinese very well (*pu-t'ung-hsiao Han-wen*). Therefore, I-ling-a, an imperial household bondservant then serving as salt commissioner for the Liang-huai district, was ordered to assist him. Sometime in March of 1781, I-ling-a was relieved of his duties as assistant drama censor and replaced by T'u-ming-a, a Mongol of the Plain Yellow banner.

Ch'üan-te and T'u-ming-a worked out between themselves the following *modus operandi*.[76] Drama scripts collected by provincial governors were first sent to Ch'üan-te, who sorted them into three categories: Those that merited complete destruction he sent to the capital for review; those which required only a few emendations to be acceptable to the regime he forwarded to T'u-ming-a; and those which contained no sedition he returned to their owners. On receipt of scripts requiring emendation, T'u-ming-a rewrote the offending passages, and forwarded his handiwork to Peking for review. Both Ch'üan-te and T'u-ming-a regularly reported to the emperor on what they had accomplished, and regularly received imperial rescripts on their reports urging them not to be too harsh or stir up trouble In his last extant memorial on drama dated 20 June 1781, Ch'üan-te reported that he had collected 424 scripts of which he had evaluated 273. Unfortunately he did not say how many he had censored, but he did report that of the forty-two scripts he had evaluated since he had last memorialized, he had ordered six banned entirely, eighteen partially amended, and eighteen returned unchanged to their owners. In a memorial dated 6 June 1781, T'u-ming-a reported that he had edited thirty-five scripts and had twenty more on hand for editing.[77]

Another board, of similar function, was also created in December 1780. Apparently in order to preserve texts which contained only a few anti-Manchu references, printers had begun to prepare special editions of works with the offensive passages excised. However, the blank spaces could be more suggestive than the actual excised wording. One of the volumes printed in this way, a text of *Ch'ing-hsia-chi* (Collected writings of the scholar of Ch'ing-hsia mountain), by Sung Lien (ca. 1600) somehow made its way into the *Ssu-k'u ch'üan-shu* itself. On the day of his appointment as director-general of the Ssu-k'u project, Ho-shen was given the task of investigating this text. The following day he memorialized, indicting Assistant Director-General Ts'ao Wen-chih and Chief Collator Tsang Sheng-yung for failing to report the blank spaces.[78] Five days later, an edict was issued on the subject of blank spaces:

> We presume that besides this work there are shops in the provinces which have printed other books, leaving spaces blank, as in the case of *Ch'ing-hsia-chi,* to which thought ought to be given and additions made. But this will probably be a difficult task for the governors-general and governors to perform satisfactorily and in like fashion. So we command the high officials of all the provinces to discover those volumes which need not be destroyed but have been printed with spaces left blank, to mark them, and to send them to the capital. They will then be turned over to the commissioners to be deliberated upon, filled in, then returned. In case there are woodblocks, the high provincial officials should obediently compare them with the revised works and engrave [the proper parts] afresh, so that they may be precisely the same.[79]

On 15 December, Ho-shen and A-kuei were appointed to direct the work of filling in the blanks. The creation of a centralized board to correct works which had been, in effect, selectively censored by Chinese literati and printers must have facilitated such prior censorship and thus relieved some of the pressure on local officials to find and destroy all anti-Manchu references. Unfortunately, as in so many of Ho-shen's activities, no records of the functioning of the committee for filling in the blanks survive.

The promulgation of indices and standards. A second procedural reform, the preparation and printing of master lists of banned books, took place in the spring of 1782. The initiative for this reform came from the group of Manchu officials surrounding the emperor, who had inherited in the late 1770s the task of supervising his various projects and campaigns, rather than from the emperor himself. Specifically, the reform was begun by Ying-lien, an uncle of Ho-shen's who served in various mostly administrative capacities at the Ssu-k'u Commission. In May 1780, Ying-lien was ordered to supervise and expedite the return of the books collected for the *Ssu-k'u ch'üan-shu* to their owners in the provinces. In order to carry out this task, the Manchu official found it necessary to prepare a master list of all the books that had been banned to insure that no work containing seditious language would be returned to its owner by mistake.[80] In April 1782, he reported to the emperor that his work was complete. Some months later the Grand Council memorialized, noting with approval the work of Ying-lien and his subordinates and observing that "Nine-tenths of the works now being submitted have already been investigated. . . . The number of different books to be destroyed, both *in toto* and *in parte,* is 780." The memorial concluded:

> We recommend that an *Index* be drawn up clearly stating title and author [or compiler of all seditious works], and that it be printed in book form for circulation among the provinces. Then even the ignorant villagers who have books hidden away will know what is against the law, and we who have the prohibition in hand will be better able to check volumes sent [to the capital]. So will provision for every book of past generations deserving suppression be made.

In such manner was the first and only known central index of books banned in the Ch'ien-lung years created.[81]

Apparently at the same time as they were preparing a master list of banned books, Ying-lien and his subordinates prepared a set of criteria for determining whether or not a book was to be judged seditious. This document consisted mostly of summaries of imperial edicts banning individual works; some of the provisions, however, were of a more general nature. The relatively limited definition of sedition adopted in the document and the repeated stress on the need for discretion and judgment were striking.[82]

Except for works with obviously rebellious wording, the commissioners assert in their preamble, in determining sedition "there will have to be a balance, which will need to be handled in various ways." Article one orders that books containing the names of military stations and outposts in Manchuria be forwarded to the capital, so that their terminology and transcriptions could be made consistent with the *Man-chou yuan-liu k'ao* then being prepared in Peking. Articles two, three, four, and five summarize edicts on the works of Ch'ien Ch'ien-i, Lü Liu-liang, Chin Pao, Chu Ta-chun and Wu Wei-yeh, and works which used the reign names of certain Ming pretenders. Article six cautions that in the censorship of encyclopedias and books of historical notes it is not necessary to destroy an entire volume simply because of a few anti-Manchu references, and orders censors to "check carefully to see in what division and what biography the treasonable part occurs and eliminate that." Article seven deals with collections of memorials. Noting the emperor's command that memorials relevant to present conditions, like those in the *Ching-shih wen-pien* (Collection on statecraft), should be preserved, it urges censors to be cautious: "whichever have words and phrases that are treasonable should be censored of that part; the balance after deliberation ought to be conserved." The concluding article orders censors to deliberate and change as necessary writings by Sung and Ming authors dealing with the Yuan, Chin and Liao dynasties.[83]

Reprimands of provincial governors for overzealousness. The new emphasis on discretion and systematization evident in the standards and memorials of Ying-lien was also apparent in the comments made by the emperor on works sent to him to be banned. While the emperor had on many occasions found that books submitted for prohibition were not seditious, he had never before 1782 reprimanded the governors who had submitted them. In the spring of that year, however, one case involving the misuse of taboo characters and some rather extravagant and clumsy historical allusions was brought to his attention by Governor Li Shih-chieh of Hunan. The emperor commented: "If in all cases, we must blow back the fur and examine the faults, picking and plucking at falsehoods, will this not cause the people to be at a loss as to what to do? This matter has come about because Governor Li Shih-chieh is not familiar with the principles of literature." In a second edict on the case issued two days later, the emperor further fulminated: "If in managing local affairs, all were as hasty and unconcerned as this, how could they be worthy of reappointment?"[84]

Later in the same year, officials presented the emperor with some rather bitter poems written by a literatus who had studied all his life but never obtained more than the first degree. The emperor commented: "There are many of this type, disappointed men singing in the wild grasses and marshlands. If we blow back the fur and examine the faults in order to punish them, then anyone expressing himself in poetry will be putting himself in danger, and the people will be at a loss as to what to do. . . . Let this edict be widely proclaimed and known."[85]

In statutory terms the censorship did not end in 1782. Provincial book bureaus continued to exist, submitting yearly consignments of books to Peking, until the end of the reign. There was even an imperial order in 1788 to provincial governors to "throw their whole energies into the execution of censorship that it may be cleaned up once and for all."[86] But there is an unmistakable decline after 1782 in the number of books submitted by the provinces to the court, in the number of litigations, and in the court's interest in the campaign. Probably, the systematization and reprimands of the early 1780s represented a signal to the literate community that the campaign had reached its conclusion, and that literati in the empire had little to gain from further censorship. The effort had grown, in large measure, because bureaucrats, expectant officials, and literati had been able to further their own interests through the pursuit of imperial goals; when this opportunity ceased to exist, the campaign declined as well.

The circumstances of these cases and the nature of imperial reprimands were widely known in the provincial bureaucracy. Four governors memorialized, acknowledging receipt of the edict reprimanding Li Shih-chieh. Governors Pa-san-t'ing of Kwangtung and T'an Shang-chung of Anhwei promised to avoid cases involving petty infractions like those of the tasteless Hunan poet;[87] Pa-san-t'ing remarked that he was particularly aware of the dangers of such cases since he had been responsible for the case of Wang Erh-yang. The governor of Kiangsi noted that the only circumstances in which an individual could be punished were ridiculing the dynasty and concealing seditious books; he agreed with the emperor that there was no need to bring to the attention of the court cases in which language had been misused, and further noted the importance of investigating all gentry accusations. In order to encourage his subordinates to act in accord with imperial wishes, the governor said, he widely publicized the outcome of all cases. Governor Li Shih himself submitted two memorials, one apologizing for his errors, and the other announcing that the imperial pardons had been conveyed to those condemned.[88]

The Eighteenth-Century Censorship in History and Historiography

This picture of the growth and decline of censorship in the eighteenth century accords with what is known of the late Ch'ien-lung era. The last quarter of the eighteenth century in China was a time when, beneath a façade of harmony, peace and prosperity, an endless array of particular interests were in competition with each other, a competition perhaps exacerbated by the silent pressure of population growth. The phenomena observed in the censorship campaign — expectant officials vying with each other to collect banned books in competition for government posts, gentry avenging grievances by turning in their fellows, and provincial governors maneuvering and dissembling to avoid imperial reprimand — were as characteristic of the era as Chu Yun's attempt to turn the Ssu-k'u project to his own ends, or the competition between government and private book holders for the best editions. That the censorship campaign grew through such phenomena may be a new finding; it should not be a startling one.

Nor should it be surprising that the campaign was directed primarily against anti-Manchu references. The Ch'ien-lung Emperor's desire to regulate the thought of his subjects through literary controls has often been suggested but never attested. On the other hand, his "strange, guilty" sensitivity to any slight of his own rule or that of his ancestors has been demonstrated.[89] Anti-Manchu historical references not only offended this sensitivity but undermined sagging Manchu morale, and quite possibly upset the delicate Manchu/Chinese balance at court. In the atmosphere of self-glorification surrounding the emperor's sixtieth birthday and the compilation of the *Ssu-k'u ch'üan-shu*, a campaign to eradicate anti-Manchu references seemed to fit.

Perhaps the most surprising aspect of the views of the censorship here presented is precisely that they are new, that previous students have failed to offer such interpretations. Partly, as was suggested in the introduction, the problem has been one of source material. But two further explanations, one ideological and the other methodological, may be offered for the character of previous interpretations.

Twentieth-century historians have seen the "literary inquisition" as an instance, perhaps the classic instance, of a clumsy and inefficient Manchu despotism which robbed China of its intellectual vigor and political vitality on the eve of its encounter with the West. Sun Yat-sen once wrote:

> Under the Manchu autocracy, those who would preserve the Chinese national spirit could not do so in writing, [they] could do so only by word of mouth. [This was because] the writings which had been transmitted were completely destroyed in the time of the Ch'ien-lung emperor. . . .Because of the proscriptions of that era which involved many litigations and condemnations, the written thought of the Chinese people was completely destroyed.[90]

Such pronouncements, of course, were part of the ideology of revolution. But the ideas behind them have been remarkably long-lived. A Republic of China book company, offering for sale a series of reprints of books suppressed during the Ch'ing dynasty, wrote in its 1977 catalog:

> Because the Chinese people have a spirit of "separating themselves from their enemies, distinguishing the Chinese from barbarians," they have long survived and prospered. After the Manchus conquered China, they came to understand this principle quite clearly, and so realized that if they were to maintain themselves in China, and cast out the Chinese, they must first destroy this element of the Chinese spirit.
>
> We know that books are the most important medium for the transmission of traditional thought and spirit. Therefore, the Ch'ing government thoroughly investigated the books that our nation had long preserved. . . . We have made every effort to locate copies of these books preserved in the country, reprint and circulate them, in order to further this company's goal of reinvigorating Chinese culture.[91]

Writing from the other side of the Taiwan straits in 1959, Chang Shun-hui blamed the literary litigations for the willingness of eighteenth-century scholars to engage in tedious and unproductive research of textual verification, and by implication, for their failure to explore the revolutionary implications of their discoveries.[92]

L. Carrington Goodrich in his English study, *The Literary Inquisition of Ch'ien-lung*, seems to have adopted his Chinese contemporaries' view of the censorship as an instance of imperial despotism, without incorporating into his work the anti-Manchu nationalist assumptions that buttressed their theories. Adding a Western legalistic metaphor to the condemnation of the Ch'ien-lung Emperor, Goodrich proclaimed that the "Ch'ien-lung Emperor stands accused at the bar of public opinion for his open interference with scholars of his day."[93]

Viewing the inquisition in this light, historians have sought to assess the extent of damage done to China's intellectual heritage, rather than explore the procedures and mechanisms through which the inquisition grew. Studies have focused on the fragmentary indices of banned books which archivists have found in central government records and in the private papers of participants in the campaign. Since 1883 three book-length essays and numerous articles have been devoted to these documents.[94] While impressive lists of condemned books have been compiled, it is unlikely that any complete list of all the works totally or partially censored by all the local censorship boards will ever be found, or indeed, was ever compiled. Furthermore, while one may assume that every book on extant lists was in fact banned, one may not assume, given the lack of centralized standards and the haphazard procedures of governors and governors-general, that the lists completely exhaust or even accurately reflect imperial purposes in the campaign. It is both dangerous and methodologically unjustified to characterize the Ch'ien-lung Emperor's policies toward scholarship and literature only on the basis of these lists.

A number of historians writing in the People's Republic in late 1979 and early 1980, men who had seen in their own lives the destruction that could be wrought by a government bent on changing the political lexicon of its people, emphasized other aspects of the Ch'ien-lung book burnings. Some argued that a potential for intellectual persecution exists whenever those with supreme power were also regarded as having supreme wisdom and the duty to reform society; and that when political authority is unrestrained by law, the ruler's will can only be constrained by circumstance. Others attributed the accusations and jealousies of the censorship campaign to the Ch'ing government's policies of "literary entrapment," and regarded both as evidence of the "feudal" character of Chinese society in the eighteenth century, remnants of which they still sadly saw about them. Ultimately, of course, these articles were probably meant to demonstrate a will to overcome "feudal remnants," but the points they made were telling. The broad pattern of censorship in the Ch'ien-lung era—a vague central initiative, growing momentum fed by officials eager to please, and finally official retreat in the face of overwhelming popular response—was one repeated many times in Chinese history. The censorship campaign manifested some of the timeless characteristics of Chinese history, but also had its roots in specific eighteenth-century realities like the evolving relationship between Chinese and their Manchu rulers, and the patterns of relations between scholars and the state. The effort to relate the campaign too closely to broader historical patterns has its truth, but inevitably oversimplifies.[95]

The analysis of this chapter has viewed the eighteenth-century censorship not solely as an instance of imperial despotism, but as an historical event, shaped by the interests and attitudes of all those who participated in it. The purpose has not been to apologize for the Ch'ien-lung Emperor; censorship of the written word is always odious, and when it involves the destruction of lives and property, it is particularly so. Without excusing the emperor in the slightest for his role in initiating the effort, one may observe that the human and literary tragedies of the campaign were not all of his making. The intellectual world of eighteenth-century China was simply too large, too complex, and too diverse to be dominated by one man. The ferocities and excesses of the era bear witness to this complexity.

The place of the censorship of the Ch'ien-lung reign in history and historiography is, therefore, doubly ironic. Not only has the campaign to secure the dynasty's virtuous name contributed to its reputation for despotism and paranoia; but the event so often cited as evidence of the Ch'ing dynasty's despotic power proves, on closer examination, to show just how tightly the exercise of imperial will in eighteenth-century China was constrained by the community of interests over which the emperor reigned.

7

Conclusion

Censorship was certainly not the only phase of the literary activities of the Ch'ien-lung court in the 1770s and 80s which reflected a wide range of interests and motives; such diversity was apparent, albeit with far less disastrous consequences, in all phases of the Ssu-k'u effort. The imperial initiative in the project was one shaped by both traditional ideals and practical considerations. The scholars, for their part, responded to the initiative in a way that reflected the complex institutional environment in which they lived, and their deeply-held beliefs about the nature of truth and the best means of pursuing it. Perhaps the interests of the bureaucracy most influenced the form of the final *Ssu-k'u ch'üan-shu* manuscript, and these were conditioned by ideological, factional and personal considerations. In its intellectual stance the *Annotated Catalog* primarily reflected the views of *k'ao-cheng* scholarship, but some of its formulations were undoubtedly constrained by Manchu ethnic sensitivities and imperial pride. Reflecting all of these interests, the Ssu-k'u project was dominated by none. The hypothesis of this research has been that the government sponsored literary activity of the late Ch'ien-lung years was the product of an interaction between scholars and the state. This conclusion will briefly explore some implications of this finding for eighteenth-century political and intellectual history.

The findings of this research together with those of other studies of eighteenth-century China suggest that government in China, like the government of most states, rested on the art of the possible. But if the Ch'ien-lung Emperor was not a despot, he was not necessarily incompetent or weak either. The capacities of the Ch'ien-lung government were great, but so were the limitations inherent in the character of imperial rule, and the weight of established social interests and habits of thought in an empire long undisturbed by foreign or domestic calamity. As committed as he was to serving as moral and symbolic leader of his empire, the Ch'ien-lung Emperor was perhaps more significant as a ruler who demonstrated the strengths and limitations of traditional rule in the face of the contrary tendencies of eighteenth-century Chinese society than as a force for the encouragement or repression of any single tendency.

The successes of the Ch'ien-lung government in the Ssu-k'u project were indeed impressive. The largest book collection of Chinese history was assembled, and a catalog that is still useful in spite of its limitations was prepared. The products of the Ssu-k'u Commission were not only monuments of government achievement, they were demonstrations of the quality and vitality of eighteenth-century scholarship. Moreover, in at least partially integrating scholars into a political hierarchy, the Ch'ien-lung government achieved a goal which had eluded many of its Ch'ing and late Ming predecessors. These accomplishments should not be underestimated. The Ch'ien-lung government held sway over an area that was, in population, geographical expanse, and social complexity, greater than that ruled by any previous Chinese government. Moreover, intellectuals were perhaps more profoundly alienated from the seventeenth and eighteenth-century Chinese government than they had ever been before. Accomplishing even the appearance of unity was no mean feat. Yet signs of dynastic decline, to the extent that they were apparent in the Ssu-k'u project, did not interfere with the public façade of harmony and achievement.

Many of the failures of the Ssu-k'u effort were, by contrast, evident in the light of critical retrospect. Few saw the manuscript errors that marred the texts of the *Ssu-k'u ch'üan-shu,* and the errors of judgment and evaluation in the *Annotated Catalog* have not yet been fully listed and documented. The most spectacular failure of the Ch'ien-lung government in the literary realm was its inability to censor expeditiously and effectively. This was, of course, a task which has frustrated many modern governments, and which was not even attempted by earlier Ch'ing emperors.

The late Ch'ien-lung era was undoubtedly a turning-point in

Ch'ing government, but it seems futile to try to prepare a scorecard of successes and failures or to attempt to characterize the age either as the beginning of the end or the end of the beginning. It is more important to observe the patterns and processes which shaped the achievements and failures of the era. These, it must be noted, were not the creation of a single man or faction. Chinese historians' explanations to the contrary, the decline of bureaucratic morale, the growth of tensions among literati, and the government's inability to utilize completely the talents of its most creative subjects did not occur overnight. If anything, these trends were related to a gradual erosion of consensus which accompanied the growth of a complex and perhaps even pluralistic society.

One evolving relationship of particular interest and importance in eighteenth-century China was that between Manchu rulers and their Chinese subjects. Much of the writing on Ch'ing history has unfortunately tended to read the concerns of seventeenth-century Manchu leaders into the policies of all Ch'ing governments. Clearly, however, the attitudes and goals of Manchus changed with changes in their socio-economic status, cultural attainments and political security. While full documentation does not yet exist of the reasons for the Ch'ien-lung Emperor's inauguration of censorship, nor can the precise nature of the relationships between Ho Shen, A-kuei and their Chinese counterparts be completely fathomed, some preliminary observations can be made about the state of Manchu-Chinese relations at mid-dynasty. The personnel list of the Ssu-k'u Commission shows that Manchus held none of the substantive offices on the Commission; they did, however, serve as proctors, proofreaders and directors-general. Manchu leaders were almost certainly aware of this fact. Their sensitivity about the ethnic differences separating them from their Chinese subjects was apparent in orders to collators to change texts of works dealing with Sino-foreign relations, in the remarkable series of publications of the 1770s and 1780s dealing with Manchu heritage and history, and most spectacularly, in their campaign of censorship. The Manchus of the eighteenth-century knew that they were a foreign minority, albeit one that had achieved a remarkable mastery of the civilizing arts of Chinese life. By contrast many Chinese of the eighteenth-century seem to have, if not precisely forgotten the foreign origins of Manchu rulers, at least accepted their leadership. Chinese scholars were willing to follow Manchu overlords. In fact, many Chinese showed themselves quite

skillful during the campaign of censorship in using Manchu sensitivities to resolve old social and economic squabbles. The apparent
incongruity of Manchu self-consciousness and Chinese tolerance
was perhaps not so strange given the relative size of the two groups:
the Manchus' small numbers and exalted social and political position
must have been very much on their minds throughout their rule.
The attitudes of Chinese and Manchus toward the ethnic issue probably led to a situation in which the difference was highly important
in the hothouse atmosphere of court politics, and of rapidly declining
significance the further away one moved from the imperial city in
Peking. Such a hypothesis would account both for the extraordinary
sensitivity of the Peking government to slights of Manchus (a sensitivity which led them to set up a committee to eliminate the character *t'u* "rabbit" from all Manchu names in the historical record) and
the relative absence of ethnically based opposition to Ch'ing rule in
the provinces.

Ch'ing rulers' consciousness of their ethnic origins probably
affected Ch'ing political and court history, but did such consciousness affect other dimensions of modern Chinese development?
What, if any, difference did it make for institutional or intellectual
history that the compilers of the *Ssu-k'u ch'üan-shu* were Manchus? To
the extent that this question implies the counterfactual query ("How
would the Chinese have done it differently?") it is, of course, unan-
swerable. Yet certain speculations may be advanced. While the
precedents for imperial book collecting were very old, in fact several
centuries had passed since any Chinese ruler had employed them.
Anxious to demonstrate, among other things, how conversant he was
with the classical Chinese past, the Ch'ien-lung Emperor awakened
long dormant traditions, probably realizing only dimly that the institutions and needs of Chinese scholars and rulers had changed. The
result was a situation in which imperial dictates were not in accord
with social and intellectual needs. Such a pattern, recurrent in
Ch'ing history, has often been seen as having a stultifying effect on
Chinese development. Yet in the case of the Ssu-k'u project, Chinese
and Manchu elites set about the task of creatively adapting traditional forms to meet new needs. The *Ssu-k'u ch'üan-shu,* offering Chinese
scholars an opportunity to collect and comment on their own intellectual heritage and consolidate their philosophical movements,
served a very different purpose from Liu Hsiang's *Pieh-lu.* As

research on Ch'ing institutions continues, such instances of institutional creativity may not prove to be uncommon responses to the changing character of Chinese society, and the unparalleled fact of prolonged and secure Ch'ing control of the Chinese imperium.

One reason for the richness, complexity, and, if the argument of the previous paragraph is credited, creativity of institutions like the Ssu-k'u Commission was that they were the work of many hands. Until recently, historians concerned with establishing the social and economic bases of elite status in traditional China have ignored the ways in which segments of the Chinese elite consciously differentiated themselves from one another. Yet one implication of the present study is that there were among China's elite many different styles of life, and conceptions of social and political duty. All those examined here were intellectuals in Edward Shils' sense of those who employed in their speech and writing with relatively greater frequency than those around them symbols of abstract scope and general reference. Yet some clearly enjoyed the political arena while others dwelt on its peripheries in institutions of image-making or recruitment. Still others like Tai Chen and Shao Chin-han preferred pure scholarship. Each of these segments of the elite had different reasons for participating in the Ssu-k'u project, and the strengths and faults of the final product reflected the diversity of the compilers.

The Ssu-k'u project may not have reflected any particularly new developments in the eighteenth-century bureaucracy, but since the tasks of collecting and evaluating books were ones for which bureaucrats' common classical education should in theory have suited them perfectly, the project may have revealed the limitations and capacities of the late eighteenth-century ruling apparatus more clearly than did other contemporary efforts. These strengths and limitations were, in a sense, built into the structure of the bureaucracy. If one were to liken Ch'ing government to a game, it would be one in which the emperor held all the cards, but the bureaucrats habitually played them. Since all legitimate authority in Chinese government derived from the emperor and his possession of a mandate from heaven, bureaucrats were in theory utterly dependent on him. In the face of an imperial order, be it one to collect and edit the best books or to destroy offensive ones, the Ch'ing emperor's servants and slaves could do nothing but obey. Surely part of the tone of anger and pent-up frustration that characterizes imperial reprimands derived from the fact that whatever officials did they did it in the emperor's name. When Hai Ch'eng allowed the *Tzu-kuan* to circulate, or when copyists made errors in the texts of Ssu-k'u books, they were in a sense misusing the authority entrusted to them from above. Misusing authority was not merely an administrative error, but a moral one.

Given the finality of the emperor's orders, however, it behooved the sensible official to prevent him from issuing them. Or, if orders were issued, the wise official made sure that they were worded as vaguely as possible so that the administrator would have the maximum scope for maneuver. Not only administrative obstacles but, as the evidence above has repeatedly demonstrated, inter-bureaucratic competition in a world largely dependent on imperial favor affected the way the emperor's policies were actually carried out. In a time when neither external challenges nor internal unrest forced priorities on the empire, and precedent and the way it was interpreted constrained the imperial will, the bureaucrats' capacity to shape central government policy in this indirect manner was great. The conventions of court-centered historiography have normally concealed the process by which bureaucrats shaped Chinese policy from view. But in the Ssu-k'u project, the outlines of bureaucrats' influence on policy making were visible, albeit through the historiographical glass dimly, for too many different sorts of officials had an interest in the project. Clearly, local educational officials, capital educational officials like Chu Yun, senior statesmen at court like Yü Min-chung and Liu T'ung-hsun, and scholars-turned-officials like Yao Nai and Shao Chin-han, all influenced the *Ssu-k'u ch'üan-shu*. The irony of Ch'ing bureaucracy was that while all of these officials would have had to bow in compliance, as many did, before direct imperial orders, the project could not have succeeded unless it served each of their interests in some way.

Probably the most interesting and historiographically important motivations to examine were those of the scholars. Often the participation of eighteenth-century scholars in the Ssu-k'u project has been characterized as an ill-advised and self-interested capitulation to a cruel and barbarous regime. This book has argued, on the contrary, that the scholars enthusiastically took part in the effort because they saw in it an opportunity to achieve goals which were beyond their private and individual resources. This finding has implications for our image of Ch'ing scholarship, and of the Ssu-k'u products. The *k'ao-cheng* scholar's concern with issues of philology and textual transmission has often been seen as a product of his fear that by addressing larger social or philosophical issues, he would court government repression. Yet, as has been shown here, the areas of Manchu sensitivity were fairly narrow and predictable. The vast majority of Ch'ien-lung editors' textual emendations and book bannings involved writings which dealt with the history of Sino-foreign relations or with the Manchus' sixteenth and seventeenth-century rise to power. No evidence has appeared in the course of this study to suggest that the Ch'ing government ever used censorship to change the fundamental directions of Chinese thought in other areas. Given this finding, one must consider the possibilities that Ch'ing intellectuals examined philological and epistemological issues either because they found it financially rewarding to do so, or because they believed such issues genuinely important. Until much more is known about the financial commitments and resources of Ch'ing scholars, the first proposition will be difficult to evaluate. It can be safely suggested, however, that hope of financial gain was not the major motive of scholarly participation in the Ssu-k'u compilation. But there is ample evidence of the truth of the second proposition in the writings of eighteenth-century intellectuals and in the *Annotated Catalog* of the *Ssu-k'u ch'üan-shu*. Chu Yun and his colleagues examined ancient texts because they believed they could recapture in them ancient truths. Shao Chin-han corrected the historical record in order to provide those who followed him with a model of conduct and political wisdom.

The products of the Ssu-k'u compilation both shaped and reflected this belief. The Ssu-k'u project afforded scholars an opportunity to review texts on a much larger scale than their private resources would permit, to inspect the *Yung-lo ta-tien,* and to pronounce their judgments from a forum which commanded the respect and attention of all literati. It is true that certain projects recommended by Chu and his circle were not approved by the Grand Council, but this was because they were impractical not because they were impolitic. Perhaps the best sign of the degree to which the book collection project reflected the concerns of eighteenth-century scholarship was the reaction it caused. Yao Nai and his followers were not opposed to the concept of a book collection project *per se,* nor did they object to the fact of scholarly collaboration. Their reaction was to the imperial endorsement of the goals and concerns of the *k'ao-cheng* movement which the Ssu-k'u project represented.

The willing participation of eighteenth-century intellectuals in the Ssu-k'u effort suggests some more general conclusions about the significance of imperial prerogatives in the Chinese scholarly world. Clearly, if truth could be certified by imperial imprimatur, it was in the interests of the opposing sides on various questions to compete to attract imperial favor. On the other hand, an emperor who perceived that the security of his empire rested on the allegiance of literate elites could hardly afford to be capricious in his endorsements. Indeed, this probably was the reason why the imperial prerogatives in the scholarly world were usually exercised by institutions of government rather than by the emperor personally. In the final analysis, was eighteenth-century scholars' competition for the favor of an emperor concerned with public opinion markedly different from the competition for public esteem which characterizes modern intellectual discourse? There were, of course, cases where emperors overstepped their role, overrode the institutions of imperial scholarship and sought to impose personal or family considerations on intellectual life. In fact, the Ch'ien-lung Emperor's effort to eradicate from the historical record unfavorable references to his Manchu ancestors probably represented such a case. Yet the chaos which followed from his orders illustrated the dangers that attended such an overstepping and the reasons why wise rulers left such tasks to the organs of government created to carry them out. To be sure, the fact that the Chinese emperor was both sage and ruler imposed constraints on the scholarly world—conventions for referring to the ruler, taboos, etc. But to a thinker who accepted the emperorship as part of the furniture of the universe, such conventions probably did not seem to be constraints, but inevitable concomitants of civilized life. Is there any evidence that these conventions impinged upon the creativity of the Chinese intellectual any more than the forms of Western scholarly discourse limit the freedom of modern thinkers? If, as the evidence of the present study suggests, the answer to this question is no, then there is no reason to suspect that the Chinese scholar in the context of his assumptions was any more restrained in his expressions of social and political opinion than his western counterpart. The challenge this poses to the historian of Chinese thought is to understand the character of constraints and, instead of decrying the depotism of Chinese rulers, assess the achievements of Chinese scholars.

NOTES
BIBLIOGRAPHY
GLOSSARY
INDEX

ABBREVIATIONS USED IN THE NOTES

CCCT:CL   National Palace Museum (Taipei) Archives, Grand Council Collection, Ch'ien-lung reign. First citations will include catalog number of memorial, date, and name of memorialist.

CSCWC   *Chang Shih-chai hsien-sheng wen-chi,* Chang I-hsuan, comp. Taipei, Wen-hai Publishing Company reprint, 1968.

ECCP   *Eminent Chinese of the Ch'ing Period,* Arthur W. Hummel, ed. Taipei, Literature House reprint, 1964.

HCCS   *Hsueh-cheng ch'üan-shu.* Taipei, Wen-hai Publishing Company reprint, 1968.

HPHSL   Yao Nai, "Hsi-pao-hsuan shu-lu," in *Hsi-pao hsuan i-shu san-chung,* Mao Yü-sheng, ed. T'ung-cheng, 1879.

KCCHLC   *Kuo-ch'ao chi-hsien lei-cheng,* Li Huan, ed. Hsiang-yin, 1884–1890.

KCT:CL   National Palace Museum Archives, Palace Collection, Ch'ien-lung reign. Form of citations same as for *CCCT:CL.*

Pan-li   *Pan-li ssu-k'u ch'üan-shu tang-an,* Wang Chung-min, ed. Peking, 1935. This work was published in two *ts'è;* I shall use roman numerals I and II to distinguish them. All official documents which appeared in *Pan-li* are cited in the notes in their *Pan-li* version.

SCHNP   Huang Yun-mei, *Shao Erh-yun (Chin-han) hsien-sheng nien-p'u.* 1933.

SYT:FP   National Palace Museum (Taipei) Archives, Edict Record Book Collection, Square series, Ch'ien-lung reign.

Shih-lu   *Ta-ch'ing kao-tsung-shun huang-ti shih-lu.* Tokyo, 1937–38.

TMTY   *Ch'in-ting ssu-k'u ch'üan shu tsung-mu t'i-yao,* Chi Hsiao-lan, ed. Taipei, Commercial Press reprint, 1971. The Commercial Press reprint is consecutively paginated, and I shall cite this pagination throughout. Since the indices to this work, and many traditional writings on it refer to *chüan* and page number (traditional style), I have provided this reference in parenthesis after each citation.

TYFTK   Shao Chin-han, "Ssu-k'u ch'üan-shu t'i-yao fen-tsuan kao," in *Shao-hsing hsien-cheng i-shu,* Ma Yung-hsi, ed. K'uei-chi, 1883.

WTYT   *Ch'ing-tai wen-tzu yü tang.* Taipei, Wen-hai Publishing Company reprint, 1975.

YMCSC   Yü Min-chung, *Yü Min-chung shou-cha,* Ch'en Yuan, ed. Taipei, Wen-hai Publishing Company reprint, 1968.

All references to the dynastic histories are to the editions published by the Chung-hua shu-chü (1959–1972). Citations include page number in the Chung-hua edition, title of article, and *chüan* number.

Where available, dates of birth and death are provided on first reference for persons not employed by the Ssu-k'u Commission or involved in a Ch'ing literary case.

Notes

Chapter 1. Introduction

1. Among the most memorable of such studies are Florian Znaniecki, *The Social Role of the Man of Knowledge*, and Karl Mannheim, *Ideology and Utopia: An Introduction to the Sociology of Knowledge*. For a survey of literature on the sociology of knowledge, see Edward A. Shils, "Intellectuals," in vol. 8 of *International Encyclopedia of the Social Sciences*, pp. 399–414.
2. Adam Yuen-chung Liu's *The Hanlin Academy: Training Ground for the Ambitious* and Miyazaki Ichisada, *China's Examination Hell*, trans. by Conrad Schirokauer are useful institutional studies, but do not really explore the significance of the government bodies they are describing. Thomas A. Metzger's *The Internal Organization of the Ch'ing Bureaucracy: Legal, Normative and Communicative Aspects* and John Watt's *The District Magistrate in Late Imperial China* explore the relationship of scholarly and political roles in Ch'ing China, but do not address the structure of policy-making institutions. For earlier periods, see John Meskill, *Academies in Ming China: A Historical Essay*, John H. Winkelman, "The Imperial Library in Southern Sung China, 1127–1279," and Howard S. Galt, *A History of Chinese Educational Institutions*.
3. See, for instance, Philip A. Kuhn and Susan Jones, "Dynastic Decline and the Roots of Rebellion," pp. 107–162; James Polachek, *The Inner History of the Opium War;* and Judith Whitbeck, "The Historical Vision of Kung Tzu-chen, 1792–1841." Susan Naquin has some interesting speculations on why this era saw significant outbreaks of millenarian rebellion in *Shantung Rebellion: the Wang Lun Uprising of 1774*, pp. 148–151, and 160–164.
4. Immanuel C. Y. Hsu emphasizes the splendors of the Ch'ien-lung reign in *The Rise of Modern China*, 3rd ed.; Alexander Woodside emphasizes the crumbling facades of the era in his draft chapter for the *Cambridge History of China*.

5. Harold L. Kahn, *Monarchy in the Emperor's Eyes: Image and Reality in the Ch'ien-lung Reign.*
6. Two of the most forceful statements of the importance of gentry allegiance for the survival of Ch'ing rule in the nineteenth and twentieth centuries are to be found in Philip A. Kuhn, *Rebellion and Its Enemies in Late Imperial China: Militarization and Social Structure, 1796–1864,* and Chuzō Ichiko, "The Role of the Gentry: An Hypothesis," pp. 297–318.
7. Yü Ying-shih, "Ts'ung Sung-Ming ju-hsueh ti fa-chan lun Ch'ing-tai ssu-hsiang shih," and "Ch'ing-tai ssu-hsiang shih ti i-ko hsin chieh-shih," and Benjamin A. Elman, *From Philosophy to Philology: Intellectual and Social Aspects of Change in Late Imperial China.* For a more complete assessment of the various theories about the origin of the textualist movement, see Chapter 2 below.
8. Some of the major areas of literati discontent with official orthodoxy in the eighteenth century are explored in Paul S. Ropp, *Dissent in Early Modern China: Ju-lin wai-shih and Ch'ing Social Criticism.* I have explored the origins of the assumption that the burden of eighteenth century scholarship was opposition to the Ch'ing regime in a paper entitled "The National Essence Movement and the Eighteenth Century," which was delivered at the March 1982 meetings of the Association for Asian Studies.
9. Luther Carrington Goodrich, *The Literary Inquisition of Ch'ien-lung,* and Wu Che-fu, *Ch'ing-tai chin-hui shu-mu yen-chiu.*

Chapter 2. The Imperial Initiative

1. Robert Frost, "For John F. Kennedy on His Inauguration," *The Poetry of Robert Frost,* p. 424.
2. *Pan-li* I, p. 1a–b. Throughout this book, I will follow the practice of referring to the emperor as the author of imperial edicts, although I am aware of the complex process by which they were composed.
3. See Yao Ming-ta, *Chung-kuo mu-lu-hsueh shih,* pp. 177–201, *passim.*
4. Hu Shih, "The Establishment of Confucianism as a State Religion During the Han Dynasty," pp. 20–21.
5. *Shih-chi,* "Basic Annals of Ch'in Shih-huang," 6.254–255. Translated in Derk Bodde, *China's First Unifier: A Study of the Ch'in Dynasty as seen in the Life of Li Ssu . . . (280?–208 B.C.)* pp. 82–83.
6. *Han-shu,* "Treatise on Art and Literature," 30.1701, trans. in T. H. Tsien, *Written on Bamboo and Silk,* p. 13.
7. On the Lius, see *Han-shu* 36.1929–1974.
8. For some sense of these conflicts, see Tjan Tjoe-som, trans. *Po Hu T'ung: The Complete Discussions in White Tiger Hall;* R. P. Kramers, "Conservatism and the Transmission of the Confucian Canon," pp. 119–172; and Jack L. Dull, "An Historical Introduction to the Apocrypha (Ch'an-wei) texts of the Han Dynasty."
9. See, for instance, K'ang Yu-wei's *Hsin-hsueh wei-ching-k'ao,* the argument of which is summarized in Liang Ch'i-ch'ao, *Intellectual Trends in the Ch'ing,* trans. by Immanuel C. Y. Hsu, pp. 92–93.

10. Robert des Rotours, vol. 2 of *Traité des fonctionaires et traité de l'armee* pp. 204–207; Charles O. Hucker, *A Dictionary of Official Titles in Imperial China,* pp. 377–378.
11. *Sui-shu,* "Treatise on Bibliography," 32.908.
12. Yao Ming-ta, *Mu-lu-hsueh,* pp. 187–188; and des Rotours, *Traité des fonctionaires* 2:207.
13. *TMTY,* p. 1775 (85.1a).
14. Yao Ming-ta, *Mu-lu-hsueh,* pp. 197–198.
15. John F. Winkelman, "The Imperial Library," p. 36.
16. Wan Ssu-t'ung, "Ming-shih i-wen-chih hsu." According to the Commercial Press editors, the passage was found in a handwritten version of the *Ming-shih-kao* in the possession of the National Library of Peking. The work is formally attributed to Wang Hung-hsu, but the librarians believe it was Wan's work. The Commercial Press editors believe that the piece might have been written as a preface to a privately compiled treatise on bibliography in the Yuan. On state publishing during the Yuan, see also Yeh Te-hui, *Shu-lin ch'ing-hua,* pp. 176–177. On the regional character of intellectual life during the Yuan, see Wing-tsit Chan, "Chu Hsi and Yuan Neo-Confucianism," pp. 197–231.
17. Ch'ien Ta-hsin, "Pu Yuan-shih i-wen-chih hsu," p. 8393.
18. Hok-lam Chan, *Control of Publishing in China, Past and Present,* pp. 5–21. See also Denis Twitchett, *Printing and Publishing in Medieval China.*
19. Kuo Po-kung, *Yung-lo ta-tien k'ao,* pp. 5–8.
20. Lynn Ann Struve, "Ambivalence and Action: Some Frustrated Scholars of the K'ang-hsi Period," p. 327.
21. Thomas S. Fisher, "Lü Liu-liang (1628–1683) and the Tseng Ching Case (1728–1733)"; Ono Kazuko, "Shincho no shisō-tōsei o megutte," pp. 99–123; and Lynn Ann Struve, "Uses of History in Traditional Chinese Society: The Southern Ming in Ch'ing Historiography."
22. William S. Atwell, "From Education to Politics: The Fu-she in Late Ming," p. 338.
23. *HCCS* 1:433–434 (24.1a–b); Ono Kazuko, "Shincho no shisō-tōsei," p. 100.
24. *HCCS* 1:461 (26.1a); Ono Kazuko, p. 100. The ambiguity of the educational intendant's position as policeman and preceptor of local students was characteristic of Ming as well as Ch'ing institutions. See Tillman Grimm, "Ming Educational Intendants."
25. *HCCS* 1:39–41 (26.1a); Ono Kazuko, p. 101. Translated in Hsiao Kung-chuan, *Rural China: Imperial Control in the Nineteenth Century,* pp. 241–242.
26. The complete text of this edict can only be found in the K'ang-hsi edition of the *Ta-Ch'ing hui-tien* 51:20a. Subsequent editions of the *Hui-tien* and *HCCS* include only abridged versions of the edict, abridgements more in accord with later versions of this regulation. See also Ono Kazuko, p. 102.
27. Robert Oxnam, *Ruling from Horseback: The Politics of the Oboi Regency,* pp. 10 and 118.
28. Lawrence D. Kessler, "Chinese Scholars and the Early Manchu State;" and the same author's *K'ang-hsi and the Consolidation of Ch'ing Rule 1661–1674,* pp. 158–166. For another perspective on the politics of the K'ang-hsi era,

see H. Lyman Miller, "Factional Conflict and the Integration of Ch'ing Politics."

29. See *ECCP*, pp. 205–206; L. Carrington Goodrich, *The Literary Inquisition of Ch'ien-lung*, pp. 75–77; and Lynn Ann Struve, "Uses of History," pp. 103–104. For Ku Yen-wu's reaction to these events, see Willard J. Peterson, "The Life of Ku Yen-wu (1613–1682) (II)."

30. Hellmut Wilhelm, "The Po-hsueh hung-ju Examination of 1679," p. 62.

31. On the Nan-shu-fang, see Silas Wu, "Nan-shu-fang chih chien-chih chi ch'i ch'ien-ch'i fa-chan," pp. 6–12. The first text published was the *Jih-chiang ssu-shu chieh-yi*, see *TMTY*, pp. 746–747 (36.5b).

32. *HCCS* 1:166–167 (7.1b–2a).

33. On the Neo-Confucian tone of K'ang-hsi official scholarship, see Wing-tsit Chan, "The *Hsing-li ching-i* and the Ch'eng-Chu School," pp. 545–546. During the course of his reign, the K'ang-hsi Emperor ordered seven collections of lecture notes printed, as follows:

Date	Text	TMTY Citation
1676	Ssu-shu	pp. 746–747 (36.5b)
1679	Shu-ching	pp. 246–247 (12.5b)
1683	I-ching	p. 83 (6.1a)
n.d.	Shih-ching	– – –
n.d.	Tzu-chih t'ung-chien	– – –
1738	Ch'un-ch'iu	p. 575 (29.1a)
1749	Li-chi	p. 421 (21.4a)

In addition to these texts, the K'ang-hsi Emperor ordered imperially commissioned (*ch'in-ting*) editions of commentaries as follows:

1715	I-ching	pp. 83–84 (6.1a)
1722	Ch'un-ch'iu	pp. 575–576 (29.1a)
1729	Shu-ching	p. 247 (12.5b)
1740	Shih-ching	p. 319 (16.5a)

The last publication of the second series, and the last two of the first series were approved (*ting*) by the K'ang-hsi Emperor, but were printed by his successors.

One of the most useful summaries of imperial publications in Ch'ing is the bibliography section of the Ch'ien-lung period work *Kuo-ch'ao kung-shih* (1769) which has been recently reprinted.

34. *Ta-Ch'ing sheng-tzu-jen huang-ti shih-lu* 39.13b, 120. 5b–6a, and 249.7b–8a.

35. Lynn Ann Struve, "Uses of History," p. 140. See also her article, "The Hsu Brothers and Semiofficial Patronage of Scholars in the K'ang-hsi Period."

36. Tai I, "Han-hsueh tan-che," p. 7.

37. *HCCS* 1:65–66 (3.1a–b).

38. Liu Shih-p'ei, "Lun chung-kuo i-ch'ien tsang-shu-lou," p. 2291.

39. Translated in Hellmut Wilhelm, "The Po-hsueh hung-ju Examination," p. 64.

40. *ECCP*, pp. 66, 93–95, 285–286; *TMTY*, pp. 2822–2823 (136.5b), 874 (47.12a). See also Silas H. L. Wu, *Passage to Power: K'ang Hsi and His Heir Apparent, 1661–1722*, p. 165. On public and private interests in the K'ang-hsi publication projects, see Jonathan D. Spence, *Ts'ao Yin and the K'ang-Hsi Emperor*, pp. 157–165.

41. Lynn Ann Struve, "Ambivalence and Action," p. 354.
42. Lynn Ann Struve, "Uses of History," pp. 193–195; Goodrich, *Literary Inquisition,* pp. 77–79; *ECCP,* p. 701.
43. *Ta-Ch'ing Shih-tsung-hsien huang-ti shih-lu* 2.20a–21a; Lionel Giles, *An Alphabetical Index to the Chinese Encyclopedia,* pp. vi–vii. Giles here draws upon the *K'uei-t'ien so-chi* by Liang Chang-chü, which was written after Liang retired in 1841. For Liang's biography, see *ECCP,* pp. 499–501.
44. The Yung-cheng Emperor's few publications reflected his interests in administration and management of the bureaucracy, and his interests in the subject of filial piety:

Date	Title	TMTY
1727	Ch'ing-han-wen hsiao-hsueh	— — —
	Hsiao-ching chi-chu	p. 654 (32.4a–6)
1727	Pa-ch'i t'ung-chih	p. 1734 (82.10a)
1736	Kung-ch'en chuan	— — —
1736	Chih-chung ch'eng-hsien	pp. 1941–1942 (92.3a)
1726	Yueh-hsin chi	— — —

45. *Ta ch'ing Shih-tsung hsien huang-ti shih-lu* 12.16a–b.
46. See Sheng Lang-hsi, *Chung-kuo shu-yuan chih-tu,* pp. 132–133, and Araki Toshikazu, "Yōsei jidai ni okeru gakushinsei no kaikaku—shu to shite sono nin yōhō o chūshin to shite," pp. 27–43.
47. *HCCS* 1:46–56 (2.5a–9b). Translated in David S. Nivison, "Ho Shen and his Accusers: Ideology and Political Behavior in the Eighteenth Century," pp. 225–226.
48. Thomas S. Fisher, "Lü Liu-liang and the Tseng Ching Case," pp. 253–279, *passim.*
49. *Lo-shan-t'ang chi* 1.9a–b. Ch'ien-lung's pre-monarchical essays have been discussed at some length in Harold Kahn, *Monarchy in the Emperor's Eyes: Image and Reality in the Ch'ien-lung Reign;* and Chang Chun-shu, "Emperorship in Eighteenth Century China."
50. *Shih-lu* 5.3b–5a.
51. *ECCP,* pp. 54–56; On Chang's appointment of friends and protégés, see Liu T'ung-hsun's accusation and the emperor's response to it in *Shih-lu* 156.7a–11b. See also R. Kent Guy, "Zhang Tingyu and Reconciliation: The Scholar and the State in the Early Qianlong Reign."
52. *HCCS* I, 92–93 (14.1b–2b); 99–106 (14.51–8b).
53. On the Yung-cheng publication of the *Hsiao-ching,* see *TMTY,* p. 654 (32.4a–b). The Yung-cheng Emperor also ordered a translation of the *Hsiao-ching* into Manchu (see *KCKS,* pp. 967–968). For the Ch'ien-lung rejection of the text, see *HCCS* I, 94–98 (14.3a–8a).
54. On the second *po-hsueh hung-ju* examinations, see Li Fu-sun, *Ho-cheng hou-lu,* and Hsiao I-shan, *Ch'ing-tai t'ung-shih* 2:33–34. On the Tseng Ching case, see Thomas Fisher, "Lü Liu-liang and the Tseng Ching Case," pp. 280–283; and L. Carrington Goodrich, *The Literary Institution of Ch'ien-lung,* pp. 84–85. For the edict condemning Tseng, see *Shih-lu* 9.10b–11a; the form of this edict suggests that the impulse for the punishment of Tseng might have come from the Board of Punishments, to which Hsu Pen (see

ECCP, p. 602), a protégé of O-erh-t'ai, had just been appointed. The source of the new policy initiatives of the early Ch'ien-lung reign is a complex subject which deserves further study.

55. See *Pan-li* I, 69b; David S. Nivison, "Ho-shen and his Accusers," p. 229; Hsiao I-shan, *Ch'ing-tai t'ung-shih* 2:14–32 *passim.*; and Goodrich, *Literary Inquisition*, pp. 94–96. Factional politics also affected the Ssu-k'u project significantly at various points. See below, chapters 4 and 5.

56. *HCCS* 1:66 (3.1b).

57. *HCCS* 1:125–126 (5.3a–b).

58. On Ch'ien-lung's trips to the south, see Sugimura Yuzo, *Kenryu katei*, pp. 16–18. On K'ang-hsi's trips, see Jonathan D. Spence, *Ts'ao Yin and the K'ang-hsi Emperor*, pp. 124–157.

59. Among the first works to appear were the *Hsieh-chi pan-fang shu* on court painters and painting, which was commissioned in 1711 and printed in 1741 (*TMTY*, p. 2270 [109.b]); and the *Ta-Ch'ing i-t'ung-chih*, printed in 1744 (*TMTY*, pp. 1460–1461 [68.3b]; a draft of the *Ming-shih*, commissioned in the K'ang-hsi reign, was presented to the court in 1739. A supplement to the *Lü-lü cheng-i*, a compendium on mathematics, music and the calendar first commissioned by K'ang-hsi in 1713, was ordered in 1741 and completed in 1746 (*TMTY*, pp. 802–803 [38.6b]).

60. The book was completed in 1740; see *TMTY*, pp. 4225–4226 (190.4b).

61. The three texts on rites were the *I-li i-shu* (*TMTY*, pp. 395–396 [20.4b]), the *Chou-kuan i-shu* (*TMTY*, p. 379 [19.7a]), and the *Li-chi i-shu* (*TMTY*, p. 422 [21.4b]); they are sometimes collectively known as the *San-li i-shu*. The *Chou-i shu-i* (*TMTY*, p. 84 [6.1b]) and *Shih-i che-chung* (*TMTY*, p. 319 [16.5a]) were both printed in 1755. The *Ch'un-ch'iu chih-chieh* (*TMTY*, p. 576 [29.1b]) was corrected and printed in 1758.

62. On the encyclopedias, see Wang Chung-han, "Ch'ing san-t'ung tsuan-hsiu k'ao," in *Ch'ing-shih tsa-k'ao;* the best introduction to the products of this project and their significance is Teng Ssu-yü and Knight Biggerstaff, *An Annotated Bibliography of Selected Chinese Reference Works*, pp. 110–115.

63. Publications on Mongolia included the *Huang-yü hsi-yü t'u-chih*, a volume of maps of Ch'ing campaigns in the area (*TMTY*, pp. 1478–1479 [68.11a]), commissioned in 1757 and completed in 1762; the *Hsi-yü t'ung-wen-chih*, a dictionary of Mongolian languages, commissioned in 1763 (*TMTY*, pp. 876–877 [41.13a]); the *Meng-ku yuan-liu*, a translation of a Mongol language work on early Mongol history (*TMTY*, pp. 1139–1140 [51.7b]); and the *Meng-ku hui-pu wang-kung kung-chi piao-chuan*, a volume of biographies of Mongol leaders issued in 1779 (*TMTY*, pp. 1283–1284 [58.4a]).

64. On the *Ch'ing-wen chien*, see *TMTY*, p. 875 (41.1a); on the *Man-chou yuan-liu-k'ao*, see *TMTY*, pp. 1478 (68.11a); the *K'ai-kuo fang-lueh* is not in *TMTY*, probably having been finished too late, but Franz Michael's *The Origins of Manchu Rule in China* is based closely on it.

65. On the *P'ing-ting liang Chin-chuan fang-lueh*, see *TMTY*, pp. 1076–1077 (49.4a); on the establishment of the *fang-lueh kuan* as a permanent office, see Li P'eng-nien, et al. editors, *Ch'ing-tai chung-yung kuo-chia chi-kuan kai-shu*, pp. 66–67.

66. On the Hsieh and Hu cases, see Hsiao I-shan, *Ch'ing-tai t'ung-shih* 2:18–21; Goodrich, *Literary Inquisition,* pp. 88–96; and *WTYT,* pp. 9–11 and 57–114.

67. *WTYT,* pp. 19–56; and Arthur Waley, *Yuan Mei, an Eighteenth Century Chinese Poet,* pp. 63–64; Waley interprets the bureaucratic significance of the Ting case slightly differently than I do.

68. *WTYT,* pp. 131–134, and 817–829.

69. *Pan-li* I, p. 1.a–b.

Chapter 3. The Scholars' Response

1. Benjamin A. Elman, *From Philosophy to Philology.*

2. The term *p'u-hsueh* was originally used in the *Han-shu* "Biographies of Confucians, Ou-yang Sheng" to describe a kind of learning pursued for its own sake by scholars unconcerned with profit or reputation: "Mao K'uan was extraordinarily talented. When he first met Han Wu-ti, he lectured on classical studies. The emperor said, 'I have always considered the *Book of Documents* to be a subject for *p'u-hsueh,* and have not favored it, but hearing K'uan speak of it, I can see its merits.' He thereupon endorsed K'uan's interpretation." See Morohashi, *Dai kanwa jiten* 6:6250; and *Han-shu* 88.3603.

The origins of the term *Han-hsueh* are fairly obvious, but its usage was complicated. Originally, the term was used to describe a certain group of eighteenth-century scholars, concentrated in Soochow, who advocated the study of Han dynasty commentaries. Many of the members of this group were appointed to the Ssu-k'u Commission, with the result that the *Han-hsueh* orientation became associated with the Ssu-k'u project. (See chapter 5.) Later, after a group of dissidents styled themselves followers of Sung learning, it became customary to call all of eighteenth-century scholarship *Han-hsueh.* One early example of this expanded usage was the title *Kuo-ch'ao Han-hsueh shih-ch'eng chi* of Chiang Fan's 1812 history of eighteenth-century scholarship. The term *k'ao-cheng* is the most neutral and perhaps least problematic description for eighteenth-century scholarship as a whole.

3. See Chang Ping-lin, "Ch'ing-ju," and "Hsueh-yin" in *Ch'ien-lun;* and Teng Shih, "Kuo-hsueh chin-lun," in *Kuo-ts'ui hsueh-pao.* I have analyzed their views at greater length in "Decadence Revisited: National Essence Views of the Eighteenth Century."

4. Hu Shih, "Ch'ing-tai hsueh-che ti chih-hsueh fang-fa," and "Chih-hsueh fang-fa yü ts'ai-liao" in *Hu Shih wen-ts'un.* Recently, Yü Ying-shih has taken issue, not so much with Hu's emphasis on *k'ao-cheng* methodology, as with his view that *k'ao-cheng* scholars completely rejected the concerns of Sung philosophy. Yü has demonstrated that it was the very concern of early Ch'ing scholars with Sung issues of metaphysics that led them to examine earlier texts. Seeing a much more gradual break between Sung thought, and Ming and Ch'ing thought than did Hu Shih, Yü has suggested that there was an "internal logic" to Chinese thought in the late imperial period, a logic which led from the "anti-intellectualism" or "intuitionism" of Ming to "intellectualism" and textual studies in the Ch'ing. See Yü

220 Notes to Pages 40–44

Ying-shih, "Ts'ung Sung-Ming ju-hsueh ti fa-chan lun Ch'ing-tai ssu-hsiang shih," and "Ch'ing-tai ssu-hsiang shih ti i-ko hsin chieh-shih."

5. Yamanoi Yū, "Minmatsu shinsho shisō ni tsuite no ichi kōsatsu." See also Benjamin Elman, "Japanese Scholarship and the Ming-Ch'ing Intellectual Transition."

6. See Yü, "Ch'ing-tai ssu-hsiang ti i-ko hsin chieh-shih," p. 147.

7. Huang Tsung-hsi's contribution to this debate was entitled *I-hsueh hsiang-shu lun* (*TMTY*, pp. 87–88 [6.3a]); Mao's was titled *T'u-shu yuan-ch'uan-pien* and Hu Wei's was titled *I-t'u ming-pien* (*TMTY*, pp. 95–96 [6.6b]). On the tradition of criticism, see Ch'ien Mu, *Chung-kuo chin-san pai-nien hsueh-shu shih* 1:229; and Liang Ch'i-ch'ao, *Intellectual Trends During the Ch'ing Period*, trans. Immanuel C. Y. Hsu, pp. 34–35. Though Liang fails to note the partisan origins of the controversy, he styles Hu Wei's book a "fatal blow to the Sung learning [of Chu Hsi]," and compares its impact in China to that of Darwin's *Origin of the Species* and Renan's *Life of Jesus* in Europe. Ku Yen-wu's note on the *Chou-i pen-i* can be found on pp. 2–4 of the modern reprint *Yuan-ch'ao-pen jih-chih-lu*. Tai Chen's *Ching-k'ao* is a rare work, but Yü Ying-shih describes Tai's comments on the *I-ching* in his "Ch'ing-tai ssu-hsiang ti i-ko hsin chieh-shih," p. 147. Fang Tung-shu's discussion of the controversy can be found in *Han-hsueh shang-tui*, pp. 1–2.

8. See Benjamin Elman, "Philosophy (I-li) Versus Philology (K'ao-cheng): The *Jen-hsin Tao-hsin* Debate." See also Yü, "Ch'ing-tai ssu-hsiang ti i-ko hsin chieh-shih," p. 148; and Tai Chün-jen, *Yen-mao ku-wen shang-shu kung-an*, pp. 79–94; and Elman, *From Philosophy to Philology*, pp. 200–212. Ku Yen-wu's remarks on the text of the *Ku-wen shang-shu* can be found in *Jih-chih-lu*, pp. 51–54; for his views on the passage on the "heart," see pp. 528–529. Huang Tsung-hsi's comments on the passage in question can be found in the preface he wrote for Yen's text, "Shang-shu ku-wen shu-cheng hsu," *Nan-lei wen-ting*, san-chi 1.1a–b. Yen's comments on the "heart" passage can be found in ch. 2, article 32 of his text.

9. Mao Ch'i-ling's responses were entitled *Ku-wen shang-shu yuan-tz'u* (*TMTY*, pp. 249–251 [12.7a], and *Shang-shu kuang-t'ing-lu* (*TMTY*, p. 251 [12.7b]). Ch'ien Mu has some interesting comments on why Mao wrote these volumes, see *Chung-kuo chin-san-pai nien hsueh-shu shih* 1:244–246. Fang Tung-shu's comments on the controversy are found in *Han-hsueh shang-tui*, pp. 25–33.

10. *TMTY*, p. 902 (42.10a).

11. For a sense of the continuity of these studies, see Benjamin Elman, "From Value to Fact: The Emergence of Phonology as a Precise Discipline in Late Imperial China"; and Elman, *From Philosophy to Philology*, pp. 212–221; as well as Chiang Yung, "Hsu," *Ku-yun piao-chun*, 3b–4a; and Tai Chen, "Ku-shih yin-lun pa," *Tai Chen wen-chi*, p. 86.

12. Ch'ien Mu, *Hsueh-shu shih* 1:307–308.

13. Tai Chen's most famous work on the rites was entitled *K'ao-kung t'u-chi*. Chiang Yung's works were entitled *Li shu kang-mu* (*TMTY*, p. 437 [22.4a]), *Li-chi hsun-i tse-yen* (*TMTY*, p. 427 [21.6b]), and *Chou-li i-i chu-yao* (*TMTY*, p. 385 [20.8b]).

14. See Ch'ien Mu, *Hsueh-shu shih* 2:495. Ling's major work on the rites texts was titled *Li-ching shih-li*.
15. On the development of this tradition, see Ch'ien Mu, *Hsueh-shu shih* 2:523–569; and Chou Yu-tung, *Ch'ing chin-ku-wen hsueh*. T'ang Chih-chün explores some of the political implications of the modern text movement in his article, "Ch'ing-tai ching-chin-wen-hsueh ti fu-hsing."
16. Tai Chen, "Yü Yao Ssu-lien [Hsi-chuan] shu," in *Tai Chen wen-chi*, pp. 141–142. This passage is translated in *Intellectual Trends During the Ch'ing Period*, p. 56.
17. To be sure, in those areas where techniques of investigation were influenced by the Jesuit presence in China, methods and conclusions came to resemble those of contemporary western scientists. For stimulating discussions of this issue, see John B. Henderson, *The Development and Decline of Chinese Cosmology*, pp. 227–256; Nathan Sivin, "Why the Scientific Revolution Did Not Take Place in China, or Didn't It?"; and Elman, *From Philosophy to Philology*, pp. 61–64, 79–85, 228.
18. Ho Ping-ti, "The Salt Merchants of Yangchow: A Study in Commercial Capitalism in Eighteenth-Century China." There is, of course, a large literature in Chinese and Japanese on the salt merchants, who commanded, in many cases, extensive networks of financial and personal connections. A thorough investigation of the acquisition and disposition of salt merchant wealth lies somewhat outside the scope of the present study.
19. Ōkubo Eiko, *Min-shin jidai shoin no kenkyū*, pp. 260–286.
20. On the Yangchow scholars, see Chang Shun-wei, *Ch'ing-tai Yang-chou hsueh-chi*.
21. Liu Shih-p'ei, "Chin-tai Han-hsueh pien-ch'ien lun," p. 3821.
22. Yeh Tě-hui, "Wu-men shu-fang chih sheng-shuai," in *Shu-lin ch'ing-hua*, pp. 254–257.
23. Nancy Lee Swann, "Seven Intimate Library Owners."
24. See chapter 5 below for a discussion of the contribution of one member of the Hangchow academic community, Shao Chin-han, to the Ssu-k'u project.
25. Li Wen-ts'ao, "Tu-men shu-ssu chih chin-hsi," reprinted in Yeh Tĕh-hui, *Shu-lin ch'ing-hua*, pp. 257–262.
26. See Cheng Chung-ying, trans. *Tai Chen's Inquiry into Goodness*, particularly "Editor's Introduction," pp. 17–30.
27. See Guy, "The Development of the Evidential Research Movement: Ku Yen-wu and the Ssu-k'u ch'üan-shu."
28. Sugimura Yuzō, *Kenryū kōtei*, pp. 140–141.
29. Sugimura Yuzō, *Kenryū kōtei*, pp. 14–15, 44; *Shih-lu* 184:7b–9a; and *ECCP* pp. 276–277.
30. Yao Ming-ta, *Chu Yun nien-p'u*, p. 3.
31. Kawata Teiichi, "Shindai gakujutsu no ichi sokumen—Shu In, Sō Shikan, Kō Ryōkitsu to shite Shō Gakusei," p. 104.
32. Yao, *Chu Yun nien-p'u*, p. 9. Chu was evidently somewhat curious about his family origins, but never expressed any desire to return to them

permanently. See "T'ung-hsing chung-huang hsien-sheng chih kuan chiao-yü ch'i fang-wen hsien-shih p'u-p'ai wei shu shu-tse," *Ssu-ho wen-chi*, pp. 147–148.

33. Yao Ming-ta, *Chu Yun nien-p'u*, p. 13.

34. William Hung, ed. *Tseng-chiao Ch'ing-ch'ao chin-shih t'i-ming pei-lu fu yin-te*, p. 101; *ECCP*, pp. 152–155, 805–807, 828.

35. For Tai Chen's reaction to this group, see "Tai Tung-yuan nien-p'u," *Tai Chen wen-chi*, p. 221. On Weng Fang-kang's connections with the group, see Chu Yun, "Ch'ü-fu yen-shih pen-tsang ch'ih-tu hsun," *Ssu-ho wen-chi*, pp. 75–76. On Yao Nai, see chapter five below. See also Yü Ying-shih, *Lun Tai Chen yü Chang Hsueh-ch'eng*, pp. 105–107.

36. Chu Kuei, "Chu-chün chu-kung shen-tao-pei," *Ssu-ho wen-chi*, p. vi; Yao Ming-ta, *Chu Yun nien-p'u*, p. 25. Chu wrote a summary of the volume, two *fu* on the campaign, and an introduction for a collection of poetry about it.

37. Chu Yun, "Pien-hsiu Chiang-chün mu-chih-lu," *Ssu-ho wen-chi*, pp. 234–236. The house was located in the southern portion of the city, just west of the Kwangtung *hui-kuan*.

38. Chu Yün, "Hsien-fu-chun hsing-chuang," *Ssu-ho wen-chi*, pp. 159–166.

39. See *Chu Yun nien-p'u*, pp. 129–141. The translation of the term *men-jen* is problematic. "Students" seems to me to convey the reality of the term in the eighteenth century more accurately than "disciples," with its biblical overtones, or "followers."

40. On Wu Lan-t'ing and Ch'a Pi-ch'ang, see Chang Hsueh-ch'eng, "Wu-fu-chun mu-pei," *Chang-shih i-shu*, 15.40a–42a, and Chu Yun, "Sung ch'a sheng pi-ch'ang chih kuan t'un-liu hsu," *Ssu-ho wen-chi*, pp. 85–87. Hsu Shen-shou was evidently an example of a student who stayed with Chu only briefly. See Chu Yun, "Yang Shu-jen mu-chih-lu," *Ssu-ho wen-chi*, p. 274.

41. Chang Hsueh-ch'eng, "Chiang Yü-ts'un p'ien-hsiu mu-chih-lu shu-hou," *CSCWC*, p. 270.

42. See chapter 2, and Araki Toshikazu, "Yōsei jidai ni okeru gakushinsei no kaikaku-shu to shite sono nin yōhō o chūshin to shite."

43. For descriptions of Chu's yamen in T'ai-p'ing, see David S. Nivison, *The Life and Thought of Chang Hsueh-ch'eng*, p. 39; and Susan L. M. Jones, "Hung Liang-chi (1746–1809): The Perception and Articulation of Political Problems in Late Eighteenth-Century China," pp. 55–56. For Chu's travels, see *Chu Yun nien-p'u*, p. 41.

44. On Pi Yuan, see *ECCP*, pp. 622–625; on Juan Yuan, see Benjamin Elman, "The Hsueh-hai-t'ang and the Rise of New Text Scholarship at Canton" and Peh-t'i Wei, "Juan Yuan: A Biographical Study with Special Reference to Mid-Ch'ing Security and Control in Southern China, 1799–1835," part I.

45. David S. Nivison, *The Life and Thought of Chang Hsueh-ch'eng*, p. 32.

46. Li Wei, "Ts'ung-yu-chi," in *Ssu-ho wen-chi*, pp. xxvi–xxvii.

47. Chang Hsueh-ch'eng, "Jen Ta-ch'un mu-chih-lu," *CSCWC*, p. 138.

48. The work is not extant.

49. Chu Yun, "Chiao-chin-fang hsiao-chi hsu," *Ssu-ho wen-chi*, pp. 83–84.

50. This personality style is vividly described by Judith Whitbeck in "The Historical Vision of Kung Tzu-chen (1792–1841)."

51. See, for instance, Chu Yun, "Han hsi-yu-hua-shan miao pei pa-wei," and "Kuan shun-t'ien-fu shu chu-pei chi," in *Ssu-ho wen-chi*, pp. 89–90, and 132–133.

53. Mao Ch'ang was a scholar of the Former Han dynasty whose recension of the *Shih-ching* became the standard text. The details of his life are sketchy; see *TMTY*, p. 293 (15.1b–2a). Chao Ch'i (ca. 110–200) is known for his commentary on the *Mencius;* see *TMTY*, p. 711 (35.1a–b). Ho Hsiu (d. ca. 180) was a specialist on the *Ch'un-ch'iu*, particularly the Kung-yang and the Ku-liang commentaries. Cheng K'ang-ch'eng (Cheng Hsuan, 127–200) was perhaps the most famous of all Han dynasty commentators; works on almost all the classics are attributed to him.

54. Hsu Shen (30–124) was the compiler of China's first etymological dictionary, *Shuo-wen chieh-tzu*, which was completed in 100.

55. Chu Yun, "Ch'üan-hsueh-p'ien," *Ssu-ho wen-chi*, pp. 78–79.

56. Chu Yun, "Shuo-wen chieh-tzu hsu," *Ssu-ho wen-chi*, p. 69.

57. Chu Yun, "Yü Chia Yun-ch'en lun shih-chi shu," *Ssu-ho wen-chi*, p. 145.

58. *Pan-li*, I, pp. 2b–3a. Punctuated in *Ssu-ho wen-chi*, p. 2. There was apparently little response to the emperor's initial edict on book collection, quoted in chapter 2. A second edict on the matter was therefore dispatched to provincial governors on 11 November. In accordance with imperial instruction, this second edict was relayed to education commissioners. Chu Yun was very likely responding to this second edict, rather than the first, with his memorial of December 1772.

59. The works which Chu Yun submitted by Tai Chen were *K'ao-kung chi-t'u* and *Ch'ü Yuan fu-chu*, neither of which were included in the imperial collection. The works submitted by Chiang Yung were the *Li-shu kang-mu* (*TMTY*, pp. 439–440 [22.a]), and the *Chou-li i i* (*TMTY*, pp. 385–386 [19.9b]). A total of sixteen works by Chiang Yung were included in the imperial collection, five of which dealt with rites. No works authored by Tai Chen were included in this collection, although many texts he reconstituted from fragments in the *Yung-lo ta-tien* were. On Chiang Yung and Tai, see *ECCP*, pp. 695–699 and Yü Ying-shih, *Lun Tai Chen yü Chang Hsueh-ch'eng*, pp. 169–178.

60. Chu submitted the *T'ung-ya* of Fang I-chih (*TMTY*, pp. 2500–2501 [119.3a]), the *Ku-shih-pi* by Fang Chung-te (*TMTY*, pp. 2877 [139.4a]), and the *Ku-chin shih-i* by Fang Chung-lü (*TMTY*, p. 2658 [126.7b]). Fang I-chih's book was printed in the imperial library, the works of his sons were only listed in the catalog, with the notation that they were not as good as the father's work. On the Fangs, see *ECCP*, pp. 232–233, and J. Willard Peterson, *Bitter Gourd: Fang I-chih and the Impetus for Intellectual Change*. The Ssu-k'u editors made some interesting remarks on the origins of *k'ao-cheng* scholarship during the Ming in the course of their review of Fang I-chih's *T'ung-ya*.

61. The imperial catalog included notices of nineteen works by Mei Ting-tso,

five works by Shih Jun-chiang, and three works by Wu Hsiao-kung, but did not mention the works Chu Yun submitted by each author, which were the *Hsuan-hsueh ch'üan-shu* by Mei, the *Yü-shan-chi* by Shih and the *Shu-nan chi* by Wu. Chu Yun submitted the *Chu-shu t'ung chien* (*TMTY*, p. 1024 [47.1b]), and the *Shan-ho liang-chieh k'ao* (*TMTY*, p. 1157 [72.3a]) by Hsu Wen-ching. on Hsu, see *ECCP*, pp. 326–327.

62. Chang Hsueh-ch'eng, "Chu Hsien-sheng pieh-chuan," *CSCWC*, p. 130.
63. Chu's memorial may be found in *Pan-li*, I, 3a–4b; there is a punctuated version in *Ssu-ho wen-chi*, pp. 2–4. No other memorials from provincial education commissioners dealing with matters of substance are extant in the National Palace Museum Archives from 1770 until the end of the reign.
64. Cheuk-woon Taam, *The Development of Chinese Libraries Under the Ch'ing Dynasty, 1644–1911*, pp. 59–61.
65. David S. Nivison, *The Life and Thought of Chang Hsueh-ch'eng*, pp. 79–80.
66. Chou Yung-nien, "Ju-ts'ang shuo," p. 2a. I have chosen to use the reprint of this essay in the *Sung-lin ts'ung-shu*. There is also a punctuated version of the text in Yang Chia-lo, comp., *Ssu-k'u ch'üan-shu kai-shu* (Nanking 1931), "Wen-hsien," pp. 9–11, but this text has quite a few misprints and variants.

   The date of the essay is uncertain. It was written as an introduction to the catalog of Chou Yung-nien's library, which was entitled *Chi-shu-yuan shu-mu*, and seems to be no longer extant. Chang Hsueh-ch'eng apparently wrote a preface for the volume when he visited Chou in 1775, in which he referred to the ideas, if not the text of the essay now known as "Ju-ts'ang-shuo." See Chang Hsueh-ch'eng, "Chou Shu-ch'eng chuan," *Chang-shih i-shu* 18.25a–26b.

   For the story of the foolish old man trying to move the mountain stone by stone, see the *Lieh-tzu*, T'ang-ho Chapter.
67. "Ju-tsang-shuo," pp. 3a–b; Nivison, *The Life and Thought of Chang Hsueh-ch'eng*, pp. 77–78.
68. Lionel Giles, "A Note on the *Yung-lo ta-tien*," pp. 137–143.
69. Ch'üan Tsu-wang, "Ch'ao *Yung-lo ta-tien* chi," *Chi-chi t'ing-chi wai-p'ien*, pp. 2757–2763.
70. Quoted in Tuan Yü-ts'ai, *Tai Tung-yuan nien-p'u*, pp. 228–229.
71. Two institutions of Han scholarship.
72. Liu Chih-chi (661–721) was a compiler of the T'ang imperial catalog. Tseng Kung (fl. 1058) worked on the Sung imperial catalog.
73. On the differences between these two systems of cataloging, see below.
74. David S. Nivison, *The Life and Thought of Chang Hsueh-ch'eng*, pp. 77–78, 156, 171–174.
75. This language is from Ku Kuang-ch'i's 1812 preface to Ch'in Ssu-fu's book catalog *Shih-yen chai shu-mu*, as quoted in Yao Ming-ta, *Chung kuo mu-lu-hsueh shih*, p. 413.
76. See, for instance, Shao Chin-han's review of the *Shih-chi chih-chieh* discussed in chapter five below. Some nineteenth-century scholars felt that the Ssu-k'u editors had not devoted enough attention to the problems of editions; one, Shao I-ch'en (1810–1860) prepared a set of notes on the various editions available of the books listed in the *Ssu-k'u ch'üan-shu chien-ming mu-lu,*

which was later published under the title *Ssu-k'u chien-ming mu-lu piao-chu.* See Teng and Biggerstaff, *An Annotated Bibliography of Selected Chinese Reference Works,* p. 25.

77. On Ou-yang Hsiu and Chao Ming-ch'eng, see note 78 below. Chu Yun appears to refer here to the *San-li t'u chih-chieh* (*TMTY,* pp. 431–432 [22.1a]) by Nieh Chung-yü (fl. 960), and the *K'ao-ku t'u* and *Hsu k'ao-ku-t'u* (see *TMTY,* pp. 2393–2395 [115.1b]) by Lü Ta-lin (fl. 1090).

78. The catalogs of Ou-yang Hsiu and Chao Ming-ch'eng are the first two entries in the section of *TMTY* devoted to stone inscriptions.

79. For convenient summaries of the history of epigraphy in China, see Ch'ien Ta-hsin's preface to his *Chin-shih wen-tzu mu-lu,* Lung Family reprint (Changsha 1884); and Benjamin Elman, *From Philosophy to Philology,* pp. 188–191.

80. Wang Ch'ang, *Chin-shih ts'ui-p'ien hsu,* 2a.

81. Weng Fang-kang, *Liang Han chin-shih chi* 1.1a.

82. Sun Hsing-yen, *Huan-yü fang-pei-lu,* p. 1.

83. On the economic position of eighteenth-century scholars, see Susan Jones, "Hung Liang-chi," pp. 53–55, and also her article, "Scholasticism and Politics in Late Eighteenth Century China," pp. 28–49.

84. David S. Nivison, *The Life and Thought of Chang Hsueh-ch'eng,* p. 45.

Chapter 4. Scholars and Bureaucrats at the Ch'ien-lung Court: The Compilation of the Ssu-k'u ch'üan-shu

1. The issue of roles in Chinese bureaucracy was first raised by Max Weber in *The Religion of China.* Weber saw office in China essentially as a reward for success in the Confucian quest for moral cultivation. The "office prebendary," he argued, was an "ethically hallowed form. It is the one position becoming a superior man because the office alone allows for the perfection of personality." (Max Weber, *The Religion of China,* trans. Hans H. Gerth, p. 160). Thomas Metzger, in his recent work on the subject, particularly chapter one of *The Internal Organization of the Ch'ing Bureaucracy,* and chapter four of *Escape from Predicament,* offers a more complex view in which intellectual training, practical considerations, and moral conceptions all play a part in shaping literati roles and orientations toward official life.

2. Some sense of the variety of people in the Hanlin Academy is conveyed in Adam Yuen-chung Lui's book, *The Hanlin Academy: Training Ground for the Ambitious, 1664–1850.*

3. Benjamin I. Schwartz, "Some Polarities in Confucian Thought," p. 5.

4. I am grateful to Professor Suzuki Chusei for this formulation (private conversation, November 1976).

5. On the origins and early functioning of the Grand Council, see Silas H. L. Wu, *Communications and Control in Imperial China,* and Fu Tsung-mao, *Ch'ing-tai chün-chi ch'u tzu-chih chi ch'i chih-chang yen-chiu,* pp. 338–418.

6. The grand councillors during the Ch'ien-lung period are listed in Fu, *Ch'ing-tai chün-chi,* pp. 535–591. On Fu-heng, see *ECCP,* pp. 252–253, *Ch'ing-shih,* pp. 4116–4121, and *KCCHLC* 29.5a–25b.

7. On Liu, see *ECCP*, pp. 533–534, and *KCCHLC* 21.22a–42a. For Liu's reputation for incorruptibility, see the comments of Chao-lien in *KCCHLC* 21.42a. Liu's accusation of Chang T'ing-yü led to Chang's resignation as president of the Board of Personnel. Although the emperor approved of Liu's indictment, he did not accept Chang's resignation. See *Shih-lu* 156.7a–11a.

8. On Yü, see *ECCP*, pp. 942–944; *Ch'ing-shih*, pp. 4242–4243; and *KCCHLC* 27.1a–7b.

9. Most discussion memorials were submitted fairly promptly after they were requested by the emperor. Of course, the New Year's holiday fell during the period Chu's memorial was being discussed, but this should not have unduly delayed Council action. On discussion memorials, see Beatrice S. Bartlett, *The Vermillion Brush: Grand Council Communications System and Central Government Decision Making in Mid-Ch'ing China*.

10. Yao Nai, *Chu-chün hsien-sheng pieh-chuan*, *KCCHLC* 128.35a.

11. Yao Ming-ta, *Chu Yün nien-p'u*, pp. 14–34.

12. Chang Hsueh-ch'eng, "Chu Yun pieh-chuan," *CSCWC*, p. 129. See also *Chu Yun nien-p'u*, p. 16.

13. Chang Hsueh-ch'eng, "Chu Yun pieh-chuan," *CSCWC*, p. 129.

14. Chu Kuei, "Chu-chün chu-kung shen-tao-pei," in *Ssu-ho wen-chi*, p. 6.

15. Chang Hsueh-ch'eng, "Chu Yun pieh-chüan, *CSCWC*, p. 136.

16. Li Wei, "Ts'ung-yu chi," in *Ssu-ho wen-chi*, p. 23. This story is often taken as evidence of Liu Tung-hsün's incorruptibility and Chu Yün's observance of the traditional rules of the Ch'ing court. (See Susan L. M. Jones, "Hung Liang-chi," p. 47). It should be contrasted with the relations that existed between Yü Min-chung and the Ssu-k'u revisers, described below.

17. *YMCSC*, p. 54.

18. Wang Lan-yin, "Chi Hsiao-lan hsien-cheng nien-p'u," p. 100.

19. *Pan-li*, I, pp. 5b–7a.

20. The preface and subject headings for Juan Hsiao-hsu's catalog of books was one of the earliest treatises on book cataloging extant in the eighteenth century. See Juan Hsiao-hsu *Ch'i-lu hsu-mu*.

21. Ma Tuan-lin, *Wen-hsien t'ung-k'ao, chüan* 1974 and 207. *TMTY*, pp. 1702–1703 (81.3b–4a).

22. Unfortunately, the Ssu-k'u project did not achieve this.

23. The allusion was to the *Han-shu*, "Treatise on Bibliography," 30.1701.

24. See above, chapter 2.

25. For the complex history of the writing of the *Ch'ün-chai tu-shu-chih*, see *TMTY*, pp. 1777–1778 (85.1b), and Ssu-yu Teng and Knight Biggerstaff, *An Annotated Bibliography*, pp. 15–16.

26. I am grateful to Beatrice S. Bartlett for this observation.

27. *TMTY*, "Hsü" (Sheng-yü), p. 2.

28. *Pan-li*, I, p. 8a–b.

29. The "meng" hexagram was traditionally listed fourth in the *I-ching;* the "Ta-tung" Ode was #203 in the Mao listing; and the "tung-kuan" section of the *Chou-li* was a part of the "K'ao-kung chi-t'u" which was traditionally appended to the end of the *Chou-li*.

30. *Pan-li,* I, p. 7b.
31. Yao Ming-ta, *Chung-kuo mu-lu hsueh-shih* p. 199.
32. *Pan-li,* I, p. 7b.
33. Tuan Yü-ts'ai, *Tai Tung-yuan nien p'u,* in *Tai Chen wen-chi,* p. 233.
34. Chang Hsueh-ch'eng, "Chu Yun pieh-chuan," *CSCWC,* p. 130.
35. *TMTY,* "Hsu" (*Chih-kuan*), pp. 1–25.
36. Kuo Po-kung, *Ssu-k'u ch'üan-shu tsuan-hsiu k'ao,* p. 60, and Yeh Te-hui, *Shu-lin ch'ing-hua,* p. 240 both interpret the personnel list in this manner.
37. Based on a sample of eleven revisers whose age can be determined.
38. Adam Yuen-chung Lui, *The Hanlin Academy,* p. 127. I am grateful to Dr. Lui for making his doctoral thesis available to me, and for showing me how this method might be applied to my study.
39. Data on the ranks attained by Hanlin bachelors is from Chu Ju-chen, comp., *Tz'u-lin chi-lueh* 4.31a–35b. The same was evidently true of Sung dynasty compilation projects. See John H. Winkelman, "The Imperial Library in Southern Sung China, 1127–1279," pp. 22–23.
40. *Pan-li,* I, pp. 54a, 86a–b.
41. Wang Tseng, comp., *Hsin-ch'a hsien-chih,* "Hsu", *1b*–2a, and Hung Liang-chi, comp., *Huai-ch'ing fu-chih* 15.6a. Wang says in his autobiographical preface to *Hsin-ch'a hsien-chih* that he was demoted after ten years' service in the Hanlin Academy. This must be an approximate figure, however. According to the *Huai-ch'ing fu-chih,* Wang was appointed assistant prefect in 1787, twelve years after his appointment to the Hanlin. Shao Chin-han briefly alludes to Wang's troubles in a letter written in 1777 (See *SCHNP* 57).
42. *Ch'ing-shih lieh-chüan* 75.37a–38a.
43. *Ch'ing-shih* 38.4453.
44. "Wu Sheng-lan kuo shih-kuan pen-chuan," *KCCHLC* 97.12b–13a. See also Susan Jones, "Hung Liang-chi," pp. 169, 192.
45. Adam Lui, *The Hanlin Academy,* pp. 102–109, *passim.*
46. I am grateful to Professor Ch'en Chieh-hsien of the National Taiwan University for this information. For some sense of Hanlin bachelors' attitudes toward these commissions, see Wu Ching-tzu, *The Scholars,* Gladys Yang and Yang Hsien-yi, translators, p. 128.
47. "Wu Sheng-lan kuo-shih-kuan pen-chuan," *KCCHLC* 97.13a.
48. *Ch'ing-shih* 363.4506.
49. Ibid., 363.4506.
50. Yao Ming-ta, *Chu Yün nien-p'u,* p. 70.
51. Yao Jung, "Yao Nai hsien-sheng hsing-chuang," *KCCHLC* 146.17a.
52. *Pan-li,* I, p. 7b.
53. *Pan-li,* I, pp. 9a, 91a.
54. *Pan-li,* I, pp. 54a, 65b, 70b.
55. Wang Tsung-yen, "Sun Ch'en-tung mu-chih-lu," *KCCHLC* 130.12b–13a.
56. *Pan-li,* I, pp. 7a, 9b–10a, 55b, 91a.
57. *Pan-li,* I, p. 7b.
58. Morohashi Tetsuji, *Dai kanwa jiten* 5:311–312.
59. *Pan-li,* I, p. 91a.
60. See *ECCP,* p. 543. Lu-fei Chih's association with the actual work of

copying texts is strengthened by the fact that when the compilation was finished, he was held accountable for errors in the manuscript. See *Pan-li*, I, 88a.

61. A-kuei, for instance, was out of the capital during most of the compilation process.

62. Ch'en Yuan, ed., *Yü Min-chung shou-cha.* I am grateful to Professor Yin Chiang-i of Tung-wu University (Taipei) for his assistance in reading the handwriting of these letters.

63. *YMCSC,* pp. 7–8.

64. On book bureaus, see *KCT:CL* 029551 (Kao-chin, 15 September 1774); *KCT:CL* 029837-1 (Ho Wei, 14 October 1774); *KCT:CL* 029453 (Sa-tsai, 7 September 1774); *KCT:CL* 029496 (San-pao, 11 September 1774); *KCT:CL* 029802 (San-pao, 12 October 1774); *KCT:CL* 029861 (Hai-ch'eng, 16 October 1774); *KCT:CL* 030000 (Yu Wen-i, 26 October 1774); *KCT:CL* 030178 (Tê-pao, 7 November 1774). In *KCT:CL* 029792 (Li Hu, 12 October 1774) Governor Li Hu reported that there was no need for a book bureau in Yunnan.

65. Actually, four offices were designated by this name: prefectural directors of schools (*chiao-shou*), rank 7A; departmental directors of schools (*hsueh-cheng**), rank 8A; district directors of schools (*chiao-yu*), rank 8A; and subdirectors of schools (*hsun-tao*), rank 8B. See Brunnert and Hagelstrom, *Present Day Political Organization of China,* pp. 431–435, passim.

   The Yung-cheng Emperor made a number of changes in the qualifications and regulations for this post. For an analysis of these, and a discussion of the office in general, see Araki Toshikazu, "Choku-sho kyogaku no sei o tsujite mitaru Yosei chika no bunkyo seisaku."

66. There are at least two sources on the number of books submitted by each province for the *Ssu-k'u ch'üan-shu.* Yang Chia-lo in *Ssu-k'u ch'üan-shu kai-shu,* "Wen-hsien," pp. 154–156 gives one set of figures, with no sources. In the Palace Museum Archives, there is a set of memorials in which provincial governors responded to an imperial order to report on the number of books they had submitted; the series of memorials is, however, incomplete. The figures given in extant provincial memorials do not always agree with Yang Chia-lo; usually Yang's figures are lower, and so may represent an earlier count. Given the difficulties of using either set of numbers, I have decided to use here the numbers of books actually reprinted in the *Ssu-k'u ch'üan-shu,* rather than those submitted for consideration by the provinces. The number of books submitted by a province which were reprinted in the compilation can be fairly easily and reliably counted, for the origin of books is noted in the *Tsung-mu t'i-yao.* I rely on Yang Chia-lo's count, based on the catalog, also presented in *Ssu-k'u ch'üan-shu kai-shu,* pp. 154–156.

67. *Pan-li*, I, p. 13a–b. On Li Chih-ying and his negotiations with salt merchants, see Preston Torbert, *The Ch'ing Imperial Household Department,* p. 108. On the numbers of books submitted by private collectors and included in the compilation, see Yang Chia-lo, *Ssu-k'u ch'üan-shu kai-shu,* "Wen-hsien," pp. 156–160.

68. *Pan-li*, I, p. 25a–26b. See also *ECCP,* pp. 230, 559, 612 and 810.

69. For the emperor's orders on the reporting and disposition of borrowed and

purchased books, see *Pan-li,* I, pp. 46a–47b. The provincial governors' reports are (Chekiang) *KCT:CL* 032228 (San-pao, 5 September 1777); (Kiangsu) *KCT:CL* 032278 (Yang Kuei, 11 September 1777); (Shantung) *KCT:CL* 032070 (Kuo-t'ai, 20 August 1777); (Honan) *KCT:CL* 032107 (Hsu Ch'ien, 23 August 1777); (Shansi) *KCT:CL* 031868 (Pa-san-t'ing, 28 July 1777): Yin Cho's memorial on behalf of the salt merchants in *KCT:CL* 031774 (Yin-cho, 19 July 1777). The books which Ma Jung submitted to the *Ssu-k'u ch'üan-shu* were returned to him, and thus not included in those to which Yin Cho referred.

70. *YMCSC,* pp. 57–58.
71. Yeh Te-hui, *Shu-lin ch'ing-hua,* p. 240. I follow Yeh in his interpretation of these documents. Note, however, that a memorial of Prince Yung Jung, quoted below, implies that copying was done before collation. The point is not a vital one; for consistency's sake, I will follow Yeh.
72. *YMCSC,* pp. 119, 38.
73. Weng Fang-kang, *Weng-shih chia-shih lueh-chi,* pp. 36b–37a.
74. Lee M. Sands, "The Liu-li Ch'ang Quarter: Potters and Booksellers in late Ch'ien-lung," pp. 35–39.
75. Yao Ming-ta, *Chu Yun nien-p'u,* pp. 69–70; Chu Yun, "Ts'ao-ch'ao hsiu-hsu," *Ssu-ho wen-chi,* pp. 84–85.
76. Chang Hsueh-ch'eng, "Chu hsien-sheng pieh-chuan," *CSCWC,* p. 130.
77. Kuei Fu, "Chou Yung-nien chuan," *KCCHLC* 130.30b. Kuei, a *chin-shih* of 1790, was one of the copyists Chou hired.
78. The *Chiu-kuo-chih* by Lu Chen was one such text. Evidently by mistake, the work was not listed in *TMTY.* See Juan Yuan, *Ssu-k'u wei-shou shu-mu,* (with *TMTY*), p. 2.
79. Two such texts were the *Nan-hu-chi* and the *Chiu wu-tai-shih* (*TMTY,* pp. 3370 [160.8a], and 1005 [46.2b]). See *SCHNP,* p. 60, which quotes a brief reminiscence by Chu Wen-tsao entitled "Shu nan-hu-chi hou," and the preface to the 1975 Chung-hua shu-chu edition of *Chiu wu-tai-shih.*
80. *Pan-li,* I, p. 27a.
81. *Pan-li,* I, p. 28a.
82. *Pan-li,* I, pp. 29b, 35b; Kuei Fu, "Chou Yung-nien chuan," *KCCHLC* 130.30b.
83. Miao Ch'üan-sun, "Yung-lo ta-tien-k'ao," in *I-feng-t'ang hsu-chi,* 4.2a–3b.
84. Sun Chieh-ti, "Lun chiao-yü-pu hsuan-yin Ssu-k'u ch'üan-shu," quoted in Kuo Po-kung, p. 230.
85. *Pan-li,* I, p. 23a.
86. Huang Fang (Colophon to the letters of Yü Min-chung), *YMCSC,* p. 126.
87. *YMCSC,* p. 86.
88. *YMCSC,* pp. 47–48.
89. One reason the emperor ordered the compilation of the *Ssu-k'u hui-yao* was that he feared he would not live to see the completion of the *Ssu-k'u ch'üan-shu.* See Kuo Po-kung, pp. 198–199.
90. *YMCSC,* pp. 86–88.
91. *TMTY,* pp. 1044–1046 (47.10a). The *t'i-yao* for this volume was evidently drafted by Shao Chin-han. See *TYFTK,* pp. 47a–48b.
92. *YMCSC,* p. 50.

93. Yao Nai, *Yao Hsi-pao hsien-sheng ch'ih-tu* 2.1b–2a. I assume Yao's time reference in this letter, *tsai kuan-chung*, refers to his service at the Ssu-k'u Commission. For Yao's biography, see chapter 4.
94. *YMCSC*, p. 34.
95. Chang Hsueh-ch'eng, "Chou Shu-ch'ang pieh-chuan," *CSCWC*, p. 151.
96. *Pan-li*, I, pp. 18b–20a.
97. Yeh Te-hui, *Shu-lin ch'ing hua*, p. 240.
98. These tendencies had their roots, of course, in the nature of the Ch'ing
emperorship. See Silas Wu, *Communications and Control in China*, pp. 107–123.
99. *Pan-li*, I, p. 23a.
100. On Ts'ao Hsiu-hsien, see *Ch'ing-shih* 321.4255; on Wang Chi-hua, see "Kuo-
shih-kuan pen-chuan," *KCCHLC* 88.38a–40a; on Ts'ai Hsin, see *Ching-shih*
321.4248.
101. *Pan-li*, I, pp. 32b, 62a.
102. *YMCSC*, p. 48.
103. *YMCSC*, p. 69. On *Yi-lin*, see *TMTY*, p. 2579 (123.3b).
104. *ECCP*, p. 228.
105. Ho Shen's biography in *ECCP* touches briefly on this episode. For more
complete documentation, see *Shih-lu*, pp. 16182–16258, *passim*, and
SYT(FP), Ch'ien-lung 45 (1781–82), Fall, pp. 117–381, *passim*. The case is
of particular interest since it is one of the few cases in which Ho Shen's
activities can be documented.
106. *SYT(FP)*, Ch'ien-lung 44 (1779–1780), Winter, p. 381.
107. *Pan-li*, I, pp. 74b, 69b–70a. *SYT(FP)*, Ch'ien-lung 46 (1781–82), Spring, p.
207.
108. *TMTY*, p. 3674 (172.6b). See also L. Carrington Goodrich and Fang
Chao-ying, eds., *Dictionary of Ming Biography* 2:1182–1885; and L. Carrington Goodrich, *The Literary Inquisition*, pp. 192–193.
109. See chapter two and David S. Nivison, "Ho Shen and his Accusers," pp.
228–229.
110. The originals of these lists are found in *SYT(FP)*. Ch'en Yuan has counted
the number of errors charged to each editor, and recorded his findings at
the end of *Pan-li*.
111. *SYT(FP)*, Ch'ien-lung 46 (1781–1782), Spring, p. 195.
112. The totals for all the years of the project are as follows:

Year	Total Errors Charged
CL 42 (1777–78)	89
CL 43 (1778–79)	221
CL 44 (1779–80)	259
CL 45 (1780–81)	2118
CL 46 (1781–82)	5006
CL 47 (1782–83)	7072
CL 48 (1783–84)	12,303
CL 49 (1784–85)	3235

113. Sun Ch'en-tung ws charged with twenty-five errors in the summer of 1781.
He died on 30 September 1780. (See Wang Tsung-yen, "Sun Ch'en-tung
mu-chih-lu," *KCCHLC* 130.12b–13a.)

114. *Pan-li*, II, pp. 2b–3a.

115. *Pan-li*, II, p. 5b.

116. *KCT:CL* 052120 (Chi Hsiao-lan, 3 December 1787).

117. *KCT:CL* 052115 (Ch'üan Tê and Tung Ch'un, 3 December 1787); *KCT:CL* 052736 (Chi Hsiao-lan, 8 January 1788). See also Huang Fang [Colophon to the letters of Yü Min-chung], *YMCSC*, p. 126.

118. See Susan Jones, "Hung Liang-chi," pp. 85–86, 156–203, *passim*. On Ch'en, see *Ch'ing-shih* 363.4506; on Yin, see David S. Nivison, "Ho Shen and his Accusers," p. 234; on Mo, see "(Mo Ch'an-lu) Kuo-shih kuan pen-chuan," *KCCHLC* 102.1a–6a

119. On the location of the manuscripts, see above. The exact number of *ts'e* per set varies slightly. See William Hung, "Preface to an *Index of the Ssu-k'u ch'üan-shu tsung-mu and Wei-shou shu-mu.*" One set is extant today in the vaults of the National Palace Museum (Taipei). Two sets are extant in the People's Republic, one in Peking, the other in Hangchow.

In 1936, the Commercial Press began a project to reprint the Ssu-k'u collection.

120. Reprinted in 1964 by the Chung-hua shu-chü (Shanghai) in a 1033 page edition.

121. Two copies of this were made, one of which is extant in the vaults of the National Palace Museum (Taipei). The table of contents of the *Ssu-k'u ch'üan-shu hui yao* is reprinted in Wu Che-fu, *Ssu-k'u ch'üan-shu hui-yao tsuan-hsiu k'ao*, pp. 131–214.

122. Reprinted by the Kuang-ya shu-chü in 700 *ts'e* in 1899.

123. The two most common editions of this work are the one cited throughout this thesis, *TMTY*, which was first produced by the Commercial Press in 1934 and has been reprinted many times since; and the lithographic edition brought out by the Ta-tung shu-chü in 1930. There are two major indices: William Hung, et al., *Ssu-k'u ch'üan-shu chi wei-shou-mu yin-te;* and Wang Yun-wu et al., *Ssu-k'u ch'üan-shu tsung-mu t'i-yao shu-ming chi tso-che so-yin*, which is printed at the end of the Commercial press edition of *TMTY*.

124. Yü Chia-hsi, *Ssu-k'u t'i-yao pien-cheng*, pp. 52–53.

125. Wang T'ai-yueh, *Ssu-k'u ch'üan-shu k'ao-cheng;* Juan Yuan, *Ssu-k'u wei-shou shu-mu t'i-yao;* and Shao I-chen, *Ssu-k'u chien-ming mu-lu piao-chu*.

126. Miao Ch'üan-sun, "Hsu," in Shao I-chen, *Ssu-k'u chien-ming mu-lu piao-chu*, p. 1.

127. Yü Yueh, "Ch'un-tsai-t'ang ch'ih-tu," quoted in Liu Chao-yu, "Min-kuo i-lai ti Ssu-k'u hsueh," p. 146.

128. L. Carrington Goodrich, *The Literary Inquisition*, pp. 5–6.

129. Ou-yang Tzu, "Ou-yang hsu," in *Ssu-k'u mu-lueh*, p. 1.

130. "Hsu," *Hsu-hsiu ssu-k'u ch'üan-shu t'i-yao*, pp. 2–3; and Wu Che-fu, "Hsien-ts'un hsu-hsiu Ssu-k'u ch'üan-shu t'i-yao mu-lu hou-chi," pp. 29–30.

131. Ch'ang P'i-te, "Why the Printing of the Entire Ssu-k'u ch'üan-shu is Significant," privately distributed, 1983.

132. *Pan-li*, I, p. 7a.

133. *YMCSC*, p. 24.

134. *YMCSC*, p. 36.

135. *YMCSC,* p. 81.

136. *Pan-li,* I, p. 1b.

137. *TMTY,* "Fan-li," p. 5.

138. Yao Ming-ta, *Chung-kuo mu-lu-hsueh shih,* p. 200.

139. *YMCSC,* p. 19.

140. *YMCSC,* p. 75.

141. *YMCSC,* p. 33.

142. Kuo Po-kung, *Ssu-k'u ch'üan-shu tsuan-hsiu-k'ao,* p. 227.

143. *TMTY,* p. 3019 (145.2a).

144. This is obviously a judgment on my part. The precise comparison of the listings of the Ssu-k'u with earlier book collections is complicated by the different purposes of the various collections, and by the vagaries of textual history.

145. Chih-hsu, editor, *Yueh-ts'ang chih-ching* (A bibliographic guide to the tripitaka).

146. See the review of *Fa-yuan chu-lin* (*TMTY,* p. 3018 [145.1b]).

147. See the review of *Cheng-teng hui-yuan* (*TMTY,* p. 3022 [145.3a]).

148. *TMTY,* p. 3034 (146.3b).

149. *TMTY,* p. 3029 (146.1a). The concluding note for the Buddhist section in Chiao Hung's *Kuo-shih ching-chi chih* (A Treatise on bibliography for the national history), pp. 145-146, makes a similar point about the disorderly character of Taoist writings, but does not stress the government's responsibility to clear up this disorder.

150. On the traditional organization of the Taoist canon, see Ofuchi Ninji, "The Formation of the Taoist Canon," pp. 253-267; and Liu Tsun-yan, "The Compilation and Historical Value of the Tao-tsang," pp. 104-119. I am grateful to Judith Boltz for her very helpful comments on this issue.

151. The *Yun-chi ch'i-chien.* See *TMTY,* p. 3055 (146.12a).

152. See the review of the *Tao-te-ching-chieh* (*TMTY,* 1.3033 [146.3a]).

153. See the review of the *Tao-te-ching-chu* (*TMTY,* p. 3036 [146.4a]).

154. *TMTY,* p. 3047 (146.8b).

155. *TMTY,* "Sheng-yü," p. 9. Translated in L. Carrington Goodrich, *The Literary Inquisition,* p. 147.

156. *YMCSC,* p. 74.

157. *YMCSC,* p. 110. The text in question was the *San-ch'ao pei-meng hui-pien* (*TMTY,* pp. 1070-1071 [49.1b]).

158. *YMCSC,* p. 26.

159. *Shih-lu,* p. 17204 (1174.8b-9a).

160. Their reports are preserved in the Ch'ing archives at the National Palace Museum (Taipei), and afford an interesting study in book circulation in the eighteenth century.

161. Chang Ping-lin, "Ai chin-shu," *Chien-lun* 4.17b.

162. Only seven individuals had more than twenty works listed in the *Annotated Catalog:* Chu Hsi (1130-1200), 28 works; Ch'en Chi-ju (1558-1639), 31 works; Lu Shen (1477-1544), 20 works; Wang Shih-chen (1634-1711), 32 works; Wei I-chieh (1616-1686), 24 works; Yang Shen (1488-1559), 35 works.

163. Chang Ping-lin, "Jih-chih-lu chiao-chi hsu," in Hsu Wen-ts'e editor, *Yuan-ch'ao-pen jih-chih-lu* (Taipei, 1958).

164. *TMTY,* pp. 1005–1007 (46.2b); For evidence of imperial interest in this text, see *YMCSC,* pp. 80, 83, and 84.
165. *SCHNP,* p. 53; For Shao's biography, see chapter 5.
166. *SCHNP,* p. 54.
167. Ch'en Yüan, *Chiu wu-tai-shih chi-pen fa-fu san-chüan, passim.*

Chapter 5. Reviewing the Reviewers: Scholarly Partisanship and the Annotated Catalog

1. Chiang Fan (1761–1831) claimed in the early nineteenth century that all the reviews "from the broadest [discussions of] classics, history, philosophy and literature, to the narrowest noted on medicine and divination" were the work of Chi Hsiao-lan himself. See Chiang Fan, "Chi Yun," *Han hsueh shih-ch'eng chi* 6.1b. Li Tz'u-ming argued in a note included in his *Yueh-man-t'ang pi-chi* (Desultory jottings from the Yueh-man Hall) that Chi's intellectual background was not broad enough for him to have written all the reviews. Li credited Chi with the reviews in the literature division only, attributing the reviews in classics, history and philosophy to Tai Chen, Shao Chin-han and Chou Yung-nien, respectively. (Li's note is quoted in Wang Lan-yin "Chi Hsiao-lan hsien-sheng nien-p'u," p. 95.)
2. Juan Yuan, "Hsu," *Ssu-k'u wei-shou shu-mu t'i-yao,* reprinted in *TMTY,* volume 5.
3. Shao Chin-han, "Ssu-k'u ch'üan-shu t'i-yao fen-tsuan-kao," in Ma Yung-hsi, comp., *Shao-hsing hsien-cheng i-shu.*
4. Yao Nai, "Hsi-pao hsuan shu-lu," in Mao Yü-sheng, *Hsi-pao-hsuan i-shu san-chung.*
5. Prefaces to the work by Miao Ch'üan-sun, Hu Ssu-ching and Weng himself are extant. See Miao Ch'üan-sun, reprinted in *I-feng-t'ang wen-chi, I-feng-t'ang-wen man ts'un,* 4:16a–b, Weng Fang-kang, *Fu-ch'u chai wen-chi,* pp. 1388–1389; and Hu Ssu-ching, *T'ui-lu ch'üan-chi* pp. 339–340. Liu Ch'eng-kan mentions seeing these drafts in a preface to the *Ssu-k'u ch'üan-shu piao-wen chien-shih,* dated summer, 1915, and quoted in Yang Chia-lo, editor, *Ssu-k'u ch'üan-shu kai-shu,* "Wen-hsien," pp. 11–25.
6. Liu Ch'üan-chih, "Chi wen-ta kung i-chi hsu," quoted in Wang Lan-yin, "Chi Hsiao-lan hsien-sheng nien-p'u," p. 95.
7. Both Liu I-an and Immanuel C. Y. Hsu, in his translator's preface to *Intellectual Trends During the Ch'ing Period,* suggest that the k'ao-cheng movement of the Ch'ien-lung and Chia-ch'ing periods represented an effort by scholars to avoid political repression by burying themselves in old classics. Hou Wai-lu in *Chung-kuo chin-tai ssu-hsiang hsueh-shuo shih* argues that Tai Chen expressed his resentment of Manchu oppression in his criticisms of Mencius.
8. Chang Hsueh-ch'eng, "Huang-Ch'ing lieh-feng shu-jen Shao-shih Yuan chu-jen mu-chih-lu," *CSCWC,* p. 59. Shao's father evidently devoted his life to the study of the *I-ching.* See Ch'ien Ta-hsin, "Tseng Shao Chin-han hsu," *Ch'ien-yen-t'ang wen-chi* 23.333–334.
9. *SCHNP,* pp. 4, 5–6.
10. *ECCP,* p. 639.
11. Chu Yün, "Shao nien-lu hsien-sheng mu-piao" in *Ssu-ho wen-chi,* 11.203–204.

12. Ch'ien Ta-hsin, "Shao-chün mu-chih-lu," in *Ch'ien-yen t'ang wen-chi* 43.686.
13. *SCHNP,* pp. 11–19, passim.
14. Wang Hui-tzu, "P'ing-ta-meng lang-lu," quoted in *SCHNP,* pp. 14–15.
15. *SCHNP,* p. 15.
16. *SCHNP,* p. 18.
17. Ch'ien Ta-hsin, "Shao-chün mu-chih-lu." One of Shao's examiners, Liu T'ung-hsun, was a close friend of Chu Yun's. According to Li Wei's biography of Chu, Shao's success in the examination was due to the fact that Chu expressed his admiration of Shao to Liu. See Li Wei, "Ts'ung-yu-chi," in *Ssu-ho wen-chi,* p. xxiv.
    Shao was ranked thirtieth in the examination. Fourteen people with ranks below Shao's were appointed to the Academy. See William Hung, ed., *Tseng-chiao Ch'ing-ch'ao chin-shih t'i-ming pei-lu fu yin-te,* p. 114.
18. Ch'ien Ta-hsin, "Tseng Shao Chih-nan hsu."
19. *SCHNP,* p. 20. Shao met Chang Hsueh-ch'eng at T'ai-p'ing, and the two formed a life-long friendship. In Chang's *I-shu,* there are eight letters to Shao and a biography of him; discussions of Shao also are found in several of Chang's "family letters." Based on these documents, Nivison comes to the conclusion that Shao was "taciturn, cautious, perhaps even dull compared with Chang, who (as he admitted) spilled out whatever strange speculations came into his head. Evidently the two needed each other." (Nivison, *Chang Hsueh-ch'eng,* p. 51). Chang's letters are useful for describing Chang, but in the absence of any responses from Shao, it may be unfair to generalize from them about Shao.
20. *SCHNP,* pp. 50–54.
21. *SCHNP,* p. 30.
22. Ch'ien Ta-hsin, "Shao-chün mu-chih-lu."
23. On the *Erh-ya* project, see below. On the project to prepare a *Hsu t'ung-chien kang-mu,* see Chang Hsueh-ch'eng, "Shao Chin-han pieh-chüan," *CSCWC,* p. 159. also Nivison, *Chang Hsueh-ch'eng,* p. 206.
24. Ch'ien Ta-hsin, "Shao-chün mu-chih-lu."
25. See above, chapter 2.
26. Chang Hsueh-ch'eng, "Che-tung hsueh-shu," *Wen-shih t'ung-i,* pp. 51–52. Chang saw himself as preceded in this tradition by Huang Tsung-hsi, the Wan brothers, and Ch'üan Tsu-wang.
27. Paul Demieville, "Chang Hsueh-ch'eng and his Historiography," p. 170.
28. Naito Kōnan, *Shina shigakushi,* pp. 356–357.
29. Ch'ien Mu, *Chung-kuo chin-san-pai-nien hsueh-shu shih* 1:31.
30. Naito Kōnan, *Shina shigakushi,* pp. 298–299. Huang and his followers believed especially in the importance of having charts as evidence of the flow of history. They also stressed the importance of being faithful to documentary evidence in history writing.
31. See Virginia Mayer Chan's dissertation on the western Chekiang tradition.
32. *TYFTK,* p. 1b.
33. This remark raises the tantalizing question of the relationship between Shao and the *chin-wen* movement. I have found no indication in Shao's

writings that he was a partisan of the movement. On the other hand, Chuang Ts'un-yü, one of the most important eighteenth-century figures in the movement, was one of Shao's *chin-shih* examiners, so there existed at least a formal relationship of disciple and teacher between the two men.

34. Ch'ien Ta-hsin, "Shao-chün mu-chih-lu."

35. Many authors trace the origins of Ch'ing phonological studies to Ku Yen-wu's *Yin-hsueh wu-shu*. See, for instance, Ch'ien Mu, *Chung-kuo chin-san-pai-nien* 1:151.

36. *TMTY*, pp. 832–834 (40.1a–b). See the text and analysis in William Hung, *Erh-ya yin-te*.

37. Shao Chin-han, *Erh-ya cheng-i*, "hsu," p. 3a.

38. Ibid., p. 4a.

39. *SCHNP*, p. 80. Huang further claims that the most famous nineteenth-century commentary on the *Erh-ya*, Hao I-hsing's *Erh-ya i-shu*, is not substantially different from Shao's. However, *Shu-mu ta-wen pu-cheng*, ed. by Chang Chih-tung, p. 42, says that Hao's work definitely superseded Shao's.

40. Hung Liang-chi, *Chüan-shih ko-shih*, chüan 8, quoted in *SCHNP*, p. 80.

41. *TMTY*, pp. 974–979 (45:2a–4b). These three commentaries have been studied and reprinted by the modern Japanese scholar Takigawa Kametarō in *Shiki kaichū kōshō*.

42. *TYFTK*, p. 2a.

43. On Mao Chin, See *ECCP*, pp. 565–566. An original of the Mao Chin edition of *Shih-chi* and commentaries currently resides in the Library of Congress.

44. *TMTY*, p. 974 (45.2a); *TYFTK*, p. 2b. The language on the Kao-yang and Kao-shih families comes from the "Basic Annals of the Five Emperors," *Shih-chi*, chüan 1 (Takigawa, 1:50). The description of Ch'in Shih-huang is from the "Basic Annals of Ch'in Shih-huang," *Shih-chi*, chüan 6 (Takigawa, 2:67). In both cases, the information in *Shih-chi chi-chieh* is not found in the other commentaries.

45. The *locus classicus* for the terms *po* and *yueh* is the *Analects* 6/25: "The Superior man, extensively studying all learning, and keeping himself under the restraint of propriety, does not exceed what is right." By Ch'ing times, the terms had come to represent two poles of scholarly activity, namely, wide reading and gathering of information (*po*) and intensive study and moral cultivation (*yueh*). Perhaps the best discussion of the terms, and the significance of having both characteristics, is Chang Hsueh-ch'eng's essay, "Po-yueh" in *Wen-shih t'ung-i*, pp. 47–51. For an analysis of this essay, see Shimada Kenji, "Shō Gakusei no ichi," pp. 519–530, particularly pp. 522–524.

46. *TYFTK*, p. 2b.

47. *TMTY*, p. 974 (45.2a).

48. Yang Lien-sheng, "The Organization of Chinese Official Historiography," p. 357.

49. Nivison, *Chang Hsueh-ch'eng*, p. 51; Chang Hsueh-ch'eng, "Shao Yü-t'ung pieh-chüan," *CSCWC*, pp. 134–135, 136.

50. *TYFTK,* p. 35a–b. The review is punctuated in *SCHNP,* p. 45.

51. *TYFTK,* p. 37a–b.

52. Naito Kōnan, *Shina shigakushi,* pp. 360–361. For Ch'ien Ta-hsin's views, see "Pa Sung-shih," *Ch'ien-yen-t'ang wen-chi* 28.432–435. Ch'ien's main complaint appears to be that the biographies of intellectual leaders of the *Sung-shih* are inadequate. Included in the *wen-chi* of Wan Ssu-t'ung and Ch'üan Tsu-wang are a number of notes revising and supplementing the biographies of military and intellectual figures of the Southern Sung dynasty. There are fourteen such notes in Ch'üan Tsu-wang's *Chi-ch'i-t'ing chi wai-p'ien,* pp. 3204–3228 (28.9a–28b), and five notes in Wan Ssu-t'ung's *Ch'un-shu i-p'ien* 11.1a–7a.

53. *SCHNP,* pp. 62–66. Chang Hsueh-ch'eng's son Chang I-hsuan was a student of Shao Chin-han's, and his comments on his father's biography of Shao (see note 48) are a major source of information on Shao's revision project. Ch'ien Ta-hsin evidently saw the manuscript, and included the table of contents in his *Shih-chia-chai yang-hsin-lu.* Judging from the table of contents, Shao set about revising the *Sung-shih* in much the same way as Wan and Ch'üan, except that while Wan and Ch'üan discussed military and intellectual figures, Shao confined himself to writers and philosophers.

54. *TYFTK,* p. 37a, *TMTY,* p. 1009 (46.3a). K'o Wei-ch'i's work was entitled *Sung-shih hsin-pien* (*TMTY,* pp. 1109–1110 [50.10a]). Shen Shih-po's was *Sung-shih ch'in cheng-p'ien.* K'o's work was given a notice in the *Annotated Catalog,* but not copied into the *Ssu-k'u,* Shen's work was not listed. There is no biography of Shen in the standard histories, and I have been unable to locate any other reference to the work.

55. *TYFTK,* p. 36a; *TMTY,* p. 1009 (46.3b).

56. *TYFTK,* p. 37a–b; *TMTY,* p. 1010 (46.3b). On the inadequate treatment of the Southern Sung resistance, see Ch'üan Tsu-wang's colophon on a biography of Yueh K'o (*Chi-ch'i-t'ing chi,* pp. 3224–3225 (28.17a–18b)).

57. *TYFTK,* p. 37a.

58. Yang Lien-sheng, "The Organization of Chinese Historiography," p. 54.

59. K'o Wei-ch'i's revision of the *Sung-shih* is criticized in the *Annotated Catalog* for praising the Chinese, and considering the "Liao and Chin dynasties as foreign, in the same class with the Hsi-hsia and Korea. How can this be fair? This type of nonsense must be distinguished from [K'o's] achievement in supplementing the *Sung-shih,*" *TMTY,* p. 1110 (50.10b).

60. Liu Han-p'ing, "Lueh lun *Ssu-k'u t'i-yao* yü *Ssu-k'u fen-tsuan-kao* ti i-t'ung ho Ch'ing-tai Han-Sung hsueh chih cheng," p. 43.

61. On K'o Wei-ch'i, see above, notes 53 and 58, and the *Dictionary of Ming Biography* 1:721–722.

62. *TMTY,* p. 1010 (46.4b).

63. See *Nan-shih,* "Biography of Liu Chih-lin," 45.1250–1251; and *Liang-shu,* "Biography of Liu Chih-lin," 50.573.

64. *TYFTK,* pp. 9b–11a; *TMTY,* pp. 980–981 (45.4b–5a).

65. Cheng Hsuan was a famous commentator on the classics; sixteen works attributed to him are listed in *TMTY.* Kan Pao lived during the Ch'in dynasty (317–420); his commentaries were extant as late as the Sui

Dynasty (589–618), but are not listed in *TMTY*. The *Han-shu* commentaries of Fu Ch'ien and Wei Chao, both of the Later Han dynasty, were also extant during the Sui, but are not available today.

66. The text was considered difficult to read even in the late Han and early Six Dynasties period, as biographies in the *Hou-han-shu* and *San-kuo-chih* attest. See *Han-shu*, "hsu," p. 9.

67. Yen Shih-ku (579–645) wrote a famous commentary on the *Han-shu*. The commentary evidently did not circulate independently, but it is included in virtually all editions of *Han-shu*.

68. *TYFTK*, p. 12a–b.

69. *TYFTK*, p. 24a–b.

70. This accusation was evidently first made by Li Yen-shou (600–680), T'ang dynasty author of the *Nan-shih* and *Pei-shih*. Though he does not base himself on the evidence Li presented, Clyde B. Sargent explores the implications of these charges in "Subsidized History: Pan Ku and the Historical Records of the Former Han Dynasty," pp. 141–142. cf. Homer H. Dubs' reply, "The Reliability of the Chinese Histories."

71. *TMTY*, p. 982 (45.5b).

72. *TMTY*, pp. 983–984 (45.5b) and 1006 (46.3a–b).

73. The works were reviewed in *TMTY* as follows: Cheng Shao, *Wu-hsueh-p'ien* (*TMTY*, p. 1110 [50.10a]); Teng Yuan-hsi, *Ming-shu* (*TMTY*, pp. 1112–1113 [50.10a–b]); Hsueh Ying-ch'i, *Hsien-chang-lu* (*TMTY*, p. 1062 [48.4a]).

74. *TMTY*, p. 1856 (90.2b).

75. *TYFTK*, p. 43a–b.

76. *TMTY*, p. 1017 (46.7b).

77. *TYFTK*, pp. 43b–44a. On Hsu, see *ECCP*, pp. 316–319.

78. *TMTY*, p. 1017 (46.7b).

79. *TYFTK*, p. 44a. Lu Wen-ch'ao also complained of this inadequacy of the *Ming-shih* in his brief note "T'i Ming-shih i-wen-chih-kao," *Pao-ching-t'ang wen-chi* 7.98.

80. See the comment of Pi Yuan quoted above, pp. 127–128.

81. *Shih-lu*, p. 14611 (995.21b–24a).

82. *TMTY*, p. 1018 (46.8a).

83. On the authorship and significance of the *Han-hsueh shang-tui*, see Hamaguchi Fujio, "Ho Tojo no *Kangaku shoda* o megutte," pp. 73–79.

84. On the authorship and provenance of the book, see above, pp. 122–123. On Mao Yü-sheng, see *Ch'ing-shih lieh-chuan* 73.29b.

85. Hamaguchi Fujio, "Ho Tojo no Kangaku hihan ni tsuite," pp. 172, and 173–178. I am grateful to Benjamin Elman for bringing the Hamaguchi articles to my attention. For an interesting discussion of the atmosphere in which the *Han-hsueh shang-tui* was written, see Benjamin A. Elman, "The Hsueh-hai-t'ang and the Rise of New Text Scholarship at Canton."

86. The question of what relationship Yao Nai's epistemological insights bore to the political stance of his disciple is a fascinating one, but lies somewhat outside the scope of the present chapter. In general, the problem of what political positions were associated with intellectual stances in eighteenth-century scholarship is one which awaits further research.

87. See Hilary J. Beattie, "The Alternative to Resistance: The Case of T'ung-ch'eng in Anhwei," pp. 256–257.
88. Yao Nai's great-great-grandfather, Yao Tuan-ko, was evidently something of a "strict constructionist." See Mao Yü-sheng," [Yao Hsi-pao] mu-chih-lu," *KCCHLC* 146.8a.
89. Yao Jung, "Yao Nai hsing-chuang," *KCCHLC* 146.16b.
90. *ECCP,* pp. 235–237. On the Tai Ming-shih case, see chapter 2 above.
91. There have been many studies of the T'ung-ch'eng School's views on writing. See, for instance, Liu Sheng-mu, ed., *T'ung-ch'eng p'ai wen-hsueh yuan-yuan k'ao.*
92. Yeh Lung, *T'ung-ch'eng-p'ai wen-hsueh-shih,* p. 126.
93. Tai Chen, "Yü Yao Ssu-lien shu," *Tai Chen wen-chi, Yuan-shan, Meng-tzu tzu-yi shu-cheng,* pp. 141–142.
94. Chang Ping-lin, *Chien-lun* 4.25a. Both Chang and Wei Yuan stress the importance of factionalism in the disputes between Han and Sung learning.
95. "[Yao Nai] kuo-shih-kuan pen-chüan," *KCCHLC* 146.6a.
96. This is reported both in Mao Yü-sheng, "Mu-chih-lu," *KCCHLC* 146.8b, and Yao Jung, "Hsing-chuang," *KCCHLC* 146.17a.
97. Yao Jung, "Hsing-chuang," *KCCHLC* 146.17a.
98. Yao Jung, "Hsing-chuang," *KCCHLC* 146.17b. In this as in other respects, Yao Jung's biography of his father is remarkably full and revealing.
99. Yao Nai, "Fu Chang-chün shu," in *Hsi-pao hsuan ch'üan-chi,* p. 65. For a further discussion of Sung learning in the late eighteenth century, see Yü Ying-shih, *Lun Tai Chen yü Chang Hsueh-ch'eng,* chapter six, note 70.
100. In addition to the biographies by Yao Jung and Mao Yü-sheng quoted above, see Ch'en Yung-kuang, "[Yao Nai] Hsing-chuang," *KCCHLC* 145.12b–16a.
101. In addition to numerous notes, letters and preface in Yao's 619-page *wen-chi.* Ch'en Yung-kuang edited a collection of Yao Nai's letters after his retirement entitled *Hsi-pao hsien-sheng ch'ih-tu.*
102. Yao Nai, "Hsiao-hsueh-k'ao hsu," *Ch'üan-chi,* pp. 47–48.
103. Yao Nai, "Shang-shu pien-hsu hsu," *Ch'üan-chi,* p. 193.
104. Yao Nai, "Shu An wen-ch'ao hsu," *Ch'üan-chi,* p. 46. The same argument can be found in many of Yao's writings.
105. Yü Ying-shih, *Lun Tai Chen yü Chang Hsueh-ch'eng* p. 111.
106. The preface cannot be found in any extant work by Ch'ien Tien, and the preface itself gives no indication of what work it was meant to introduce. However, one of Ch'ien's works was entitled *Lun-yü hou-lu.* This work was completed in 1776, just as Yao was on the verge of leaving Peking. In a letter to Chiang Sung-ju (see *Ch'üan-chi,* p. 73). Yao speaks of discussing learning with Tai Chen and his colleagues, and writing a preface to Ch'ien Tien's works to illustrate his point. If the preface quoted here and the preface Yao speaks of in his letter to Chiang Sung-ju are the same, then it is quite possible that these lines reflect Yao's thought just as he was leaving the capital. Since the preface quoted here begins and ends with references to Confucius and the preservation of his teachings, it may quite possibly have been written to introduce *Lun-yü hou-lu.* On Ch'ien Tien, see *ECCP,* p. 156.

107. Yao Nai, "Tseng Ch'ien Hsien-chih hsun," *Ch'üan-chi,* p. 84.

108. See pp. 59–60 above.

109. Yao Nai, "Tseng Ch'ien Hsien-chih hsun," *Ch'üan-chi,* p. 84.

110. Ibid., p. 84.

111. Some of the books Yao reviewed were not listed in the *Catalog* at all. Of the ninety-nine books Yao reviewed, sixty-six were included in the *Ssu-k'u ch'üan-shu* collection, sixteen were described in the *Annotated Catalog* but not included in the collection and seventeen were not listed in the *Annotated Catalog.*

Yao Nai's draft reviews are so numerous that they would justify a monograph by themselves. I have selected the reviews of *Ku-shih, Shan-hai-ching, Mo-tzu,* and *Chung-yung chi-lueh* because they seem to illustrate as well as any the differences in scholarly styles and substances between Yao and the Ssu-k'u editors.

112. The *Ku-shih* was completed in 1065. It dealt only with the portions of the *Shih-chi* which describe events prior to the time of Confucius. It revised six of the ten basic annals in *Shih-chi,* sixteen of the thirty biographies of hereditary and noble houses, and thirty-seven of seventy individual biographies.

113. Naito Kōnan, *Shina Shigakushi,* p. 219.

114. Chu Hsi, "Ku-shih yü-lun," in *Chu-tzu chi* 71.33a.

115. *HPHSL* 2.1a–b.

116. *TMTY,* pp. 1903–1904 (92.6a–b).

117. *TMTY,* pp. 1092–1093 (50.3a–b).

118. For an excellent summary of the contents of *Shan-hai-ching* and the various interpretations of it, see Naito Kōnan, *Shina shigakushi,* pp. 93–95.

119. See Naito, *Shina shigakushi,* p. 93, and note 125 below.

120. Yang Shen's views are expressed most clearly in "Shan-hai-ching hou-hsu" *Sheng-an ch'üan-chi* 2.17; Yang also wrote a short commentary entitled *Shan-hai-ching pu-chu.* For Yang's biography, see *Dictionary of Ming Biography* 2:1531–1534.

121. Hui Tung (1697–1758) evidently wrote a commentary on the *Shan-hai-ching* stressing its use as a geography book, but the commentary is not extant. Hsu Wen-ching (1667–after 1756) used information from the *Shan-hai-ching* extensively for verification of geographical information in his commentary on the *Shu-ching.* In 1783, Pi Yuan published an edition of the *Shan-hai-ching* with commentary, and as Sun Hsing-yen's preface to that work makes clear (see *Sun Yuan-ju wai-chi,* 3.5a–6a) those who participated in that project saw the *Shan-hai-ching* principally as a geography book. The most extensive commentary on the *Shan-hai-ching* was done by a younger contemporary of Yao Nai, Hao I-hsing (1757–1825), and was entitled *Shan-hai-ching chien-shu.*

122. Chu Hsi, "Ch'u-tz'u pien-cheng," B.1b–3b, in *Ch'u-tz'u chi-chu.* Chu made the comments in the course of refuting an earlier attempt, by Hung Hsing-tzu (1090–1155) to use *Shan-hai-ching* to comment on the "T'ien-wen" section of the *Ch'u-tz'u.*

123. *HPHSL* 2.10a–b.

124. *TMTY,* pp. 2938–2939 (142.1a–b).

125. *TMTY,* p. 2939 (142.1b).
126. The *Han-shu* "I-wen-chih" lists *Mo-tzu* as a work in 71 *p'ien.* For some reason, Sung bibliographies only record 61 *p'ien.*
127. The suspect portions are *p'ien* 40–43, and 52–71.
128. This episode is related in the *Han Fei-tzu,* section 32. Burton Watson, a modern translator of *Mo-tzu* remarks that "The *Mo-tzu,* whatever the interest of its ideas, is seldom a delight to read. . . . The style as a whole is marked by a singular monotony of sentence patterns, and a lack of wit or grace that is atypical of Chinese literature." See Burton Watson, *Basic Writings of Mo-tzu, Hsun-tzu and Han-Fei-tzu,* pp. 14–15.
129. *HPHSL* 3.5b.
130. Ibid., 3.5b.
131. See Mencius, 3B/9.
132. *TMTY,* pp. 2452–2453 (117.16).
133. Augustinius Tseu, *The Moral Philosophy of Mo-tzu,* pp. 35–36, footnote.
134. Wang Chung, "Mo-tzu hsu," and "Mo-tzu hou-hsu" in *Shu-hsueh, nei-p'ien,* 3.1a–4b.
135. On Weng Fang-kang, see *ECCP,* pp. 856–858. On his note of sympathy to Yao Nai, see Yao Jung, "Yao Nai hsing-chuang," *KCCHLC* 146.17b.
136. Weng Fang-kang, "Shu Mo-tzu," *Fu-ch'u-chai wen-chi,* pp. 618–620. On the significance of this controversy, see Hu Shih, "Weng Fang-kang yü Mo-tzu," in *Hu Shih wen-ts'un,* pp. 931–933.
137. Wing-tsit Chan, *A Source Book in Chinese Philosophy,* p. 96.
138. Tu Wei-ming, *Centrality and Commonality: An Essay on the Chung-yung,* p. 3.
139. For the difference in the two editions, which appeared in 1183 and 1189, see below. In the process of reprinting Shih Tun's work, Chu Hsi apparently changed the name from *Chung-yung chi-chieh* (Collected interpretations of the *Chung-yung*) to *Chung yung chi-lueh* (roughly, Collected approximations of the *Chung-yung*).
140. *TMTY,* pp. 721–723 (35.5a–6b).
141. *TMTY,* p. 724 (35.6b).
142. *HPHSL* 1.6a.
143. *TMTY,* p. 725 (35.6b).
144. The editors made these remarks in their review of the *Ssu-shu huo-wen TMTY,* p. 723 (35.5a).
145. *TMTY,* p. 725 (35.6b).
146. *ECCP,* p. 237.
147. Liang Ch'i-ch'ao, *Intellectual Trends During the Ch'ing Period,* p. 78.
148. Although Yao Nai referred to some of his ancestors as the founders of the T'ung-ch'eng school, this may have involved more filial piety than historical accuracy. Most authors agree that the Sung learning, as here portrayed, began with Yao.
149. Yü Ying-shih, *Lun Tai-Chen yü Chang Hsueh-ch'eng,* pp. 3–4. It might be added that with the changing political circumstances of the early nineteenth-century, the intellectual and political significance of the Sung learning changed slightly, producing the phenomenon that Hamaguchi Fumio described in the articles cited in note 88.

Chapter 6. Ch'ui-mao ch'iu-tz'u: *Blowing Back the Fur and Examining the Faults*

1. Kuo Po-kung, *Ssu-k'u ch'üan-shu tsuan-hsiu-k'ao,* p. 3.
2. Yang Chia-lo, *Ssu-k'u ch'üan-shu kai-shu,* pp. 17–19.
3. L. Carrington Goodrich, *The Literary Inquisition of Ch'ien-lung,* p. 30.
4. *Shih-lu,* p. 13465 (919.19a–b).
5. See Hsieh Kuo-chen, *Wan-Ming shih-chi k'ao,* and Lynn Ann Struve, "Uses of History in Traditional Chinese Society: The Southern Ming in Ch'ing Historiography."
6. *Shih-lu,* p. 13029 (904.29a–30a).
7. Both L. Carrington Goodrich, *Literary Inquisition* p. 53, and Wu Che-fu, in his *Ch'ing-tai chin-hui shu-mu yen-chiu,* pp. 27–62, aver that the destruction of anti-Manchu literature was the main goal of the Ch'ien-lung censorship, and this category fits the vast majority of works they list as censored.
8. *Shih-lu,* pp. 14077–14078 (964.9b–11b). I follow the translation in Goodrich, *Literary Inquisition,* p. 110 with some modifications. One problem with the Goodrich translation is that it does not make clear that the edict begins with summaries of two previous edicts.
9. *KCT:CL* 029802 (San-pao, 12 October 1774); *KCT:CL* 029880 (Kao-chin, 17 October 1774).
10. *KCT:CL* 029909 (P'ei Tsung-hsi, 19 October 1774). As cited in chapter three, the two sources on the number of books submitted by the provinces for *Ssu-k'u ch'üan-shu* do not always agree; in the case of Anhwei province, however, both sources give 516 (Yang Chia-lo, p. 154; *KCT:CL* 032624 (Min Ao-yun, 22 October 1776)).
11. *KCT:CL* 030000 (Yü Wen-i, 26 October 1774). Yang Chia-lo, p. 154, gives a total of 200 books for Fukien; *KCT:CL* 032364 (Chung-yin, 23 September 1776) gives 216.
12. On Kwangtang, see *KCT:CL* 030178 (Teh-pao, 7 November 1774); on Kwangsi see *KCT:CL* 029861 (Hai-ch'eng, 16 October 1774). Hai-ch'eng reported that 503 books had been submitted for the *Ssu-k'u ch'üan-shu* by October 1774. Yang Chia-lo, p. 154, gives a total of 664, *KCT:CL* 032205 (Hai-ch'eng, 3 September 1776) gives 1038.
13. The major concerns of the court in the late summer and early fall of 1774, aside from the suppression of the Wang Lun rebellion in Shantung, and the campaign against the Chin-chuan Miao in the southwest (see below), were the prosecution of a case against the eunuch Kao Yun-ts'ung who had been found guilty of divulging the contents of secret imperial rescripts, and the provision of relief for victims of a flood of the Yellow River. None of these had much to do with the issue of Sino-Manchu relations or with the suppression of anti-Manchu literature. There is no mention of the issue of Sino-Manchu relations in the *Shih-lu* for the two months preceding or following the 10 September edict (*Shih-lu,* ch. 960–967 passim.) or in the Ch'ing basic annals for the years 1774–1776 (*Ch'ing-shih,* pp. 171–189 passim.). Neither the *Shang-yü tang (fang-pen)* nor a bundle of draft edicts from the summer and fall of 1772, which I was fortunate enough to have

the opportunity to inspect in the First Historical Archives in Beijing, shed much light on the origins of censorship. The edict of September, 1774 was issued at a time when the Ch'ien-lung Emperor was on his summer retreat in Jehol, but it was drafted by Yü Min-chung in the capital. It may be significant that when he drafted the edict Yü himself was under a cloud, having been held responsible and demoted in the case of Kao Yun-ts'ung. On this case, see Preston Torbert, *The Ch'ing Imperial Household Department*, pp. 131–135.

14. I am grateful to Professor Jonathan D. Spence of Yale for directing my attention to this "clue." For a more detailed study of the Ch'ien-lung Emperor's literary projects, see chapter 2 above. For a discussion of how these projects changed the character of Manchu studies, see Ch'en Chieh-hsien's preface to the Hsueh-sheng shu-chü edition of *Pa-ch'i t'ung-chih*, particularly pp. 3–7.

15. *TMTY,* p. 875 (41.11a).

16. *TMTY,* p. 1016 (46.7a).

17. These were the *Sheng-ch'ao hsun-chieh chu-chen (lu)* (*TMTY,* p. 1284 [58.4b]), the *K'ai-kuo fang-lueh,* the *Man-chou yuan-liu-k'ao,* (See *Shih-lu,* pp. 15260–15262 [1319.4b–8a]). The last two works were completed too late for inclusion in *TMTY.*

18. These were the *Shih-tsung wang-kung kung-chi-piao, Ming-ch'en tsou-yi* (See *Shih-lu,* pp. 16759–16760 (1043.25b–26a)), and a Chinese translation of *Man-chou ch'a-shen ch'a-t'ien tien-li,* all completed too late for inclusion in *TMTY.*

19. *Shih-lu,* pp. 14425–14426 (983.6b–7b).

20. *Shih-lu,* p. 15261 (1319.5a–b).

21. A-kuei et al. eds., *K'ai-kuo fang-lueh,* preface, 4a.

22. *ECCP,* p. 7.

23. *Shih-lu,* pp. 14152–14153, (967.32a–34b). Quoted in Richard Lu-kuen Jung, "The Ch'ien-lung Emperor's Suppression of Rebellion," p. 218.

24. Gertraude Roth Li, "The Rise of the Early Manchu State," p. 4. These documents have since been published twice; once by the Mambun Rōtō Kenkyukai of the Toyo Bunko led by Kanda Nobuo: *Mambun Rōtō* (Tokyo, 1955–1962); and once by the National Palace Museum, Taipei: *Man-chou chiu-tang* (Taipei, 1969). The two publications are based on different versions of the originals; the Palace Museum version is generally believed to be more complete and reliable.

25. A recent article on the eighteenth century has termed the labor of reconstructing inquisition procedures "Sisyphean." Actually thanks to the clarity of Ch'ing documents and the classification procedures of the Palace Museum in Taipei, the labor can be accomplished relatively easily. It may be useful here to indicate briefly the size and significance of the archival data base that supports the statements about inquisition procedures in this section. For the years 1774 to 1779, there are 1225 memorials in the *KCT* and *CCCT* collections dealing with the literary inquisition. I believe that these represent between one-half and one-third of the original documents

on the event, based on a study of the gaps in the collection, internal evidence in some memorials which indicates that they were a part of a
numbered sequence, and a careful count of the numbers of responses by
provincial governors to selected imperial edicts. This evidence is summarized below.

a. *Gaps in archives.* There are five gaps for the years 1774 to 1779 in the
two memorial collections of the Museum, *KCT* and *CCT:*

1. Gap in *KCT*–23 January 1773 to 16 October 1773.
2. Gap in *KCT*–1 February 1775 to 7 February 1777.
3. No *CCCT* archives before 28 January 1778.
4. Gap in *KCT*–28 January 1778 to 23 June 1778.
5. Gap in *CCT*–27 February 1778 to 21 September 1778.

Such gaps exist throughout the Palace Museum Collection. Archivists
speculate that they were caused when historians, working in the collection
before it was moved during the war, pulled out the documents of certain
time-periods in which they were interested. Such documents were stored
in a separate place and never returned to the original collection. Perhaps
they still exist in one of the several wartime locations of the collection.

b. *Numbered Sequences.* A number of provincial governors had the habit
of recording the number of times the provinces they governed had submitted books for banning in the submission memorials. For this practice, historians can be grateful to San-pao, who originated it when he was
governor of Chekiang and carried it with him to Hupei. It is possible to
compare the number of submission memorials extant with the number of
such memorials which theoretically existed. The results of such a comparison are summarized below.

Province	Number of Submissions in Sequence	Number of Documents Extant	Percentage
Chekiang	18	9	50%
Fukien	7	2	28%
Hunan	5	2	40%
Hupei	8	6	75%
Kueichow	6	4	66%
Yunnan	5	1	20%
Kwanghsi	4	1	25%
		Average:	43%

c. *Response to imperial edicts.* It appears to have been de rigeur for all of
the provincial governors to submit memorials responding to certain sorts
of imperial edicts. It is therefore possible to compare the number of
gubernatorial responses extant with the number that once theoretically
existed. This measure is, of course, not as reliable as the previous one
since the possibility that a governor responded to a given edict, in the
absence of any written evidence of his response, can only be a presumption.

Edict	Number of Responses
5 August 1774	9
10 September 1774	6
18 May, 12 August 1777*	17
27 August 1778	6
26 January 1779	8 (9)**
6 September	10 (11)**

* This edict was issued twice, apparently because so few governors responded to it the first time.

**In both these cases, a newly appointed governor repeated acknowledgements made by his predecessor.

This data base is, admittedly, far from complete. However, for two reasons I believe it to be strong enough to support the statements made about inquisition procedures in this section. First, much of the language in the extant documents is repetitious; governors decided on a form to be used in submitting seditious books and repeated it over and over.

Second, any major changes in inquisition procedures were announced in imperial edicts. Since the record of imperial edicts is fairly complete (no edict about the inquisition has been found which is not in either *Pan-li* or *Shih-lu*), I believe I have not missed any major shifts of inquisition procedure.

26. *KCT:CL* 032031 (Ch'en Hui-tzu, 15 August 1777), *KCT:CL* 029802, *KCT:CL* 029907 (P'ei Tsung-hsi, 20 October 1774), and *KCT:CL* 029880.

27. The Yung-cheng Emperor made a number of changes in the qualifications and regulations for this post. For an analysis of these, and a discussion of the office in general, see Araki Toshikazu, "Choku-shō kyōgaku no sei o tsūjite mitaru Yōsei chika no bunkyō seisaku." On the Ch'ien-lung Emperor's image, see the rescript of 1750 and edict of 1753 on the importance of regular inspection of local educational officials' ability and vigor, *Hsueh-cheng ch'üan-shu* 1:419 (23.8a) and 1:423–425 (23.9b–11a). On local educational officials' control of educational lands, see the edict of 1721 ordering local educational officials to report clearly at the time of their assumption of office on the amount of land, number of books, and number of implements pertaining to the office they are assuming. *Hsueh-cheng ch'üan-shu* 1:417–418 (23.7a–b). On local education receipt of examination fees, see Miyazaki Ichisada, *China's Examination Hell*, p. 31.

28. *KCT:CL* 030466 (San-pao, 20 February 1777) *CCCT* 021115 (Kuei-lin, 9 December 1778). Hsiao Kung-chuan, *Rural China: Imperial Control in the Nineteenth Century*, pp. 72–83.

29. *KCT:CL* 029861 (Hai-ch'eng, 16 October 1774); *KCT::CL* 032946 (Pi-yuan, 24 November 1777). On book purchasing, see *KCT:CL* 030695 (Yang-kuei, 11 March 1776); *KCT:CL* 032376 Kao-chin, 23 September 1777).

30. *KCT:CL* 036136 (Li Hu, 12 October 1778). Actually Governor Li refers to the appointment of one man, expectant magistrate Ch'en Ching-li, (*chin-shih* 1756) whose duty it was to receive and evaluate the books. Whether or not Ch'en was head of a bureau, the strictures below about the difficulty

of evaluating seditious books would apply. (Shantung) *KCT:CL* 038326 (Kuo-T'ai, 4 June 1778); (Szechwan) *KCT:CL* 037783 (Wen-teh, 2 May 1779). The four Szechwan bureaus were in Ch'eng-tu, Ch'ung-ching, Ya-chou and Sheng-ching. (Shensi) *CCT:CL* 026623 (Lo Erh-chin, 2 May 1780); (Chihli) *CCCT:CL* 029202 (Yuan Shou-t'ung, 7 January 1781).

31. For samples of memorials accompanying books submitted to the emperor, see *SYTFP:CL* 45 (Fall) 159; *SYTFP:CL* 45 (Winter) 317; *SYTFP:CL* 46 (Summer) 335. See a memorial from the Grand Council which describes books as having "been turned over to the historiographers Chi Yun and Lu Hsi-hsiung for editing." (*Pan-li*, I, p. 89a–b, translated in Goodrich, *Literary Inquisition*, p. 214) and a letter from Yü Min-chung to Lu Hsi-hsiung which orders Chi and Lu to search for treasonous books (*YMCSC*, p. 97).

32. *Pan-li*, I, p. 55b.

33. *Pan-li*, I, p. 70a.

34. I am grateful to Beatrice S. Bartlett for sharing this information from her exhaustive research on the Grand Council with me.

35. *Shih-lu*, pp. 15032–15033 (1022.20a–21a), translated in Goodrich, *Literary Inquisition*, pp. 157–158.

36. *KCT:CL* 030466.

37. *KCT:CL* 030695.

38. *KCT:CL* 030925 (San-pao, 28 April 1777). San-pao used similar language in many memorials. See Goodrich, *Literary Inquisition*, pp. 159–160.

39. Cited in Araki, "Choku-shō Kyōgaku," pp. 77–78.

40. *KCT:CL* 030466. I obtain this figure by adding the numbers of books submitted in the seventh through thirteenth, sixteenth and eighteenth consignments for Chekiang. The figures and sources are as follows:

Number in Sequence	Source			Volumes Submitted
7–8	*KCT:CL*	03095	(17 April 1777)	120
9	*KCT:CL*	03125	(24 June)	454
10	*KCT:CL*	031808	(28 July)	549
11	*KCT:CL*	032229	(5 September)	1957
12	*KCT:CL*	032446	(3 October)	298
13	*CCCT:CL*	019448	(20 March 1778)	307
16	*KCT:CL*	036007	(2 October)	105
18	*CCCT:CL*	023848	(22 July 1779)	456

Reports from the fourteenth, fifteenth and seventeenth consignments are missing, but undoubtedly they would add to the totals. A version of the 2 October 1778 memorial with a list of the titles being submitted is also available (*CCCT: CL* 036342).

41. *KCT:CL* 030695; *KCT:CL* 032872 (Yang-kuei, 18 November 1777).

42. I obtained the figure for submission prior to 1778 by taking San-pao's report on the total number of books submitted from Hupei in the five submissions prior to 11 September 1778 and subtracting from it the three submissions San-pao made himself. San-pao's 11 September report is

CCCT:CL 020733; the other submissions are listed below. San-pao's five submissions were as follows:

Number in Sequence	Source			Volumes Submitted
3	*CCCT:CL*	019402	(1 March)	524
(4)	(One document missing, number			
	obtained from subsequent report)			1341
5	*CCCT:CL*	021724	(29 July)	613
6	*CCCT:CL*	021227	(2 December)	1817
7	*CCCT:CL*	022661	(12 February 1779)	667
8	*KCT:CL*	039290	(21 May)	26
9	*CCT:CL*	o25117	(25 January)	725

43. *KCT:CL* 033091 (Kao-chin, 6 December 1777); *KCT:CL* 032872 (Yang-kuei, 18 November 1777). In September of 1777, the emperor issued an edict commending the procedure, and ordering that it be employed in all the provinces. See *Shih-lu*, p. 15263 (1319.10a–b).
44. For Wang's biography, see *ECCP*, pp. 819–820; and Meng Shen, *Hsin-shih ts'ung-k'an*, pp. 581–595. The documents condemning Wang include *KCT:CL* 033054 (Hai-ch'eng, 3 December 1777), and the materials reprinted in *Chang-ku ts'ung-p'ien*, Wang Hsi-hou case, Chi 1.1a–9b. Wang was supposed to have said in his preface that the problem with the *K'ang-hsi tzu-tien* was "the difficulty of penetrating it" (*ch'uan-kuan chih-nan*). On interrogation, Wang claimed that he had meant that the Chinese writing system was difficult, rather than that the *K'ang-hsi tzu-tien* was poorly organized. He also subsequently claimed that he had written the tabooed imperial names merely to show people what they were. He observed the taboos properly, however, in the second edition of his dictionary.
45. For the court's attitude toward accusations, see below.
46. *Chang-ku ts'ung-p'ien*, Wang Hsi-hou case, Chi 2.20a.
47. These figures are based on Chuang Chi-fa's summary of the Palace Museum Archives on the Wang Hsi-hou case. Since not all governors specifically identified Wang's books in their consignments, and some merely referred to "books by Wang," it is likely that these figures understate, perhaps considerably, the number of Wang's works found in the empire. See Chuang Chi-fa, "Wang Hsi-hou Tzu-kuan-an ch'u-t'an," pp. 144–147.
48. *Chang-ku ts'ung-p'ien*, Wang Hsi-hou case 2.12a, 2.22a.
49. *WTYT*, pp. 305–306.
50. Ch'en Yuan, "Shih-hui chü-lieh," p. 552.
51. *WTYT*, p. 307.
52. *WTYT*, pp. 349–351. Governor Li holds the record for the number of inquisition cases prosecuted in a single year (five). He also seems to have been responsible for the creation of a book bureau in Hunan, and for an increase in the number of books submitted for burning from the province.
53. *WTYT*, pp. 363–364.
54. See the cases of Fang Kuo-t'ai (*WTYT*, pp. 767–770); Lou Sheng (*WTYT*, pp. 789–792); and Kao Chih-ch'ing (*WTYT*, pp. 761–765).

55. See the cases of Yü Teng-chiao (*WTYT,* pp. 817–829); Chu Ssu-tsao (*WTYT,* pp. 803–804); Cho Chang-ling (*WTYT,* pp. 567–580); Chu T'ing-cheng (*WTYT,* pp. 443–449); and Yeh T'ing-t'ui (*WTYT,* pp. 541–547).

56. For a stimulating discussion of the limitations of Freedman's model of South China landholding lineages, see Patricia Ebrey, "Types of Lineages in Ch'ing China: A Reexamination of the Chang Lineage of T'ung-cheng."

57. See the cases of Ch'eng Ming-yin (*WTYT,* pp. 549–566); Ch'en Hsi-sheng (*WTYT,* pp. 387–393); Li Ta-pen (*WTYT,* pp. 363–370); and Liang San-ch'uan (*WTYT,* pp. 751–755).

58. Ho Ping-ti, *Studies on the Population of China: 1386–1953,* p. 270.

59. On minor marriages, the reason why Chinese in recent times have chosen them, and the financial stakes involved, see Arthur P. Wolf and Huang Chieh-shan, *Marriage and Adoption in China, 1845–1945,* pp. 82–93, 261–271, and 272–281.

60. *WTYT,* pp. 387–393.

61. *WTYT,* p. 734.

62. Yang Ching-ya and Meng Ch'ing-yu, editors, *Te-hsing t'ung-chih,* 2.2b, 7.21a–b, 8:16b, 9:9b, 11a–b, 12b.

63. Chu's given name is sometimes written with the roof radical, sometimes without.

64. See Evelyn S. Rawski, *Education and Popular Literacy in Ch'ing China,* pp. 47–48. The subject headings are taken from Chiang Ker-chiu's translation: *The Three Character Classic.*

65. *WTYT,* pp. 443–447. A translation of the passage from Confucius can be found in Arthur Waley, trans., *Analects of Confucius,* Book 14, Chapter 18. On the queue, see Robert Entenmann, "De Tonsura Sino-Tartarica: The Queue in Early Ch'ing China." For an instance of resistance to the queue, see Frederick Wakeman, "Localism and Loyalism During the Ching Conquest of Kiangnan."

66. See Denis Twitchett, "The Fan Clan's Charitable Estate, 1050–1760."

67. William Theodore de Bary, *Sources of the Chinese Tradition,* p. 267.

68. Over fifty *mou* of land, including twenty *mou* of farmland, worth 166 taels were seized.

69. *Te-hsing t'ung-chih* 5.22b.

70. *ECCP,* p. 279.

71. Maurice Freedman, *Lineage Organization in Southeastern China,* pp. 46–76.

72. Jack H. Potter, "Land and Lineage in Traditional China," p. 126.

73. Feng Han-chi, "The Chinese Kinship System," pp. 180–181.

74. *Shih-lu,* p. 16375 (1118.17b–18b).

75. "Ch'üan-teh kuo-shih-kuan pen-chuan," *KCCHSC* 300.10b–11a.

76. This plan of operations was not set forth in any single document. I have pieced it together from the memorials of Ch'üan-te and T'u-ming-a in the Archives. In addition to the documents cited in notes below, these include: *CCT:CL* 029314 (Ch'üan-te, 14 December 1780); *CCCT:CL* 030030 (T'u-ming-a, 9 April 1781); *CCT:CL* 031109 (T'u-ming-a, 21 July 1781); and *KCT:CL* 039197 (T'u-ming-a, 17 October 1781).

77. *CCT:CL* 030717 (Ch'üan-te, 20 June 1781); and *CCCT:CL* 030718 (T'u-ming-a, 6 June 1781).
78. On this case, see *Pan-li*, I, pp. 69b–70a; the *Ch'ing-hsia-chi* was reviewed in *TMTY*, p. 3674 (172.6b). Also see chapter 3 above.
79. *Shih-lu*, p. 16381 (1119.5b–6a). Translated in Goodrich, *Literary Inquisition*, p. 192.
80. *TMTY* (Addendum number 1), p. 83. Translated in Goodrich, *Literary Inquisition*, pp. 211–212.
81. I have not seen an original of this document. This translation is from Goodrich, *Literary Inquisition*, pp. 214–215. This index was first printed by Yao Chin-yuan in 1883. Reprinted in *TMTY* (Addendum number 1), pp. 81–116.
82. Goodrich, *Literary Inquisition*, pp. 216–218. The original language, somewhat abridged, can be found in Wu Che-fu, *Ch'ing-tai chin-hui shu-mu yen-chiu*, pp. 87–88. I have not seen the original of the document. According to Goodrich, however, it had no date. Goodrich dates it after the memorials of Ying-lien, which seems plausible for three reasons.
    a. Its location within the sequence of documents in which it was found. See Goodrich, *Literary Inquisition*, p. 218.
    b. Since the Ssu-k'u Commissioners who submitted the memorial were at work through these years on a master list of banned books, it seems reasonable to assume that they prepared a statement of the reasons for banning books at the same time.
    c. The document mentions the work of Wu Wei-yeh, which was described by San-pao as "never previously submitted" for censorship in a memorial of 15 October 1779. (*Wen-hsien ts'ung-pien*, coll. 8, "Wei-ai-shu ch'ing-tan" 3b).
83. For the edicts condemning these works, see (Ch'ien Ch'ien-i) Goodrich, *Literary Inquisition*, pp. 100–107; (Lü Liu-liang) Goodrich, p. 85, note 5; (Chü Ta-chun) Goodrich, pp. 112–136; (Chin Pao) Goodrich, pp. 144, 149; (Edict on reign names) Goodrich, pp. 138–140.
84. *WTYT*, pp. 764–765.
85. *WTYT*, p. 770.
86. Goodrich, *Literary Inquisition*, pp. 227–228.
87. (Anhui) *KCT:CL* 041209 (T'an Shang-chung, 15 May 1782); (Kwangtung) *KCT:CL* 041330 (Pa-san-ting, 31 May 1782).
88. *KCT:CL* 041191 and *KCT:CL* 041364 (Li Shih-hao, 3 June 1783).
89. David S. Nivison, "Ho-shen and His Accusers," p. 236.
90. Quoted in Kuei Cheng-hsien, *Ch'ing-tai wen-hsien chi-lüeh*, pp. 4–5.
91. 1977 catalog of the Wei-wen Publishing Company, Taipei, p. 1.
92. Chang Shun-hui, *Ch'ing-tai Yang-chou hsueh-chi*, pp. 1–2.
93. Goodrich, *Literary Inquisition*, p. 6.
94. The first index of banned books was assembled in 1883 by Yao Chin-yuan, and included in his collectanea *Chih-chin-chai ts'ung-shu;* it was subsequently reprinted as an addendum to the Commercial Press Reprint of the *Annotated Catalog*. Yao's work has been supplemented twice, first by Teng Shih's publication in 1907 of an index kept at the Chiang-ning bureau

(*Ch'ing-tai chin-hui shu-mu pu-i*), and again in 1925 with the publication in *Pei-ching ta-hsueh kuo-hsueh-men chou-k'an* of a Grand Council memorial dated 1783, which summarized seven submissions of banned books to the Emperor. These two supplements, together with Yao Chin-yuan's original work, have recently been reprinted in a one-volume edition, *Ch'ing-tai chin-hui shu-mu; Ch'ing-tai chin-shu chih-chien-lu,* (Shanghai, 1957). As research in the Ch'ing Archives continued in the twenties and thirties, other lists, mostly those which accompanied individual consignments of books from provinical capitals to Peking, were found and published. In 1932, Ch'en nai-ch'ien published an index to all the known lists of banned books entitled *So-yin-shih-ti chin-shu tsung-lu* (Shanghai, 1932). Since the 1930s, there have been numerous articles tabulating and analyzing the various entries on these lists. In 1969, Wu Che-fu of the staff of the National Palace Museum, Taipei, published a study of banned books indices which summarized all previous work, and such additional materials as he could find in the Palace Museum Archives, *Ch'ing-tai chin-hui shu-mu yen-chiu.* Pages 99–117 of this work contain a convenient summary of the extant indices.

95. In the fall of 1979 and the winter of 1980, several articles on the censorship of the Ch'ien-lung era appeared in the Chinese press. Among these were: K'ung Li, "Lun Ch'ing-tai ti wen-tzu-yü"; Tso Pu-ch'ing, "Ch'ien-lung fen-shu"; Wang Ssu-chih, "Ming-Ch'ing wen-tzu-yü chien-lun"; and Wei Ch'ing-yuan, "Ch'ung-tu Ch'ing-tai wen-tzu-yü-tang." I am grateful to Thomas Fisher for sharing his observations on these articles with me.

Bibliography

Araki Toshikazu 荒木敏一. "Choku-shō kyōgaku no sei o tsūjite mitaru Yōsei chika no bunkyō seisaku" 直省教學の制を通じて觀たる 雍正治下の文教政策 (The provincial school inspectors as an example of Yung-cheng period educational policy), *Tōyōshi kenkyū* 東洋史 研究 16.4:70–94 (March 1958).

——. Yōsei jidai ni okeru gakushinsei no kaikaku—shu toshite sono nin yōhō chūshin toshite" 雍正時代に於學政の改革一主として 其人任田を中心として (The reform of provincial education commissioners in the Yung-cheng period—with special emphasis on the utilization of personnel), *Tōyōshi kenkyū* 東洋史研究 18.3:27–41 (December 1959).

Atwell, William S. "From Education to Politics: The Fu-she," in *The Unfolding of Neo-Confucianism*, ed. Wm. Theodore de Bary. New York, Columbia University Press, 1975.

Bartlett, Beatrice S. *The Vermillion Brush: Grand Council Communications System and Central Government Decision Making in Mid-Ch'ing China*. Yale University Press, forthcoming.

Beattie, Hilary J. "The Alternative to Resistance: The Case of T'ung-ch'eng in Anhwei," in *From Ming to Ch'ing*, ed. Jonathan D. Spence and John E. Wills. New Haven, Yale University Press, 1979.

Bodde, Derk. *China's First Unifier: A Study of the Ch'in Dynasty as seen in the Life of Li Ssu . . . (280?–208 B.C.)*. Leiden, E. J. Brill, 1938.

Brunnert, H. S. and V. V. Hagelstrom. *Present Day Political Organization in China*, tr. A. Beltchenko and E. E. Moran. Shanghai, Kelly and Walsh, 1912.

Carter, Thomas Francis. *The Invention of Printing in China and Its Spread Westward*. New York, Columbia University Press, revised edition, 1931.

Chan, Hok-lam. *Control of Publishing in China, Past and Present*. Canberra, The Australian National University, 1983.

Chan, Wing-tsit. *A Source Book in Chinese Philosophy*. Princeton, Princeton University Press, 1963.

———. "The *Hsing-li ching-i* and the Ch'eng-Chu School of the Sixteenth Century," in *The Unfolding of Neo-Confucianism*, ed. Wm. Theodore deBary. New York, Columbia University Press, 1975.

———. "Chu Hsi and Yuan Neo-Confucianism," in *Chinese Thought and Religion Under the Mongols*, ed. Wm. Theodore deBary and Hok-lam Chan. New York, Columbia University Press, 1982.

Chang-ku ts'ung-pien 掌故叢編 (Collected historical documents), Nos. 1–10. Pei-p'ing, Palace Museum, Jan. 1928–Oct. 1929.

Chang Chih-tung 張之洞, supplemented by Fan Hsi-tseng 范希曾 *Shu-mu ta-wen pu-cheng* 書目答問補正 (A primer on bibliography, with supplements). Chang's Preface, 1875; Fan's Preface, 1931. Taipei, Hsin-hua shu-chü reprint, 1975.

Chang Chun-shu. "Emperorship in Eighteenth Century China," *Journal of the Institute of Chinese Studies of the Chinese University of Hong Kong* 7.2:551–569 (1974).

Chang Hsueh-ch'eng 章學誠. *Wen-shih t'ung-i* 文史通義 (Principles of literature and history). 1771. Hong Kong, T'ai-p'ing shu-chü reprint, 1973.

———. *Chiao-ch'ou t'ung-i* 校讐通義 (Principles of textual collation). 1779. Peking, Ku-chi ch'u-pan-she reprint, 1956.

———. *Chang-shih i-shu* 章氏遺書 (The remaining writings of Chang Hsueh-ch'eng), ed. Liu Ch'eng-kan 劉承幹. Wu-hsing, Chia-yeh-t'ang, 1922.

Chang I-hsuan 章貽選, comp. *Chang Shih-chai hsien-sheng wen-chi* 章實齋先生文集 (The collected writings of Chang Hsueh-ch'eng). Taipei, Wen-hai Publishing Company reprint, 1968.

Chang Ping-lin 章炳麟. *Chien-lun* 檢論 (Collected essays). Taipei, Kuang-wen shu-chü reprint, 1970.

———. "Jih-chih-lu chiao-chi" 日知錄校記 (On editing the Record of Knowledge Accumulated Day by Day), in *Yuan ch'ao-pen Jih-chih-lu* 原抄本日知錄 (The original hand-copied edition of the Record of Knowledge Accumulated Day by Day), ed. Hsu Wen-ts'e 徐文冊. Taipei, Ming-lun ch'u-pan-she, 1958.

Chang Shun-wei 張舜徽. *Ch'ing-tai yang-chou hsueh-chi* 清代揚州學記 (Record of Yangchow scholarship in the Ch'ing period). Shanghai, Jen-min ch'u-pan-she, 1962.

Ch'ang Pi-te 昌彼德. *Chung-kuo mu-lu-hsueh chiang-i* 中國目錄學講義 (Lectures on Chinese bibliography). Taipei, Wen-hsueh shih-chi ch'u-pan-she, 1973.

Ch'en Chieh-hsien 陳捷先. "Lun Pa-ch'i t'ung-chih" 論八旗通志 (On the general history of the eight banners), in *Pa-ch'i t'ung-chih* (General history of the eight banners). Taipei, Taiwan hsueh-sheng shu-chü reprint, 1968.

Ch'en Yuan 陳垣. "Shih-hui chü-lieh" 史諱舉例 (On the omission of sacred and imperial names in Chinese writings). *Yen-ching hsueh-pao* 燕京學報 4:537–651 (December, 1924).

————. *Chiu Wu-tai-shih chi-pen fa-fu san-chüan* 舊五代史輯本發覆三卷 (Three chapters on the editing of the Old History of the Five Dynasties). Peking, Fu-jen ta-hsueh, 1937.

————, ed. *Ch'ing-tai wen-chi p'ien-mu fen-lei so-yin* 清代文集篇目分類索引 (A subject index to collected writings of the Ch'ing period). 1935. Taipei, Kuo-feng ch'u-pan-she reprint, 1965.

Cheng Chung-ying, ed. and tr. *Tai Chen's Inquiry into Goodness.* Honolulu, East-West Center Press, 1971.

Chi Yun (Hsiao-lan) 紀昀 (曉嵐). *Yueh-wei-ts'ao-t'ang pi-chi* 閱微草堂筆記 (Desultory jottings from Yueh-wei Hall). 1800. Taipei, Ta chung-kuo t'u-shu kung-ssu reprint, 1974.

————. *Chi Wen-ta kung i-chi* 紀文達公遺集 (Remaining writings of Mr. Chi Wen-ta [Hsiao-lan, Yun]), comp. Chi Shu-hsing 紀樹馨. Peking, Hung-po shu-chü, 1812.

Chia I-chün 賈逸君. "Ch'ing-tai wen-tzu-yü k'ao-lueh" 清代文字獄考略 (On the literary cases of the Ch'ing period), *Révue de l'université franco-chinoise* 10.5:65–94 (March 1937).

Chiang Fan 江藩. *Han-hsueh shih-ch'eng chi* 漢學師承記 (Notes on the transmission of Han learning from teacher to pupil). 1818. Shanghai, Commercial Press reprint, 1934.

Chiang Yung 江永. *Ku-yun piao-chun* 古韻標準 (Ancient phonology). Taipei, Kuang-wen shu-chü reprint, 1966.

Chiao Hung 焦竑. *Kuo-shih ching-chi chih* 國史經籍志 (A treatise on bibliography for the national history). 1590. Changsha, Commercial Press reprint, 1939.

Ch'ien Mu 錢穆. *Chung-kuo chin-san-pai-nien hsueh-shu shih* 中國近三百年學術史 (Chinese intellectual history in the last three hundred years). 1938. Taipei, Commercial Press reprint, 1972.

Ch'ien Ta-hsin 錢大昕. *Chin-shih-wen pa-wei* 金石文跋尾 (A catalog of stone inscriptions). 1787. Changsha, Lung family reprint, 1888.

————. *Ch'ien-yen-t'ang wen-chi* 潛研堂文集 (Collected works of the Ch'ien-yen Studio). 1806. Shanghai, Commercial Press reprint, 1935.

————. "Pu Yuan-shih i-wen-chih hsu," 補元史藝文志序 (Preface to a treatise on art and literature prepared to supplement the Yuan history), in *Erh-shih-wu-shih pu-pien* 二十五史補編 (Supplements to the twenty-five standard histories) 4:8393–8394. Shanghai, K'ai-ming shu-chü, 1937.

Ch'ien Tseng 錢曾. *Shu-ku-t'ang ts'ang-shu-mu* 述古堂藏書目 (Catalog of books stored at Shu-ku Hall), in *Yueh-ya-t'ang ts'ung-shu* 粵雅堂叢書 (Collectanea from the Yueh-ya studio), Wu Ch'ung-yao 伍崇曜 comp. Taipei, Hua-wen shu-chü reprint, 1965.

Chih Wei-ch'eng 支偉成. *Ch'ing-tai p'u-hsueh ta-shih lieh-chuan* 清代樸學大師列傳 (Biographies of the great masters of the unadorned learning movement during the Ch'ing dynasty). 1925. Taipei, I-wen Publishing Company reprint, 1970.

Chin Chien 金簡, ed. *Ch'in-ting Wu-ying-tien chu-chen pan ch'eng-shih* 欽定武英殿聚珍版程式 (Procedures for printing with assembled pearls in the Wu-ying throne hall, authorized by the emperor), in *Chu-chen ts'ung-shu*

聚珍叢書 (Collectanea printed with assembled pearls). Peking, 1777.

Ch'in-ting hsueh-cheng ch'üan-shu 欽定学政全書 (A complete manual for education commissioners, authorized by the emperor), ed. Yü Min-chung 于敏中 . Revised edition presented to the Emperor in 1782. Taipei, Wen-hai Publishing Company reprint, 1968.

Ch'in-ting ssu-k'u ch'üan-shu chien-ming mu-lu 欽定四庫全書簡明目錄 (Abbreviated catalog of the Complete Library of the Four Treasuries, authorized by the emperor), ed. Yung-jung 永瑢 et al. 1782. Shanghai, Ku-tien wen-hsüeh reprint, 1957.

Ch'in-ting ssu-k'u ch'üan-shu tsung-mu t'i-yao 欽定四庫全書總目提要 (Annotated general catalog of the Complete Library of the Four Treasuries, authorized by the emperor), ed. Chi Hsiao-lan 紀曉嵐 et al., 1782. Taipei, Commercial Press reprint, 1971.

Ch'ing-shih 清史 (History of the Ch'ing dynasty). Taipei, National Defense Research Institute, 1961.

Ch'ing-shih lieh-chuan 清史列傳 (Historical biographies of the Ch'ing dynasty). Shanghai, Chung-hua Book Company, 1928.

Ch'ing-tai chin-hui shu-mu (pu-yi), Ch'ing-tai chin-hui chih-chien-lu 清代禁燬書目 (補遺), 清代禁書知見錄 (An index to books banned during the Ch'ing (with supplements), a list of books known to have been banned during the Ch'ing). Shanghai, Commercial Press reprint, 1957.

Ch'ing-tai wen-tzu-yü tang 清代文字獄檔 (Archives of literary cases during the Ch'ing). 1931. Taipei, Wen-hai Publishing Company reprint, 1975.

Chou Yü-tung 周予同 . *Ching chin-ku wen-hsueh* 經今古文學 (Studies of the old and new texts of the classics). Shanghai, Commercial Press, 1936.

Chou Yung-nien 周用年 . "Ju-ts'ang-shuo" 儒藏説 (A proposal for a Confucial tripitaka), in *Sung-lin ts'ung-shu* 松隣叢書 (Master Sung-lin's collectanea), comp. Wu Ch'ang-fu 吳昌夫 . 1917.

Chu Hsi 朱熹 . "Ch'u-tz'u pien-cheng" 楚辭辯證 (Correct readings of the Ch'u-tz'u), in *Ch'u-tz'u chi-chu* 楚辭集注 (Interpretations of the Ch'u-tz'u). Peking, Chung-hua shu-chü reprint, 1963.

———. "Ku-shih yü-lun" 古史語論 (Additional discussion of the Ku-shih), in *Chu-tzu chi* 朱子集 (A collection of the writings of Master Chu), ed. Hsu Shu-ming 徐樹銘 . 1810.

Chu Ju-chen 朱汝珍 . *Tz'u-lin chi-lueh* 詞林輯略 (A list of Hanlin scholars). Peking, 1929.

Chu Yun 朱筠 . *Ssu-ho wen-chi* 筍河文集 (Collected writings of Master Ssu-ho) ed., Li Wei 李威 . 1815. Shanghai, Commercial Press reprint, 1936.

Ch'üan Tsu-wang 全祖望 . *Chi-ch'i-t'ing chi wai-p'ien* 鮚埼亭集外篇 (A supplement to the collected writings of the Master of Chi-ch'i Pavilion), comp. Tung Ping-ch'un 董秉純 . 1776. Taipei, Wen-hai Publishing Company reprint, 1969.

———. *Chi-ch'i-t'ing chi* 鮚埼亭集 (Collected writings of the Master of Chi-ch'i Pavilion), comp. Shih Meng-chiao 史夢蛟 . 1804. Shanghai, Commercial Press reprint, 1929.

Chuang Chi-fa 莊吉發 . "Wang Hsi-hou *Tzu-kuan* an ch'u-t'an" 王錫侯

字貫案初探 (A preliminary investigation of the case of Wang Hsi-hou's dictionary), *Shih-yuan* 史原 10.4:137–156 (October 1973).

Chuzo Ichiko. "The Role of the Gentry: An Hypothesis," in *China in Revolution: The First Phase,* ed. Mary C. Wright. New Haven, Yale University Press, 1968.

Demieville, Paul. "Chang Hsueh-ch'eng and his Historiography," in *Historians of China and Japan,* ed. by Edwin Pulleyblank and Wm. G. Beasley. London, Oxford University Press, 1961.

Dubs, Homer H. "The Reliability of the Chinese Histories," *Far Eastern Quarterly* 6:23–43 (1946–47).

Dull, Jack L. "An Historical Introduction to the Apocrypha (Ch'an-wei) Texts of the Han Dynasty." Ph.D. dissertation, University of Washington, 1966.

Ebrey, Patricia, "Types of Lineages in Ch'ing China: A Re-examination of the Chang Lineage of T'ung-ch'eng," *Ch'ing-shih wen-t'i* 4.9:1–20 (June 1983).

Elman, Benjamin A. "Japanese Scholarship and the Ming-Ch'ing Intellectual Transition," *Ch'ing-shih wen-t'i* 4.1:1–22 (June 1979).

——— . "The Hsueh-hai-t'ang and the Rise of New Text Scholarship at Canton," *Ch'ing-shih wen-t'i* 4.2:51–81 (December 1979).

——— . "From Value to Fact: The Emergence of Phonology as a Precise Discipline in Late Imperial China," *Journal of the American Oriental Society* 102.3:493–500 (July–October 1982).

——— . "Philosophy (*I-li*) Versus Philology (*K'ao-cheng*): The Jen-hsin Tao-hsin Debate," *Toung Pao* 69.4 & 5: 175–222 (1983).

——— . *From Philosophy to Philology: Intellectual and Social Aspects of Change in Late Imperial China.* Cambridge, Mass., Council on East Asian Studies, Harvard University, 1984.

Entenmann, Robert E. "De Tonsura Sino-Tartarica: A Study of the Queue in Early Ch'ing China." Seminar Paper, Harvard University, 1974–1975.

Fang Tung-shu 方東樹. *Han-hsueh shang-tui* 漢學商兌 (A polemic against Han learning). Taipei, Commercial Press reprint, 1974.

Feng Han-chi. "The Chinese Kinship System," *Harvard Journal of Asiatic Studies* 2.2:141–269 (July 1937).

Fisher, Thomas S. "Lü Liu-liang (1628–1683) and the Tseng Ching Case (1728–1733)." Ph.D. dissertation, Princeton University, 1974.

Fishman, Olga. *Tszi Yun Zametki iz Khizhni Velikoe v Malom.* Moscow, Izdatelstvo Nauka, 1974.

Freedman, Maurice. *Chinese Lineage and Society: Fukien and Kwangtung.* London, Athlone Press, 1966.

——— . *Lineage Organization in Southeast China.* London, Athlone Press, 1958.

Frost, Robert. *The Poetry of Robert Frost.* New York, Holt, Rinehart and Winston, 1969.

Fu Tsung-mao 傅宗懋. *Ch'ing-tai chün-chi-ch'u tsu-chih chi chih-ch'ang chih yen-chiu* 清代軍機處組織及職掌之研究 (Research on the organization and official duties of the Grand Council during the Ch'ing dynasty). Taipei, Chia-hsin shui-ni kung-ssu wen-hua chi-chin-hui.

Galt, Howard S. *A History of Chinese Educational Institutions*. London, A. Probsthain, 1951.

Giles, Lionel. *An Alphabetical Index to the Chinese Encyclopedia*. 1911. Taipei, Ch'eng-wen shu-chü reprint, 1970.

——— . "A note on the Yung-lo ta-tien," *New China Review* 2:137–153 (1920).

Goodrich, L. Carrington. *The Literary Inquisition of Ch'ien-lung*. Baltimore, American Council of Learned Societies Studies of China and Related Societies, 1935. Reprinted by Paragon Book Co., New York, 1966.

——— and Fang Chao-ying, eds. *Dictionary of Ming Biography*. New York, Columbia University Press, 1976.

Grimm, Tillman. "Ming Educational Intendents," in *Chinese Government in Ming Times: Seven Studies*, ed. Charles O. Hucker. New York, Columbia University Press, 1969.

Guy, R. Kent. "Decadence Revisited: National Essence Views of the Eighteenth Century." Unpublished paper, presented at the annual meetings of the Association for Asian Studies, March 1983.

——— . "The Development of the Evidential Research Movement: Ku Yen-wu and the *Ssu-k'u ch'üan-shu*," *Tsing-hua Journal of Chinese Studies*, New series, 16.1 and 2: 97–118 (Dec. 1984).

——— . "Zhang Tingyu and Reconciliation: Scholars and the State in the Early Qianlong Period." *Journal of Late Imperial China* 7.1 (June, 1986).

Hamaguchi Fujio 濱口富士雄 . "Hō Tōju no *Kangaku shōda* o megutte" 方東樹の漢學商兌を繞って (On Fang Tung-shu's Han-hsueh shang-tui), *Taitō bunka daigaku kangaku kaishi* 大東文化大學漢學會誌 15:73–79 (1976).

——— . "Hō Tōju no *Kangaku hihan ni tsuite*" 方東樹の漢學批評について (Fang Tung-Shu's criticisms of Han learning), *Nihon Chūgoku gakkaiho* 日本中國學會報 30.17:165–178 (1978).

Henderson, John B. *The Development and Decline of Chinese Cosmology*. New York, Columbia University Press, 1984.

Ho Ping-ti. "The Salt Merchants of Yangchow: A Study of Commercial Capitalism in Eighteenth Century China," *Harvard Journal of Asiatic Studies* 17:130–168 (1954).

——— . *Studies on the Population of China, 1368–1953*. Cambridge, Mass., Harvard University Press, 1959.

——— . *The Ladder of Success in Imperial China*. New York, Columbia University Press, 1962.

Hou Wai-lu 侯外廬 *Chin-tai Chung-kuo ssu-hsiang hsueh-shuo* 近代中國思想學說史 (History of modern Chinese thought and scholarship). Shanghai, Sheng-huo shu-tien, 1947.

Hsiao I-shan 蕭一山 . *Ch'ing-tai t'ung-shih* 清代通史 (General history of the Ch'ing dynasty). Revised edition. Taipei, Commercial Press, 1976.

Hsiao Kung-chuan. *Rural China: Imperial Control in the Nineteenth Century*. Seattle, University of Washington Press, 1960.

Hsieh Ch'i-k'un 謝啟昆. *Hsiao-hsueh-k'ao* 小學考 (On the lesser learning). 1802. Hangchow, Chekiang Book Bureau, 1888.

Hsieh Kuo-chen 謝國楨. *Wan-Ming shih-chi k'ao* 晚明史籍考 (Annotated catalog of late Ming historical works). Pei-ping, Kuo-li Pei-p'ing t'u-shu-kuan, 1933.

Hsu Chi-ying 許霽英. "Ch'ing Ch'ien-lung ch'ao wen-tzu-yü chien-piao" 清乾隆朝文字獄簡表 (A simplified chart of the literary cases during the Ch'ien-lung reign of the Ch'ing dynasty), *Jen-wen yüeh-k'an* 人文月刊 8.4:1–13 (June 1937).

Hsu-hsiu ssu-k'u ch'üan-shu t'i-yao 續修四庫全書提要 (Enlarged and revised catalogue of the four treasuries). Taipei, Commercial Press, 1972.

Hsu Shih-ying 許時英. *Chung-kuo mu-lu-hsueh shih* 中國目錄學史 (History of Chinese bibliography). Taipei, Chung-hua wen-hua ch'u-pan-shih-yeh wei-yuan-hui, 1954.

Hsueh Chü-cheng 薛居正. *Chiu Wu-tai shih* 舊五代史 (Old history of the Five Dynasties). Peking, Chung-hua shu-chü, 1976.

Hu Shih. "The Establishment of Confucianism as a State Religion during the Han Dynasty," *Journal of the North China Branch of the Royal Asiatic Society* 60:20–41 (1929).

——— 胡適. *Hu Shih wen-ts'un* 胡適文存 (Extant writings of Hu Shih). Shanghai, Ya-tung t'u-shu-kuan, 1929–1930.

———. *The Chinese Renaissance.* Chicago, University of Chicago Press, 1934.

Hu Ssu-ching 胡思敬. *T'ui-lu ch'üan-chi* 退廬全集 (Collected works of Hu Ssu-ching). 1924. Taipei, Wen-hai Publishing Company reprint, 1973.

Huang-Ch'ing k'ai-kuo fang-lueh. 皇清開國方略 (An account of the founding of our dynasty), ed. A Kuei 阿桂 et al. Peking, 1789.

Huang Tsung-hsi 黃宗羲. *Nan-lei wen-ting* 南雷文定 (Collected prose of Huang Tsung-hsi). 1688. Shanghai, Chung-hua shu-chü reprint, 1936.

Huang Yun-mei 黃雲眉. *Shao Erh-yun (Chin-han) hsien-sheng nien-p'u* 邵二雲 (晉涵) 先生年譜 (A chronological biography of Mr. Shao Erh-yun (Chin-han). 1933. Hong Kong, Ch'ung-wen shu-tien reprint, 1972.

Hucker, Charles O. *A Dictionary of Official Titles in Imperial China.* Stanford, Stanford University Press, 1985.

Hummel, Arthur W. *Eminent Chinese of the Ch'ing Period.* 1943–1944. Taipei, Literature House reprint, 1964.

Hung, William (Hung Yeh 洪業). "Preface to an Index to *Ssu-k'u ch'üan-shu tsung-mu* and *Wei-shou shu-mu*," *Harvard Journal of Asiatic Studies* 4.1:47–58 (1939).

———, ed. *Ssu-k'u ch'üan-shu chi wei-shou shu-mu yin-te* 四庫全書及未收書目引得 (Index to the annotated catalog of the Four Treasuries and the bibliography of books not included). Harvard-Yenching Institute Sinological Index Series, no. 7, 1932. Taipei, Chinese Materials and Research Aids Service Center reprint, 1966.

———, ed. *Erh-ya yin-te* 爾雅引得 (An index to the Erh-ya). Harvard-Yenching Institute Sinological Index Series, Supplement no. 18, 1941. Taipei, Chinese Materials and Research Aids Service Center reprint, 1966.

———, ed. *Tseng-chiao Ch'ing-ch'ao chin-shih t'i-ming pei-lu fu yin-te* 曾校清朝進士提名碑錄附引得 (Collated lists of *chin-shih*, as inscribed in stone, of the Ch'ing dynasty, with index). Harvard-Yenching Institute Sinological Index Series, no. 9, 1944. Taipei, Chinese Materials and Research Aids Service Center reprint, 1966.

Jones, Susan L. M. "Scholasticism and Politics in Late Eighteenth-Century China," *Ch'ing-shih wen-t'i* 3.4:28–49 (December 1975).

────── . "Hung Liang-chi (1746–1809): and the Perception and Articulation of Political Problems in Late Eighteenth-Century China." Ph.D. dissertation, Stanford University, 1971.

Juan Hsiao-shu 阮孝緒 . "Ch'i-lu hsu-mu" 七錄序目 (Introduction to the Seven Summaries), in *Chin-shih-han ts'ung-shu* 晉石厂叢書 (Mr. Chin-shih-han's collectanea), ed. Yao Wei-tsu 姚慰祖 . Kuei-an, 1881.

Juan Yuan 阮元 . *Ssu-k'u wei-shou shu-mu t'i-yao* 四庫未收書目提要 (Annotated catalog of books not included in the annotated catalog of the Complete Library of the Four Treasuries). 1876. Reprinted in Commercial Press edition of *Ch'in-ting ssu-k'u ch'üan-shu tsung-mu t'i-yao* 欽定四庫全書 總目提要 (Annotated general catalog of the Complete Library of the Four Treasuries, commissioned by the emperor), vol. 5. Taipei, 1971.

Jung, Richard Lu-kuen. "The Ch'ien-lung Emperor's Suppression of Rebellion." Ph.D. dissertation, Harvard University, 1979.

Kahn, Harold L. *Monarchy in the Emperor's Eyes: Image and Reality in the Ch'ien-lung Reign.* Cambridge, Mass., Harvard University Press, 1971.

Kawata Teiichi 河田悌一 . "Shindai gakujutsu no ichi sokumen—Shu In, So Shinkan, Ko Ryokitsu to shite Shō Gakusei" 清代學術の一側面— 朱筠.邵晉涵.洪亮吉として章學誠 (One view of scholarship during the Ch'ing dynasty—Chu Yun, Shao Chin-han, Hung Liang-chi and Chang Hsueh-ch'eng), *Tōhō gakuhō* 東方學報 57:84–105 (January 1979).

Kessler, Lawrence D. "Chinese Scholars and the Early Manchu State," *Harvard Journal of Asiatic Studies* 31:179–200 (1971).

────── . *K'ang-hsi and the Consolidation of Ch'ing Rule.* Chicago, University of Chicago Press, 1976.

Kramers, R. P. "Conservatism and the Transmission of the Confucian Canon," *Journal of Oriental Studies* 11:119–132 (1955).

Ku Chieh-kang. "A Study of Literary Persecutions During the Ming," tr. L. Carrington Goodrich, *Harvard Journal of Asiatic Studies* 3:3–4:254–311 (1938).

Ku Yen-wu 顧炎武 . *Yuan-ch'ao-pen Jih-chih-lu* 原抄本日知錄 (The original manuscript of Knowledge Acquired Day by Day), ed. Hsu Wen-ts'e 徐文冊 . Taipei, Ming-lun ch'u-pan-she reprint, 1958.

Kuei Cheng-hsien 歸諮先 . *Ch'ing-tai wen-hsien chi-lueh* 清代文獻紀略 (Sketches of literary cases during the Ch'ing dynasty). 1944. Taipei, Wen-hai Publishing Company reprint, 1971.

Kuhn, Philip A. *Rebellion and Its Enemies in Late Imperial China: Militarization and Social Structure, 1796–1864.* Cambridge, Mass., Harvard University Press, 1970.

────── and Susan Mann Jones. "Dynastic Decline and the Roots of Rebellion," *Cambridge History of China,* Vol. 10, Pt. 1. New York, Cambridge University Press, 1978.

K'ung Li 孔立 . "Lun Ch'ing-tai ti wen-tzu-yü" 論清代的文字獄 (On literary cases of the Ch'ing dynasty), *Chung-kuo-shih yen-chiu* 中國史 研究 3:129–140 (March 1979).

Kuo-ch'ao kung-shih 國朝宮史 (History of the palace in our times). Presented to the Emperor in 1769. Taipei, Wen-hai Publishing Company reprint, 1970.

Kuo Po-kung 郭佰恭. *Yung-lo ta-tien k'ao* 永樂大典考 (A study of the Grand Encyclopedia of the Yung-lo Emperor). Shanghai, Commercial Press, 1937.

————. *Ssu-k'u ch'üan-shu tsuan-hsiu k'ao* 四庫全書篡修考 (A study of the compilation of the Complete Library of the Four Treasuries). Shanghai, Commercial Press, 1937.

Kuo-ts'ui hsueh-pao 國粹學報 (National essence journal). Shanghai, 1905–1911. Taipei, Wen-hai Publishing Company reprint, n.d.

Lasswell, Harold D. "Censorship," in *International Encyclopedia of the Social Sciences*, ed. R. A. Seligman. New York, Macmillan, 1930.

Liang Ch'i-ch'ao. *Intellectual Trends During the Ch'ing Period*, tr. Immanual C. Y. Hsu. Cambridge, Mass., Harvard University Press, 1959.

Li Chin-hua 李晉華. *Ming-shih tsuan-hsiu k'ao* 明史篡修考 (A study of the compilation of the Ming dynastic history). Peking, Yenching Journal of Chinese Studies Monographs, no. 3, 1933.

Li Fu-sun 李富孫. *Ho-cheng hou-lu* 鶴徵後錄 (A later record of the summoning of the cranes). Chia-hsing, 1807.

Li, Gertraude Roth. "The Rise of the Early Manchu State." Ph.D. dissertation, Harvard University, 1975.

Li P'eng-nien 李鵬年, et al. ed., *Ch'ing-tai chung-yang kuo-chia chi-kuan kai-shu* 清代中央國家機關概述 (The organs of central government during the Ch'ing). Harbin, Heilungjiang Jen-min ch'u-pan-she, 1984.

Li Wen-tsao 李文藻. "Tu-men shu-ssu chih chin-hsi" 都門書肆之今昔 (The evolution of bookstores in the capital). Reprinted in Yeh Te-hui, *Shu-lin ch'ing-hua*, pp. 257–262.

Li Yen-shou 李延壽. *Nan-shih* 南史 (History of the Southern Dynasties). Peking, Chung-hua shu-chü, 1975.

Li Yuan 李垣, ed. *Kuo-ch'ao ch'i-hsien lei-cheng ch'u-pien* 國朝耆獻類徵初編 (Classified collection of biographies of famous men of our dynasty). Hsiang-yin, Li family, 1884–1890.

Li Yuan-tu 李元度. *Kuo-ch'ao hsien-cheng shih-lueh* 國朝先正事略 (Brief accounts of worthies of our dynasty). 1866. Taipei, Wen-hai Publishing Company reprint, 1967.

Liu Chao-yu 劉兆祐. "Min-kuo i-lai ti Ssu-k'u hsueh" 民國以來的四庫學 (Studies on the Four Treasuries since the Republican era), *Han-hsueh yen-chiu t'ung-hsun* 漢學研究通訊 2.3:146–151 (July 1983).

Liu Han-p'ing 劉漢屏. "Lueh lun Ssu-k'u t'i-yao yü Ssu-k'u fen-tsuan-kao ti i-t'ung ho Ch'ing-tai Han-Sung hsueh chih cheng" 略論四庫提要與四庫分篡稿的異同和清代漢宋學之爭 (On the differences between Shao Chin-han's drafts and the final Ssu-k'u reviews and the competition between Han and Sung learning in the Ch'ing). *Li-shih chiao-hsueh* 1979.7: 40–44.

Liu I-an 劉益安. "Lun Ch'ien-chia k'ao-chü-hsueh-p'ai ti li-shih tso-yung chi p'i-p'an chi-cheng wen-t'i" 論乾嘉考據學派的歷史作用及

批判繼承問題 (Some questions on the historical role, function and transmission of Han learning of the Ch'ien-chia period), in *Chung-kuo chin-san-pai-nien hsueh-shu ssu-hsiang lun-chi* 中國近三百年學術思想論集 (Collected articles on Chinese intellectual history of the last three hundred years), ed. Chou K'ang-hsieh 周康燮. Hong Kong, Ch'ung-wen shu-chü, 1974.

Liu Sheng-mu 劉聲木 ed. *T'ung-ch'eng wen-hsueh yuan-yuan k'ao* 桐城文學淵流考 (A background study of T'ung-ch'eng literature). Taipei, Shih-chieh shu-chü, 1962.

Liu Shih-p'ei 劉師培 (Kuang-han 光漢). "Lun chung-kuo i-chien tsang-shu-lou" 論中國宜建藏書樓 (On the need to build libraries in China). *Kuo-ts'ui hsueh-pao* 國粹學報 19 (1906). Taipei, Wen-hai Publishing Company reprint, pp. 2287–2292.

——. "Chin-tai Han-hsueh pien-ch'ien lun" 今代漢学變遷論 (A discussion of the evolution of contemporary Han learning), in *Kuo-ts'ui hsueh-pao* 國粹學報 31 (1907). Taipei, Wen-hai Publishing Company reprint, pp. 3819–3822.

Liu Tsun-yan. "The Compilation and Historical Value of the Tao-tsang," in *Essays on the Sources for Chinese History*, ed. by Donald D. Leslie, et al. Columbia, S.C., South Carolina University Press, 1979.

Lo Chi-tsu 羅繼祖. *Chu Ssu-ho hsien-sheng nien-p'u* 朱筠河先生年譜 (Chronological biography of Mr. Chu Ssu-ho). Shanghai, Commercial Press, 1931. Taipei, Wen-hai Publishing Company reprint, 1969.

Lo-shan-t'ang chi 樂善堂集 (Writings from the Lo-shan Studio). Second edition. Peking, 1750.

Lu Wen-ch'ao 盧文弨. *Pao-ching-t'ang wen-chi* 抱經堂文集 (Collected writings from the Pao-ching Studio). 1797. Shanghai, Commercial Press reprint, 1937.

Liu, Adam Yuen-chung. "The Practical Training of Government Officials under the Early Ch'ing, 1644–1795," *Asia Major* 16.1–2:82–95 (1971).

——. *The Hanlin Academy: Training Ground for the Ambitious, 1644–1850*. Hamden, CT., Archon Books, 1981.

Ma Tuan-lin 馬端臨. "Ching-chi k'ao" 經籍考 (On bibliography), in *Wen-hsien t'ung-kao* 文獻通考 (Comprehensive examination of source materials), pp. 1513–1967. Shanghai, Commercial Press reprint, 1936.

Mannheim, Karl. *Ideology and Utopia: An Introduction to the Sociology of Knowledge.* New York, Harcourt Brace & Co., 1946.

Meng Sen 孟森. *Ming-Ch'ing shih-lun chu chi-k'an* 明清史論著集刊 (Collection of articles on the Ming and Ch'ing periods), ed. Yang Chia-lo 楊家駱. Taipei, Shih-chieh shu-chü, 1961.

Meskill, John. *Academies in Ming China: A Historical Essay.* Tucson, University of Arizona Press, 1982.

Metzger, Thomas A. *The Internal Organization of the Ch'ing Bureaucracy: Legal, Normative and Communicative Aspects.* Cambridge, Mass., Harvard University Press, 1973.

————. *Escape From Predicament: Neo-Confucianism and China's Evolving Political Culture.* New York, Columbia University Press, 1977.

Miao Ch'üan-sun 繆荃孫. *I-feng-t'ang wen-chi* 藝風堂文集 (Collected writings from the I-feng Studio). Taipei, Wen-hai Publishing Company reprint, 1973.

Michael, Franz. *The Origins of Manchu Rule in China.* Baltimore, The Johns Hopkins Press, 1942.

Miller, H. Lyman. "Factional Conflict and the Integration of Ch'ing Politics." Ph.D. dissertation, Georgetown University, 1974.

Miyazaki, Ichisada. *China's Examination Hell,* tr. Conrad Shirokauer. New York, Weatherhill, 1976.

Morohashi, Tetsuji 諸橋轍次. *Dai kanwa jiten* 大漢和辞典. Tokyo, Taishukan shoten, 1955–1960.

Naitō Konan 內藤湖南. "Shina mokurokugaku" 支那目錄學 (Chinese bibliography), 1926, in *Naitō Konan zenshu* 內藤湖南全書 (The complete works of Naitō Konan), vol. 12. Tokyo, Chikuma shobō, 1969.

————. *Shina shigakushi* 支那史学史 (A history of Chinese historiography), in *Naitō Konan zenshu,* vol. 11. Tokyo, Chikuma shobō, 1969.

Naquin, Susan. *Shantung Rebellion: The Wang Lun Uprising of 1774.* New Haven, Yale University Press, 1981.

National Palace Museum (Taipei) Archives. Palace Memorial Collection, Ch'ien-lung reign.

————. Grand Council Memorial Collection, Ch'ien-lung reign.

————. Grand Council, Palace and Edict Record Book (Square series) Collections.

Nishino Teijin 西野直一. "Kenryu-katei no soshu soshoka o noberu" 乾隆嘉慶の蘇州の藏書家をのじる (Book owning families of Soochow during the Ch'ien-lung and Chia-ch'ing periods), *Chūgoku kankei ronsō jirō* 中國關係論説資料 5.2:182–189 (January–June 1966).

Nivison, David S. "Ho Shen and his Accusers: Ideology and Political Behavior in the Eighteenth Century," in *Confucianism in Action* ed., David S. Nivison and Arthur F. Wright. Stanford, Stanford University Press, 1959.

————. *The Life and Thought of Chang Hsueh-ch'eng.* Stanford, Stanford University Press, 1966.

Ofuchi, Ninji. "The Formation of the Taoist Canon," in *Facets of Taoism,* ed. by Holmes Welch and Anna Seidel. New Haven, Yale University Press, 1979.

Ōkubo Eiko 大久保英子. *Min-Shin jidai shoin no kenkyū* 明清時代書院の研究 (Research on academies in the Ming-Ch'ing period). Tokyo, Kokusho kankōkai, 1976.

Ono Kazuko 小野和子. "Shinsho no shisō tōsei o megutte" 清初の思想統制をめぐって (Thought control during the early Ch'ing period), *Tōyōshi kenkyū* 18.3:99–123 (December 1959).

Oxnam, Robert. *Ruling from Horseback: The Politics of the Oboi Regency.* Chicago, University of Chicago Press, 1976.

Peterson, Willard J. "The Life of Ku Yen-wu," *Harvard Journal of Asiatic Studies* 28:114-156 (1968) (Part One), and 29:201-247 (1969) (Part Two).
——. *Bitter Gourd: Fang I-chih and the Impetus for Intellectual Change.* New Haven, Yale University Press, 1979.
Pa-ch'ao sheng-hsun 八朝聖訓 (Instructions of the sage during the eight reigns). Peking, latest preface, 1856.
Pan Ku 班固. *Han-shu* 漢書 (History of the Han). Peking, Chung-hua shu-chü, 1962.
Pan-li See *Pan-li ssu-k'u ch'üan-shu tang-an.*
Pan-li ssu-k'u chüan-shu tang-an 辦理四庫全書檔案 (Archives of the compilation of the Complete Library of the Four Treasuries), ed., Wang Chung-min 王重民. Pei-p'ing, Kuo-li pei-p'ing tu-shu-kuan, 1934.
Polachek, James. *The Inner Opium War.* Forthcoming.
Potter, Jack H. "Land and Lineage in Traditional China," in Maurice Freedman, ed., *Family and Kinship in Traditional Chinese Society.* Stanford, Stanford University Press, 1970.

Quarterly Journal of Bibliography. Peip'ing, National Library of Peip'ing, 1934-1944.

Rawski, Evelyn. *Education and Popular Literacy in Ch'ing China.* Ann Arbor, University of Michigan Press, 1979.
Ropp, Paul S. *Dissent in Early Modern China: Ju-lin wai-shih and Ch'ing Social Criticism.* Ann Arbor, University of Michigan Press, 1981.
des Rotours, Robert. *Traité des fonctionaires et traité de l'armée.* Leiden, E. J. Brill, 1947-48.

Sands, Lee M. "The Liu-li-ch'ang Quarter: Potters and Booksellers in late Ch'ien-lung." Seminar paper, Yale University, 1977.
Sargent, Clyde B. "Subsidized History: Pan Ku and the Historical Records of the Former Han Dynasty," *Far Eastern Quarterly* 3:119-143 (1943-1944).
Schierlitz, Ernst. "Zur Technik der Holtztypendruck aus dem Wu-ying-tien in Peking," *Monumenta Serica* 1:17-38 (1937).
Schwartz, Benjamin I. "Some Polarities in Confucian Thought," in *Confucianism and Chinese Civilization,* ed. Arthur F. Wright. New York, Atheneum, 1964.
Shao Chin-han 邵晉涵. *Erh-ya cheng-i* 爾雅正義 (An index to the Erh-ya). Yü-yao, 1788.
——. "Ssu-k'u ch'üan-shu t'i-yao fen-tsuan kao" 四庫全書提要分纂稿 (Draft reviews prepared for the annotated catalog of the Complete Library of the Four Treasuries), in *Shao-hsing hsien-cheng i-shu* 紹興先正遺書 (Writings by masters of Shao-hsing), ed. Ma Yung-hsi 馬用錫. K'uei-chi, Hsu Shou-hsueh, 1883.
Shao I-chen 邵懿辰. *Ssu-k'u chien-ming mu-lu piao-chu* 四庫簡明目錄標注 (Marginal notes on the annotated catalogue of the imperial library). Jen-ho, Shao-shih pan-yeh-lu, 1911.

Sheng Lang-hsi 盛朗西. *Chung-kuo shu-yuan chih-tu* 中國書院制度 (The system of academies in China). 1934. Taipei, Hua-shih Publishing Company reprint, 1977.

Shih-lu See *Ta-ch'ing li-chao shih-lu.*

Shils, Edward F. "Intellectuals," in *International Encyclopedia of the Social Sciences* ed. David A. Sills. New York, Macmillan, 1968.

Shimada Kenji 島田虔次. "Shō Gakusei no ichi" 章學誠の位置 (Chang Hsueh-ch'eng's position in the history of Chinese thought), *Tōhō gakuhō* 41:519-530 (1968).

Sivin, Nathan. "Why the Scientific Revolution Did Not Take Place in China, or Didn't It?," *Chinese Science* 5:45-66 (1982).

Spence, Jonathan D. *Ts'ao Yin and the Kang-hsi Emperor, Bondservant and Master.* New Haven, Yale University Press, 1966.

Ssu-ma Ch'ien 司馬遷. *Shih-chi* 史記 (Records of the historian). Peking, Chung-hua shu-chü, 1962.

Stone, Lawrence D. "Prosopography," *Daedalus* 100.1:46-79 (Winter 1971).

Struve, Lynn A. "Uses of History in Traditional Chinese Society: The Southern Ming in Ch'ing Historiography." Ph.D. dissertation, University of Michigan, 1974.

————. "Ambivalence and Action: Some Frustrated Scholars of the K'ang-hsi Period," in *From Ming to Ch'ing,* ed. Jonathan D. Spence and John E. Wills, Jr. New Haven, Yale University Press, 1979.

————. "The Hsu Brothers and Semiofficial Patronage of Scholars in the K'ang-hsi Period," *Harvard Journal of Asiatic Studies* 42:231-266 (1982).

Sugimura Yūzō 杉村勇造. *Kenryū kotei* 乾隆皇帝. Tokyo, Sangen Sha, 1961.

Sun Hsing-yen 孫星衍. *Sun-shih tz'u-t'ang shu-mu* 孫氏祠堂書目 (Catalog of the library of the Sun family at the literary hall). 1810. Shanghai, Commercial Press reprint, 1935.

————. *Huan-yü fang-pei-lu* 寰宇訪碑錄 (Stone inscriptions of the empire). 1820. Shanghai, Commercial Press reprint, 1935.

————. *Sun Yuan-ju wai-chi* 孫淵如外集 (Additional writings of Sun Yuan-ju [Hsing-yen]), ed. Wang Chung-min 王重民. Shanghai, Commercial Press, 1931.

Swann, Nancy Lee. "Seven Intimate Library Owners," *Harvard Journal of Asiatic Studies* 1.3:363-390 (1936).

Taam, Cheuk-woon. *The Development of Libraries under the Ch'ing Dynasty.* Shanghai, Commercial Press, 1934.

Ta-Ch'ing li-ch'ao shih-lu 大清歷朝實錄 (Veritable records of successive reigns of the Ch'ing dynasty). Tokyo, Okura shuppan kabushiki kaisha, 1937-1938.

Tai Chen 戴震. *Tai Chen wen-chi, yuan-shan, meng-tzu tzu-i shu-cheng* 戴震文集原善孟子字義疏證 (The collected writings of Tai Chen, the original goodness, and a philological commentary on the Mencius). Taipei, Ho-lo t'u-shu ch'u-pan-she reprint, 1975.

————. *Kao-kung chi-t'u* 考工記圖 (Illustration of the implements described in the "Record of Technology" chapter of the Rites of Chou). 1755.

Reprinted in Juan Yuan comp., *Huan-ch'ao ching-chieh* 皇朝經解 (Classical commentaries of our times). Kuangchou, 1860.

Tai Chün-jen 戴君仁. *Yen-Mao ku-wen Shang-shu kung an* 閻毛古文尚書公案 (The case of the Yen Jo-chu versus Mao Ch'i-ling on the Old Text Documents). Hong Kong, Chi Sheng Book Co., 1963.

Tai Yi 戴逸. "Han-hsueh tan-che" 汉学探析 (A preliminary analysis of the Han learning), *Ch'ing-shih yen-chiu chi* 清史研究集 (A collection of research on Ch'ing history) 2:1-45 (1982).

Takigawa, Kametarō 瀧川龜太郎. *Shiki kaichū kōshō* 史記會注考證 (A study of assembled commentaries on the Shih-chi). Tokyo, Tōhō bunka gakuin, 1932-1934.

T'ang Chih-chün 湯志均. "Ch'ing-tai ching-chin-wen-hsueh ti fu-hsing" 清代經今文學的復興 (The revival of new text learning in the Ch'ing), *Chung-kuo-shih yen-chiu* 1980.2:145-156 (June, 1980).

Te-hsing t'ung-chih 德興通志 (Gazetteer of Te-hsing county), ed. Yang Chung-ya 楊重雅 and Meng Ch'ing-yun 孟慶雲. Te-hsing, 1872. Taipei, Ch'eng-wen ch'u-pan-she reprint, 1975.

Teng Shih 鄧實. "Kuo-hsueh chin-lun" 國學今論 (National scholarly traditions in contemporary perspective), *Kuo-ts'ui hsueh-pao* 國粹學報 4, 5 (1904). Taipei, Wen-hai Publishing Company reprint, pp. 392-496, 517-526.

Teng Ssu-yü and Knight Biggerstaff. *An Annotated Bibliography of Selected Chinese Reference Works.* Third revised edition. Cambridge, Mass., Harvard-Yenching Institute, 1971.

Tjan Tjoe-som. *Po Hu-t'ung: The Comprehensive Discussions in the White Tiger Hall.* Leiden, E. J. Brill, 1949.

Torbert, Preston M. *The Ch'ing Imperial Household Department: A Study of its Organization and Principal Functions, 1662-1796.* Cambridge, Mass., Council on East Asian Studies, Harvard University, 1977.

Tseu, Augustinius A. *The Moral Philosophy of Mo-tzu.* Taipei, China Printing Ltd., 1965.

Tsien, T. H. "A History of Bibliographic Classification in China," *The Library Quarterly* 22.4:307-324.

——. *Written on Bamboo and Silk.* Chicago, University of Chicago Press, 1962.

Tso Pu-ch'ing 左步青. "Ch'ien-lung fen-shu" 乾隆焚書 (Ch'ien-lung's burning of the books), *Ku-kung po-wu-yuan yuan-k'an* 故宮博物院院刊 1:28-37 (February 1980).

Twitchett, Dennis. "The Fan Clan's Charitable Estate, 1050-1760," in *Confucianism in Action,* ed. David S. Nivison and Arthur F. Wright. Stanford, Stanford University Press, 1959.

Tu Wei-ming. *Centrality and Commonality: An Essay on the Chung-yung.* Honolulu, University of Hawaii Press, 1976.

Tuan Yü-ts'ai 段玉裁. "Tai Tung-yuan hsien-sheng nien-p'u" 戴東原先生年譜 (A chronological biography of Mr. Tai Tung-yuan), in *Tai Chen wen-chi* 戴震文集. Hong Kong, Chung-hua shu-chü, 1974.

Wakeman, Frederic, Jr. "Localism and Loyalism during the Ch'ing Conquest of Kiangnan," in *Conflict and Control in Late Imperial China,* ed. Frederic

Wakeman, Jr., and Carolyn Grant. Berkeley, Univ. of California Press, 1975.

Waley, Arthur. *Yuan Mei: An Eighteenth Century Chinese Poet*. London, Allen and Unwin, 1956.

Wan Ssu-t'ung 萬斯同. *Ch'ün-shu i-pien* 羣書疑辨 (Resolving doubts about various books). 1816. Taipei; Kuang-wen shu-chü reprint.

——. "Ming-shih i-wen-chih hsu" 明史藝文志序 (Preface to the treatise on art and literature prepared for the Ming dynastic history), in *Ming-shih i-wen-chih pu-pien, fu-pien* 明史藝文志補編續編 (Essays on art and literature prepared for the Ming dynastic history, with supplements). Shanghai, Commercial Press, 1959.

Wang Ch'ang 王昶. *Chin-shih ts'ui-pien* 金石萃編 (A compendium of stone inscriptions). 1805. Taipei, I-wen Publishing Co., 1966.

Wang Chung 汪中. *Shu-hsueh* 述學 (On scholarship). 1792. Taipei, Kuang-wen shu-chü reprint, 1970.

Wang Chung-han 王鍾翰. *Ch'ing-shih tsa-k'ao* 清史雜考 (Studies in Ch'ing history). Beijing, Jen-min ch'u-pan-she, 1957.

Wang Lan-yin 王蘭蔭. "Chi Hsiao-lan hsien-sheng nien-p'u" 紀曉嵐先生年譜 (A chronological biography of Mr. Chi Hsiao-lan [Yun]), *Shih-ta yueh-k'an* 師大月刊 1.6:77–106 (September 1932).

Wang Po-hou. *The Three Character Classic*, tr. Chiang Ker-chiu. Singapore, Chung-hua Mandarin Institute, 1941.

Wang Ssu-chih 王思治. "Ming Ch'ing wen-tzu-yü chien-lun" 明清文字獄簡论 (A brief discussion of literary cases during the Ming and Ch'ing), *Jen-min jih-pao* 人民日報, August 24, 1979.

Wang T'ai-yueh 王太岳. *Ssu-k'u ch'üan-shu k'ao-cheng* 四庫全書考證 (An examination of the Complete Library of the Four Treasuries). 1895. Commercial Press reprint, Shanghai, 1936.

Wang Tseng 王增, comp. *Hsin-ch'a hsien-chih* 新蔡縣志 (Gazetteer of Hsin-ch'a County). Hsin-ch'a, 1796.

Watson, Burton. *Basic Writings of Mo-tzu, Hsun-tzu and Han Fei-tzu*. New York, Columbia University Press, 1967.

Watson, James L. "Chinese Kinship Reconsidered: Anthropological Perspectives on Historical Research," *The China Quarterly* 92:589–622 (December 1982).

Watt, John R. *The District Magistrate in Late Imperial China*. New York, Columbia University Press, 1972.

Weber, Max. *The Religion of China*. New York, Free Press paperback, 1968.

Wei Cheng 魏徵 *Sui-shu* 隋書 (History of the Sui). Peking, Chung-hua shu-chü, 1973.

Wei Ch'ing-yuan 韦庆远. "Ch'ung-tu Ch'ing-tai wen-tzu-yü tang" 重读清代文字獄档 (On rereading archives of literary cases during the Ch'ing), *Tu-shu* 读书 3:90–100 (June 1979).

Wei, Peh-t'i, "Juan Yuan: A Biographical Study with Special Reference to Mid-Ch'ing Security and Control in Southern China, 1799–1835." Ph.D. dissertation, University of Hong Kong, 1981.

Welch, Holmes, and Anna Seidel, eds. *Facets of Taoism*. New Haven, Yale University Press, 1979.

Wen-hsien ts'ung-pien 文獻叢編 (Collectanea from the historical records office). Peip'ing, Palace Museum, 1930–1936.

Weng Fang-kang 翁方綱. *Liang Han chin-shih-chi* 兩漢金石集 (Collection of Han stone inscriptions). 1789. Taipei, I-wen Publishing Company reprint, 1966.

——. *Weng-shih chia-shih lueh-chi* 翁氏家事略記 (A history of the Weng family). Peking, 1818.

——. *Fu-ch'u-chai wen-chi* 復出齋文集 (Collected writings from the Fu-ch'u Studio), comp. Li Yen-chang 李彦章. 1836. Taipei, Wen-hai Publishing Company reprint, 1969.

Whitbeck, Judith. "The Historical Vision of Kung Tzu-chen (1792–1841)." Ph.D. dissertation, University of California, Berkeley, 1980.

Wiens, Mi-chu. "Anti-Manchu Thought During the Early Ch'ing," *Harvard University Papers on China* 22A:1–24 (May 1969).

Wilhelm, Hellmut. "The Po-hsueh hung-ju Examination of 1679," *Journal of the American Oriental Society* 71.1:60–66 (March 1951).

Wilkinson, Endymion. *The History of Imperial China: A Research Guide.* Cambridge, Mass., East Asian Research Center, Harvard University, 1974.

Winkelman, John H. "The Imperial Library of Southern Sung China, 1127–1279," *Transactions of the American Philosophical Society* 64.8 (1974).

Wolf, Arthur P. and Huang Chieh-shan. *Marriage and Adoption in China, 1845–1945.* Stanford, Stanford University Press, 1980.

Woodside, Alexander B. "The Ch'ien-lung Reign." Draft Chapter for *The Cambridge History of China,* Volume 9. Forthcoming.

Wu Che-fu 吳哲夫. *Ch'ing-tai chin-hui shu-mu yen-chiu* 清代禁燬書目研究 (A study of the indices of banned books during the Ch'ing). Taipei, Chia-hsin shui-ni kung-ssu wen-hua chi-chin-hui, 1969.

——. "Hsien-ts'un hsu-hsiu ssu-k'u ch'üan-shu t'i-yao mu-lu cheng-li hou-chi" 現存續修四庫全書提要目錄整理後記 (Remarks after having arranged the content of the extant enlarged and revised catalog on the Four Treasuries), *Ku-kung wen-hsien* 故宮文獻 1.3:29–43 (June 1970).

——. *Ssu-k'u ch'üan-shu hui-yao tsung-hsiu k'ao* 四庫全書薈要纂修考 (The compilation of the essentials of the Complete Library of the Four Treasuries). Taipei, Ku-kung ts'ung-k'an, 1976.

Wu Ching-tzu. *The Scholars,* tr. Gladys Yang and Yang Hsien-yi. New York, Grosset and Dunlap, 1972.

Wu-ying-tien chü-chen pan ts'ung-shu 武英殿聚珍版叢書 (Collectanea printed from assembled pearls in the Wu-ying Throne Hall). 1794. Canton, Kuang-ya shu-chü reprint, 1899.

Wu, Silas. (Wu Hsiu-liang 吳秀良). "Nan-shu-fang chih chien-chih chi ch'i ch'ien-ch'i fa-chan" 南書房之建置及其前期之發展 (The founding of the southern library and its early development), *Ssu yü yen* 思與言 5.6:6–12 (1968).

——. *Communications and Control in Imperial China.* Cambridge, Mass., Harvard University Press, 1970.

——. *Passage to Power: The K'ang-hsi Emperor and his Heir Apparent.* Cambridge, Mass., Harvard University Press, 1979.

Yamanoi Yū 山井湧. "Minmatsu Shinsho shisō ni tsuite no ichi kōsatsu" 明末清初思想について の一考察 (An overview of the late Ming and early Ch'ing thought), *Tōkyō Shinagaku hō* 東京支那學報 11:37–54 (1965).

Yang Chia-lo 楊家駱. *Ssu-k'u ch'üan-shu kai-shu* 四庫全書概述 (A compilation of source materials on the Complete Library of the Four Treasuries). 1930. Taipei, Chung-kuo hsueh-tien kuan reprint, 1970.

Yang, L. S. "The Organization of Chinese Official Historiography," in *Historians of China and Japan,* Edwin Pulleyblank and Wm. G. Beasley, editors. London, Oxford University Press, 1961.

Yang Li-ch'eng 楊立誠, comp. *Ssu-k'u mu-lueh* 四庫目略 (The Four Treasuries revisited). Hangchow, Chekiang Publishing Co., 1929.

Yang Shen 楊慎. *Sheng-an ch'üan-chi* 升菴全集 (The complete works of Sheng-an). 1795. Shanghai, Commercial Press reprint, 1936.

Yang Shih-ch'i 楊士奇, ed. *Wen-yuan-ko shu-mu* 文淵閣書目 (Catalog of books stored at the Wen-yuan Pavilion). 1441. Shanghai, Commercial Press reprint, 1937.

Yao Ch'a 姚察 and Yao Ssu-lien 姚思廉. *Liang-shu* 梁書 (History of the Liang). Peking, Chung-hua shu-chü, reprint, 1973.

Yao Ming-ta 姚名達. *Chu Yun nien-p'u* 朱筠年譜 (Chronological biography of Chu Yun). Shanghai, Commercial Press, 1933.

——. *Chung-kuo mu-lu-hsueh shih* 中國目錄學史 (A history of Chinese bibliography). Changsha, Commercial Press, 1938. Taipei, Commercial Press reprint, 1974.

Yao Nai 姚鼐. "Hsi-pao-hsuan shu-lu" 惜抱軒書錄 (Book reviews of [Yao Nai]), in *Hsi-pao-hsuan i-shu san-chung* 惜抱軒遺書三種 (Three unpublished manuscripts by Yao Nai) ed. Mao Yü-sheng 毛嶽生. T'ung-ch'eng. 1879.

——. *Hsi-pao hsien-sheng ch'ih-tu* 惜抱先生尺牘 (Personal letters of Yao Nai), ed. Ch'en Yung-kuang 陳用光. Wu-hsi, Hsiao-wan-ying t'ang, 1909.

——. *Hsi-pao hsuan ch'üan-chi* 惜抱軒全集 (Complete works of Yao Nai). 1921. Shanghai, Shih-chieh shu-chü reprint, 1936.

Yeh Lung 葉龍. *T'ung-ch'eng p'ai wen-hsueh shih* 桐城派文學史 (History of the T'ung-ch'eng school of literature). Taipei, Wen-chin ch'u-pan-she, 1975.

Yeh Te-hui 葉德輝. *Shu-lin ch'ing-hua* 書林清話 (Notes on bibliography). 1920. Peking, Chung-hua shu-chü reprint, 1957.

Yü Chia-hsi 余嘉錫. *Mu-lu-hsueh fa-wei* 目錄學發微 (Discoveries in bibliography). Peking, Chung-hua shu-chü, 1963.

——. *Ssu-k'u t'i-yao pien-cheng* 四庫提要辯證 (Revisions of the reviews in the Complete Library of Four Treasuries). N.p., 1937.

Yü Min-chung 于敏中. *Yü Wen-kung (Min-chung) shou-cha* 于文公 (敏中) 手扎 (Letters of Yü Min-chung), ed. Ch'en Yuan 陳垣. Taipei, Wen-hai Publishing Company reprint, 1968.

Yü Ying-shih 余英時. "Ch'ing-tai ssu-hsiang shih ti i-ko hsin chieh-shih" 近代思想史的一個新解釋 (A new explanation of Ch'ing intellectual history), in *Li-shih yü ssu-hsiang* 歷史與思想. (History and thought). Taipei, Lien-ching ch'u-pan-she, 1976.

———. "Ts'ung Sung-Ming ju-hsueh ti fa-chan lun Ch'ing-tai ssu-hsiang-shih" 從宋明儒学的發展論清代思想史 (A discussion of Ch'ing intellectual history from the point of view of the development of Sung and Ming Confucianism), in *Li-shih yü ssu-hsiang*. Taipei, Lien-ching ch'u-pan-she, 1976.

———. *Lun Tai Chen yü Chang Hsueh-ch'eng* 論戴震與章學誠 (A study of Tai Chen and Chang Hsueh-ch'eng). Hong Kong, Lung-men shu-tien, 1976.

Yueh-ts'ang chih-chin 閱藏知津 (A bibliographic guide to the tripitaka), ed. .Chih-hsu 智旭. First Preface, 1654. Taipei, Hsin-wen-feng ch'u-pan-she reprint, 1973.

Znaniecki, Florian. *The Social Role of the Man of Knowledge*. New York, Columbia University Press, 1940.

Glossary

A-kuei 阿桂

Ch'a Pi-ch'ang 查必昌
Ch'a Shih-kuei 查世桂
ch'an 諂
Chang Hsi-nien 長羲年
Chang Hsueh-ch'eng 章學誠
Chang Jo-kuei 張若桂
Chang Ping-lin 章炳麟
Chang Shou-chieh 張守節
Chang Shun-hui 張舜徽
Chang T'ing-yü 張廷玉
Chang Yü-shu 張玉書
Ch'ang-shu 常熟
Chao Ch'i 趙岐
Chao Fu 趙復
Chao Ming-ch'eng 趙明誠
Chao-ming t'ai-tzu 昭明太子
Ch'ao Kung-wu 晁公武
Ch'en An-p'ing 陳安平
Ch'en Ch'ang-ch'i 陳昌齊
Ch'en Chi-ju 陳繼儒
Ch'en Ching-li 陳經禮
Ch'en Hsi-sheng 陳希聖

Ch'en Hui-tsu 陳輝祖
Ch'en Meng-lei 陳夢雷
Ch'en Nai-ch'ien 陳乃乾
Ch'en Ti 陳第
Ch'en Yuan 陳垣
Ch'en Yueh-wen 陳燿文
Ch'en Yung-kuang 陳用光
cheng 政
Cheng Chiao 鄭樵
Cheng Hsuan 鄭玄
Cheng K'ang-ch'eng 鄭康成
Cheng-teng hui-yuan 正燈會元
Ch'eng Chin-fang 程晉芳
Ch'eng-kung-ts'e 成工冊
Ch'eng Ming-yin 程明諲
Chi Hsiao-lan 紀曉嵐
Chi-hsien shu-mu 集賢書目
Chi-ku-lu 集古錄
chi-kuo-chi 計過記
Chi-shu-yuan shu-mu 籍書園書目
Ch'i-lu hsu-mu 七錄序目
Ch'i-lueh 七略
Ch'i-tan 契丹
Chiang-che 江浙

269

Chiang Fan 江藩
Chiang Kuang-ta 江廣達
Chiang Sung-ju 蔣松如
Chiang Yü-ts'un 蔣漁村
Chiang Yung 江永
chiao 教
Chiao-hua chin-fang hsiao-chi 椒花吟舫小集
Chiao Hsun 焦循
Chiao Hung 焦竑
Chiao-ch'ou t'ung-i 校讎通義
chieh 羯
chien-sheng 監生
Ch'ien Ch'ien-i 錢謙益
Ch'ien Mu 錢穆
Ch'ien Ta-chao 錢大昭
Ch'ien Ta-hsin 錢大昕
Ch'ien Tien 錢坫
Ch'ien Tung-yuan 錢東垣
Chih-chin-chai ts'ung-shu 咫進齋叢書
chih-chung ch'eng-hsien 執中成憲
Chin Chien 金簡
Chin-ch'uan 金川
Chin Pao 金堡
chin-shih 進士
Chin-shih 金史
Chin-shih-lu 金石錄
Ch'in Hui-t'ien 秦蕙田
Ch'in Shih-huang-ti 秦始皇帝
Ch'in Ssu-fu 秦思復
Ch'in-ting ssu-shu-wen 欽定四書文
Ching-shih wen-pien 經世文編
Ch'ing 清
Ch'ing-shih 清史
Ch'ing Han-wen hsiao-hsueh 清漢文小學
Ch'ing-hsia-chi 青霞集
Ch'ing-tai chin-hui shu-mu pu-i 清代禁燬書目補遺
Ch'ing-wen-chien 清文鑑
Chiu-chang hsuan-shu 九章算書
Chiu-kuo-chih 九國志
Chiu Man-wen tang 舊滿文檔
Chiu Wu-tai-shih 舊五代史
Ch'iu Yueh-hsiu 裘日修
cho 濁
Cho Ch'ang-ling 卓長齡
Chou-i pen-i 周易本義
Chou-i shu-i 周易述義
Chou-kuan i-shu 周官義疏
Chou Ping-t'ai 鄒炳泰
Chou Yung-nien 周用年
Chu Chieh 祝浹
Chu Hsi 朱熹
Chu Huang-fan 祝煌燔
Chu Hui 祝洄
Chu Kuei 朱珪
Chu-shu t'ung-chien 竹書統箋
Chu Ssu-tsao 朱思藻
Chu P'ing-chang 祝平章
Chu Ti 朱棣
Chu T'ing-cheng 祝廷諍
Chu-tzu chu-i 朱子諸義
Chu Wen-tsao 朱文藻
Chu Yun 朱筠
Chu Yun-wen 朱允炆
ch'u-ch'i-men 出其門
Ch'u-tz'u 楚辭
Ch'u-tz'u pien-cheng 楚辭編正
chü-jen 舉人
Ch'ü Ta-chun 屈大均
chuan-tzu fen-jao 轉滋紛擾
ch'uan-kuan chih-nan 穿貫之難
chüan 卷
Ch'üan-hsueh-pien 勸學篇

Ch'üan-te 全德
Ch'üan Tsu-wang 全祖望
Chuang T'ing-lung 莊廷鑨
Chuang Ts'un-yü 莊存與
Chuang-tzu 莊子
Chuang Yun-ch'eng 莊允城
ch'ui-mao ch'iu-tz'u 吹毛求疵
chün-tzu 君子
Ch'ün-chih-chai shu-mu 郡直齋書目

Ch'un-ch'iu 春秋
Ch'un-ch'iu chih-chieh 春秋直解
Ch'ün-shu ssu-lu 群書四錄
Chung-hsing kuan-ko shu-mu 中興
館閣書目
Chung-shan shu-yuan 鍾山
書院
Chung-yin 鍾音
Chung-yung 中庸
Chung-yung chi-chieh 中庸集解
Chung-yung chi-lueh 中庸輯略
Ch'ung-wen tsung-mu 崇文總目

Erh-ya 爾雅
Erh-ya i-shu 爾雅義疏

Fa-yuan chu-lin 法苑珠林
fan-lieh 凡例
Fan Mou-chu 范懋柱
Fang Chao-ying 房兆楹
Fang Chung-lü 方中履
Fang Chung-te 方中德
Fang I-chih 方以智
Fang Kuo-t'ai 方國泰
Fang-lueh-kuan 方略館
Fang Pao 方苞
Fang Tung-shu 方東樹
Feng T'ing-cheng 馮廷正
fu-chiao-kuan 副校官

Fu Ch'ien 服虔
Fu-heng 傅恆
Fu-lung-an 福隆安

Hai-ch'eng 海成
Han Ch'eng-ti 漢成帝
Han Ching-ti 漢景帝
Han Fei-tzu 韓非子
Han-hsueh 漢學
Han-hsueh shang-tui 漢學商對
Han-shu 漢書
Han Wu-ti 漢武帝
Hang Shih-chün 杭世駿
Hao I-hsing 郝懿行
Hao-shuo 郝碩
Ho Cho 何焯
Ho-hsiao 和孝
Ho Hsiu 何休
Ho-shen 和珅
Hou-han-shu 後漢書
Hsi-hsia 西夏
Hsi-yü t'ung-wen-chih 西域通文志

Hsiao-ching chi-chu 孝經集注
Hsiao-hsueh-k'ao 小學考
Hsieh-chi pan-fang-shu 協紀辨方書
Hsieh Chi-shih 謝濟世
Hsieh Ch'i-k'un 謝啟昆
Hsieh Chin 解縉
Hsien-chang lu 憲章錄
Hsin-ch'ang 新昌
Hsin-hsueh wei-ching k'ao 新學偽經

Hsin Wu-tai-shih 新五代史
Hsing-li ta-ch'üan 性理大全
Hsiu-ning 休寧
hsiu-ts'ai 秀才
Hsiung Tz'u-li 熊賜履
hsu 序

Hsu Ch'ien-hsueh 徐乾學
Hsu k'ao-ku-t'u 續考古圖
Hsu Kuang 徐廣
Hsu Kuang-ch'i 徐光啟
Hsu Pen 徐本
Hsu san-tzu-ching 續三字經
Hsu Shen 許慎
Hsu Ta-ch'un 徐大椿
Hsu Wen-ching 徐文靖
Hsuan-hsueh ch'üan-shu 算學全書
hsueh-cheng* 學政
hsueh-cheng 學正
Hsueh Chü-cheng 薛居正
Hsueh-hai-t'ang 學海堂
Hsueh Ying-ch'i 薛應旂
hsun-ku 訓詁
hu 胡
Hu Chung-tsao 胡中藻
Hu Wei 胡渭
Huan-yü fang-pei lu 寰宇訪碑錄

Huang Ching-jen 黃景仁
Huang-Ch'ing ching-chieh 皇清
　　經解
Huang Fang 黃芳
Huang-lao 黃老
Huang P'ei-lieh 黃丕烈
Huang Shou-ling 黃壽齡
Huang-ti 黃帝
Huang Tsung-hsi 黃宗羲
Huang yü hsi-yü t'u-chih 皇輿
　　西域圖志
Huang Yun-mei 黃雲眉
Hui-chou 徽州
Hui Tung 惠棟
Hung Liang-chi 洪亮吉

i 夷

i 義
I-ching 易經
i-hsu 議敘
I-hsueh hsiang-shu lun 易學象數論
I-li 儀禮
I-li i-shu 儀禮義疏
I-lin 意林
I-ling-a 伊齡阿
I-lun 義論
I-t'u ming-pien 易圖明辨

Jao-chou 饒州
jen-hsin 人心
Jen Ta-ch'un 任大春
Jih-chiang ssu-shu chieh-i 日講四書
　　解義
Jih-chih-lu 日知錄
Jih-hsia chiu-wen-k'ao 日下舊問考
Ju-ts'ang-shuo 儒藏說
Juan Hsiao-hsu 阮孝緒
Juan Yuan 阮元

K'ai-kuo fang-lueh 開國方略
Kan Pao 干寶
Kanda Nobuo 神田信夫
kang-mu 綱目
K'ang-hsi tzu-tien 康熙字典
Kao Chih-ch'ing 高治清
Kao-chin 高晉
Kao-hsing 高辛
Kao-seng chuan 高僧傳
Kao-yang 高陽
Kao Yun-ts'ung 高雲從
k'ao-cheng 考正
K'ao ku-t'u 考古圖
K'ao-kung chi-t'u 考工記圖
k'ao Li 考禮
ko-wu 格物

K'o Wei-ch'i 柯維祺
Ku-chin shih-i 古今釋疑
Ku K'ai-chih 顧愷之
Ku Kuang-chi 顧廣圻
Ku-liang chuan 穀梁傳
Ku-shih 故史
Ku-shih-pi 古事比
ku-wang t'ing-chih 姑妄聽之
ku-wen 故文
Ku-wen shang-shu k'ao 故文尚書考

Ku-wen shang-shu yuan-tz'u 古文尚書寃詞
Ku Yen-wu 顧炎武
Ku-yun piao-chun 古韻標準
Kuei Fu 桂馥
Kuei-t'ien so-chi 歸田瑣記
Kung-chen chuan 功臣傳
Kung-fei 公非
Kuo Po-kung 郭佰恭
kung-shih 公是
Kuo-t'ai 國泰
Kung-yang chuan 公羊傳

lei-yao 類要
li 理
li 禮
li 里
Li-chi 禮記
Li-chi i-shu 禮記義疏
Li Chih-ying 李質穎
Li Fu 李紱
Li Hu 李湖
Li Kuang-ti 李光地
Li Lun-yuan 李掄元
Li Po 李白
Li-sao 離騷
Li Shih-chieh 李世傑
Li Shih-yao 李侍堯

Li Shou-ch'ien 勵守謙
Li Ta-pen 黎大本
Li-tai ming-ch'en tsou-i 歷代明臣奏義
Li Tz'u-ming 李慈銘
Li Wei 李威
Li Yen-shou 李延壽
Li Yung 李顒
Liang-ch'ao kang-mu pei-yao 兩朝綱目備要
Liang Chang-chü 梁章鉅
Liang Ch'i-ch'ao 梁啟超
Liang-Han chin-shih chi 兩漢金石記

Liang-huai 兩淮
Liang San-ch'uan 梁三川
Liang Shang-kuo 梁上國
Liang-shu 梁書
Liang Wu-ti 梁武帝
Liao-Chin-Yuan shih 遼金元史
Liao-Chin-Yuan shih kuo-yü chieh 遼金元史國語解
Liao-shih 遼史
Lieh-tzu 列子
Ling T'ing-kan 凌廷堪
Liu Chen-yü 劉震宇
Liu Chih-chi 劉知幾
Liu Chih-lin 劉之遴
Liu Ch'üan-chih 劉權之
Liu Feng-lu 劉逢祿
Liu Han-p'ing 劉漢屏
Liu Hsiang 劉向
Liu Hsin 劉歆
Liu Jo-yü 劉若愚
Liu-li-ch'ang 琉璃廠
Liu Shih-p'ei 劉師培
Liu-shu yin-yun piao 六書音均表
Liu Tsung-chou 劉宗周
Liu T'ung-hsun 劉統勳

Lo-erh-chin 勒爾謹
Lou Sheng 樓繩
Lu 魯
Lu Ch'ang 盧昶
Lu-fei chih 陸費墀
Lu Hsi-hsiung 陸錫熊
Lu Shen 陸深
Lu Wen-ch'ao 盧文弨
Lü-li yuan-yuan 律歷淵源
Lü Liu-liang 呂留良
Lü-lü cheng-i 律呂正義
Lü Ta-lin 呂大臨
Lun-yü hou-lu 論語後錄
Lung-kang chien-piao 瀧岡阡表

Ma Tuan-lin 馬端臨
Ma Yü 馬裕
Ma Yung-hsi 馬用錫
Mambun Roto 滿文老檔
Man-chou ch'a-shen ch'a-t'ien tien-li
　　滿洲祭神祭天典禮
Man-chou chiu-tang 滿洲舊檔
Man-chou yuan-liu kao 滿洲源流考
Mao Ch'ang 毛亨
Mao Ch'i-ling 毛奇齡
Mao Chin 毛晉
Mao-shih ku-yin-k'ao 毛詩古音考
Mao Yü-sheng 毛齮生
Mei Ting-tso 梅鼎祚
men-jen 門人
meng 蒙
Meng-ku wang-kung kung-chi piao-chuan 蒙古王公功績表傳
Meng-ku yuan-liu 蒙古源流
Miao 苗
Min Ao-yuan 閔鶚元
ming 命
Ming-ch'en tsou-i 名臣奏義
Ming-shih 明史

Ming-shu 明書
Ming T'ai-tsu 明太祖
Mo Chan-lu 莫瞻菉
Mo-tzu 墨子
Mu-lan 木蘭
mu-yu 幕友

Nan-hu-chi 南湖集
Nan-shih 南史
Nan-shu-fang 南書房
Nieh Ch'ung-i 聶崇義
Ning-kuo 寧國

O-erh-t'ai 鄂爾泰
O-erh-teng-pu 額爾登部
Ou-yang Hsiu 歐陽修

Pa-ch'i t'ung-chih 八旗通志
pa-ku 八股
Pa-san-t'ing 巴三廷
Pai-Sung i-ch'an 百宋一廛
Pan Ku 班固
Pan Piao 班彪
P'an Lei 潘耒
pao-chia 保甲
Pao Shih-kung 鮑士恭
P'ei Tsung-hsi 裴宗錫
P'ei-wen yun-fu 佩文韻府
P'ei Yin 裴駰
P'eng Yuan-tuan 彭元端
pi 比
pi-shu-chien 秘書監
Pi-yuan 畢沅
pi-yung 辟雍
Pieh-lu 別錄
p'ien 篇
P'ing-ting liang Chin-ch'uan fang-lueh
　　平定兩金川方略
P'ing-ting Tsun-ko-erh fang-lueh
　　平定準噶爾方略

po 博
Po-hsueh hung-ju 博學鴻儒
Po-i 伯夷
Po Lü Liu-liang ssu-shu chiang-i 駁
呂留良四書講義
pu-t'ung-hsiao Han-wen 不通曉
漢文
Pu Yuan-shih i-wen-chih 補元史藝
文志
p'u-hsueh 樸學

San-ch'ao pei-meng hui-pien 三朝北
盟會編
San-kuo-chih 三國志
San-li i-shu 三禮義疏
San li-t'u chi-chieh 三禮圖集解
San-pao 三寶
Shan-hai-ching 山海經
Shan-hai-ching pu-chu 山海經補
注
Shan-ho liang-chieh k'ao 山河兩戒考

Shang-shu 尚書
Shang-shu kuang-t'ing lu 尚書廣聽
錄
Shang-shu ku-wen shu-cheng 尚書古文
疏正
Shang-shu pien-wei 尚書辨偽
Shao Chin-han 邵晉涵
Shao Hsiang-jung 邵向榮
Shao T'ing-ts'ai 邵廷采
Shen-hsien chuan 神仙傳
Shen-nung 神農
Shen Shih-po 沈世泊
Shen-tzu 申子
Sheng-ching-chih 盛京志
sheng-yuan 生員
shih 時
Shih-ch'eng k'ao-wu 史乘考誤
Shih-chi chi-chieh 史記集解

Shih-chia-chai yang-hsin-lu 十駕齋養新
錄
Shih-ching 詩經
Shih-i che-chung 詩義折中
Shih Jun-chang 施閏章
shih-shih ch'iu-shih 實事求是
Shih Tun 石敦
shih-wen 時文
Shih-yen-chai shu-mu 石研齋書目
shu-chü 書局
Shu-ching 書經
Shu-ho-te 舒赫德
Shu-nan-chi 術南集
Shun-t'ien 順天
Shuo-wen 說文
Shuo-wen chieh-tzu 說文解字
So-yin-shih-ti chin-shu tsung-lu 索引式
的禁書總錄
Ssu-k'u ch'üan-shu 四庫全書
Ssu-k'u ch'üan-shu chien-ming mu-lu
四庫全書簡明目錄
Ssu-k'u ch'üan-shu hui-yao 四庫全書
薈要
Ssu-k'u shu-mu 四庫書目
Ssu-ma Chen 司馬貞
Ssu-pu pei-yao 四部備要
Ssu-shu huo-wen 四書或問
Su Ch'e 蘇轍
Su Tung-p'o 蘇東坡
Sun Ch'en-tung 孫辰東
Sun Chieh-ti 孫楷第
Sun-tzu 孫子
Sung-chih 宋志
Sung Lien 宋濂
Sung-shih 宋史
Sung-shih hsin-pien 宋史新編

Ta-Ch'ing i-t'ung-chih 大清一總志
Ta-Hsia ta-Ming hsin-shu 大夏
大明新書

Ta-hsueh 大學
Ta-hsueh yen-i 大學衍義
Ta-i chueh-mi-lu 大義覺迷錄
Ta-t'ung 大同
Ta-yeh cheng-yü shu-mu 大業正御書目
Tai Chen 戴震
Tai Ming-shih 戴名世
T'ai-p'ing 太平
T'an Shang-chung 譚尚忠
t'an-to wu-te 貪多無得
T'ang Hsuan-tsung 唐玄宗
T'ang T'ai-tsung 唐太宗
tao 道
Tao-hsin 道心
Tao-te ching 道德經
Tao-te ching-chieh 道德經解
Tao-te ching-chu 道德經注
Tao-tsang 道藏
t'ao 謟
Te-hsing 德興
Teh-pao 德保
Teng Hsi-yuan 鄧錫元
Teng Hui 鄧譓
Teng Shih 鄧實
Teng Yuan-hsi 鄧元錫
t'i-hsueh-kuan 提學官
t'i-tiao-kuan 提調官
T'i-yao 提要
T'ien Wen-ching 田文鏡
Ting-tzu 丁子
Ting Wen-pin 丁文彬
T'o-pa 拓跋
T'o-t'o 托托
Tsa-hsueh-pien 雜學辨
tsai kuan-chung 在館中
Ts'ai Hsin 蔡新
Tsang Sheng-mo 倉聖脈
Ts'ao Wen-chih 曹文埴

ts'e 冊
Tseng Ching 曾靜
Tseng Kung 曾鞏
tso 左
Tso-li wan-shih chih-p'ing-shu 佐理萬世治平書
tsuan-hsiu 纂修
tsung-ts'ai 總裁
tsung yueh-kuan 總閱官
Tu Fu 杜甫
Tu-shu-chih 讀書志
Tu Yü 杜預
Tuan Yü-ts'ai 段玉裁
Tung Ch'un 董椿
Tung-fang wen-hua shih-yeh wei-yuan-hui 東方文化事業委員會
Tung kuan 東關
t'u 圖
t'u 兔
T'u-ming-a 圖明阿
T'u-shu chi-ch'eng 圖書集成
T'ung-ch'eng 桐城
T'ung-chien kang-mu hsu-pien 通鑑綱目續編
T'ung-chih 通志
T'ung-chih-t'ang ching-chieh 通志堂經解
t'ung-lun 通論
T'ung-ya 通雅
Tzu-chih t'ung-chien 資治通鑑
Tzu-chih t'ung-chien kang-mu 資治通鑑綱目
Tzu-hsiao-chi 資孝集
Tzu-kuan 字貫
Tzu-ssu 子思
Tzu-yang 紫陽
Tz'u-t'ang pi-chi 祠堂筆記

wan 萬
Wan Ssu-t'ung 萬斯同
Wang Ch'ang 王昶
Wang Chi-hua 王際華
Wang Ch'i-shu 王啟淑
Wang Chung 汪中
Wang Erh-yang 王爾陽
Wang Hsi-chih 王羲之
Wang Hsi-hou 王錫侯
Wang Hui-tsu 汪輝祖
Wang Hung-hsu 王鴻緒
Wang Lun 王倫
Wang Lung-nan 王瀧南
Wang Ming-sheng 王鳴盛
Wang Nien-sun 王念孫
Wang Po-hou 王伯候
Wang Shih-chen 王士禎
Wang Tsung-yen 王宗炎
Wang Yang-ming 王陽明
Wang Yao-ch'en 王堯臣
Wei Chao 韋昭
Wei Cheng 魏徵
Wei I-chieh 魏裔介
Wei-shih hsien-shu 偽時憲書
Wei Yuan 魏源
Wen-chang cheng-tsung 文章正宗

wen-chi 文集
Wen-hsien-ch'u 文獻處
Wen-hsien ta-ch'eng 文獻大成

Wen-hsien t'ung-k'ao 文獻通考
Wen-hsuan 文選
Wen-chin-ke 文津閣
Wen-she 文社
Wen-te 文德
Weng Fang-kang 翁方綱
Wu Hsiao-kung 吳蕭公
Wu-hsueh-pien 吾學編

Wu-hu 蕪湖
Wu Lan-t'ing 吳蘭庭
Wu-li t'ung-k'ao 五禮通考
Wu San-kuei 吳三桂
Wu Sheng-lan 吳省蘭
Wu Wei-yeh 吳偉業
Wu-ying-tien 武英殿

Yang Ch'ang-lin 楊昌霖
Yang Chu 楊朱
Yang Shen 楊慎
Yao Chin-yuan 姚覲元
Yao Jung 姚瑩
Yao Ming-ta 姚名達
Yao Nai 姚鼐
Yao Tuan-ko 姚端恪
yeh-shih 野史
Yeh T'ing-t'ui 葉廷推
Yen Hsi-shen 顏希深
Yen Jo-chü 閻若璩
Yen Shih-ku 顏師古
Yin Chiang-i 尹講義
Yin-chih 胤祉
Yin Cho 寅著
Yin-fu-ching 陰符經
Yin-hsueh wu-shu 音學五書
Ying-lien 英廉
yu 友
Yü 禹
Yü Chi 余集
Yü Chia-hsi 余嘉錫
Yü Min-chung 于敏中
Yü Shan-chi 愚山集
Yü T'eng-chiao 余騰蛟
Yü Wen-i 余文儀
Yü-yao 餘姚
Yü Yueh 俞樾
Yuan Mei 袁枚
Yuan-shou-t'ung 袁守侗

yueh 約
Yueh-hsin-chi 悅心集
Yueh-man-t'ang pi-chi 越縵堂
　　　　筆記

Yun-ch'i ch'i-chien 雲笈七籤
Yung-jung 永瑢
Yung-lo 永樂
Yung-lo ta-tien 永樂大典

Index

Ssu-k'u Commission *(continued)*
punishment for negligence, 99;
Ho-shen as director-general, 103;
evaluation of work, 105; Shao
Chin-han's work at, 126, 129;
appointment and resignation of
Yao Nai, 143
Ssu-k'u shu-mu: reference to in edict,
37
Stone inscriptions: interest in of Peking scholars, 55; catalogs of, 64,
65; listing of considered by
Grand Council, 73
Su Ch'e: claim of common source of
Taoism and Buddhism, 115; *Ku-
shih* by, 146, 147
Sui dynasty: book collection during,
13
Sung dynasty: intellectuals of, impact of views in Ch'ing period, 5,
6; book catalogs compiled during,
13; collection of books produced
during, 58
Sung learning: divergence from Han
learning, 140, 145; adverse comment on, 154; as reaction to Han
learning movement, 155; hypothesis as to Yao Nai's role in origins, 156
Sung-shih: record of book borrowing,
13; view of eighteenth century
scholars on, 131; criticism of by
Shao Chin-han, 131, 132, 133

Ta-ch'ing i-t'ung-chih: compilation, 61
Ta-hsueh yen-i: Manchu edict directing study, 18
Ta-yeh cheng-yü shu-mu: compilation, 13
Tai Chen: interpretation of classical
texts, 43, 44, 45; association with
class of 1754 group, 50; inclusion
of works in *Ssu-k'u ch'üan-shu,* 57;
use of *Yung-lo ta-tien,* 62; appointment to Ssu-k'u Commission, 80;
association with Shao Chin-han,
126, 128; Yao Nai's request to be
a student of, 142

Tai Ming-shih: prosecution of, 23,
24, 142
T'ang dynasty book catalogs, 13
Tao-te-ching, 114, 115
Taoism: books included in Ssu-k'u
catalog, 111–115; works on magic
and divination classified as Taoist, 115, 149
Teng Hui: censorship case, 180
Ting Wen-pin: censorship and condemnation, 32
T'ien Wen-ching: report of expectant
educational officials, 173
Ts'ai Hsin: punishment for negligence, 99
Tseng Ching: sedition trial, 24; execution, 27
T'u-shu chi-ch'eng: publication, 22;
reference to by Ch'ien-lung Emperor, 35; comparison to *Yung-lo
ta-tien,* 77; presentation to book
owners contributing to Ssu-k'u
project, 91
Tuan Yü-ts'ai: phonology of, 43
T'ung-chien kang-mu hsu-pien: emendations ordered by emperor, 116
T'ung-chih-t'ang ching-chieh: praise of in
edict by Ch'ien-lung Emperor, 28
Tzu-chih t'ung-chien: Manchu edict
directing study, 18

Wan Ssu-t'ung: report of intellectual
life under Yuan rulers, 13; opinions of *Sung-shih,* 131
Wang Ch'ang: in class of 1754, 50;
use of stone inscriptions, 65
Wang Chung: as member of Chu
Yun's circle, 52; prefaces to *Mo-
tzu,* 151
Wang Erh-yang: censorship case,
177, 196
Wang Hsi-hou: censorship case,
174–179, 181
Wang Hung-hsu: role in *Ming-shih,*
137
Wang Ming-sheng: as *k'ao-cheng*

Harvard East Asian Monographs

69. Eric Widmer, *The Russian Ecclesiastical Mission in Peking during the Eighteenth Century*

70. Charlton M. Lewis, *Prologue to the Chinese Revolution: The Transformation of Ideas and Institutions in Hunan Province, 1891–1907*

71. Preston Torbert, *The Ch'ing Imperial Household Department: A Study of its Organization and Principal Functions, 1662–1796*

72. Paul A. Cohen and John E. Schrecker, eds., *Reform in Nineteenth-Century China*

73. Jon Sigurdson, *Rural Industrialization in China*

74. Kang Chao, *The Development of Cotton Textile Production in China*

75. Valentin Rabe, *The Home Base of American China Missions, 1880–1920*

76. Sarasin Viraphol, *Tribute and Profit: Sino-Siamese Trade, 1652–1853*

77. Ch'i-ch'ing Hsiao, *The Military Establishment of the Yuan Dynasty*

78. Meishi Tsai, *Contemporary Chinese Novels and Short Stories, 1949–1974: An Annotated Bibliography*

79. Wellington K. K. Chan, *Merchants, Mandarins, and Modern Enterprise in Late Ch'ing China*

80. Endymion Wilkinson, *Landlord and Labor in Late Imperial China: Case Studies from Shandong by Jing Su and Luo Lun*

81. Barry Keenan, *The Dewey Experiment in China: Educational Reform and Political Power in the Early Republic*

82. George A. Hayden, *Crime and Punishment in Medieval Chinese Drama: Three Judge Pao Plays*

83. Sang-Chul Suh, *Growth and Structural Changes in the Korean Economy, 1910–1940*

84. J. W. Dower, *Empire and Aftermath: Yoshida Shigeru and the Japanese Experience, 1878–1954*

85. Martin Collcutt, *Five Mountains: The Rinzai Zen Monastic Institution in Medieval Japan*

STUDIES IN THE MODERNIZATION OF THE
REPUBLIC OF KOREA: 1945–1975

86. Kwang Suk Kim and Michael Roemer, *Growth and Structural Transformation*

87. Anne O. Krueger, *The Developmental Role of the Foreign Sector and Aid*

88. Edwin S. Mills and Byung-Nak Song, *Urbanization and Urban Problems*

89. Sung Hwan Ban, Pal Yong Moon, and Dwight H. Perkins, *Rural Development*

90. Noel F. McGinn, Donald R. Snodgrass, Yung Bong Kim, Shin-Bok Kim, and Quee-Young Kim, *Education and Development in Korea*

91. Leroy P. Jones and Il SaKong, *Government, Business and Entrepreneurship in Economic Development: The Korean Case*

92. Edward S. Mason, Dwight H. Perkins, Kwang Suk Kim, David C. Cole, Mahn Je Kim, et al., *The Economic and Social Modernization of the Republic of Korea*

93. Robert Repetto, Tai Hwan Kwon, Son-Ung Kim, Dae Young Kim, John E. Sloboda, and Peter J. Donaldson, *Economic Development, Population Policy, and Demographic Transition in the Republic of Korea*

106. David C. Cole and Yung Chul Park, *Financial Development in Korea, 1945-1978*

107. Roy Bahl, Chuk Kyo Kim, and Chong Kee Park, *Public Finances during the Korean Modernization Process*

94. Parks M. Coble, *The Shanghai Capitalists and the Nationalist Government, 1927-1937*

95. Noriko Kamachi, *Reform in China: Huang Tsun-hsien and the Japanese Model*

96. Richard Wich, *Sino-Soviet Crisis Politics: A Study of Political Change and Communication*

97. Lillian M. Li, *China's Silk Trade: Traditional Industry in the Modern World, 1842-1937*

98. R. David Arkush, *Fei Xiaotong and Sociology in Revolutionary China*

99. Kenneth Alan Grossberg, *Japan's Renaissance: The Politics of the Muromachi Bakufu*

100. James Reeve Pusey, *China and Charles Darwin*

101. Hoyt Cleveland Tillman, *Utilitarian Confucianism: Ch'en Liang's Challenge to Chu Hsi*

102. Thomas A. Stanley, *Ōsugi Sakae, Anarchist in Taishō Japan: The Creativity of the Ego*

103. Jonathan K. Ocko, *Bureaucratic Reform in Provincial China: Ting Jih-ch'ang in Restoration Kiangsu, 1867-1870*

104. James Reed, *The Missionary Mind and American East Asia Policy, 1911-1915*

105. Neil L. Waters, *Japan's Local Pragmatists: The Transition from Bakumatsu to Meiji in the Kawasaki Region*

108. William D. Wray, *Mitsubishi and the N.Y.K., 1870-1914: Business Strategy in the Japanese Shipping Industry*

109. Ralph William Huenemann, *The Dragon and the Iron Horse: The Economics of Railroads in China, 1876-1937*